George Minchin Minchin

Hydrostatics and elementary Hydrokinetics

George Minchin Minchin

Hydrostatics and elementary Hydrokinetics

ISBN/EAN: 9783743349766

Manufactured in Europe, USA, Canada, Australia, Japa

Cover: Foto ©berggeist007 / pixelio.de

Manufactured and distributed by brebook publishing software (www.brebook.com)

George Minchin Minchin

Hydrostatics and elementary Hydrokinetics

PREFACE

IN this work no previous acquaintance with the nature and properties of a fluid is assumed. As in my treatise on Statics, I have begun with the very elements, and, assuming that the student's reading in pure mathematics is advancing simultaneously with his study of Hydrostatics, I have endeavoured to lead him into the advanced portions of the subject. It will be noticed, however, that the way in which the reader is introduced to the notion of a perfect fluid is very different from that which is usually adopted in similar treatises. A definition of a perfect fluid founded upon the elementary facts and principles of the theory of strain and stress is not calculated to produce the impression of simplicity, more especially when the symbols of the Differential Calculus are employed in the process.

I maintain, however, that in such a presentation of the basis of the subject there is really nothing which a beginner who is familiar with the elements of Geometry, Algebra, and Trigonometry cannot readily understand. The prevalent view that the fundamental notions of the Differential Calculus are a mystery which the beginner should not dare to approach, and which cannot be unveiled until great experience in mathematics has been attained, has long seemed to me to be a most unfortunate fallacy. Indeed, I think that, with the aid of a few simple illustrations borrowed from ordinary experience, the process of *differentiation* and the notion of a *limiting value* towards which a ratio tends will be found much more easy of attainment and comprehension than are the initial simple processes of Algebra themselves—namely, those of subtraction, multiplication, and transformation.

Hence I have not hesitated to define the particular kind of body which forms the ideal basis of Hydrodynamics—viz., a *perfect fluid*—with reference to the nature of its stress and strain; and such a definition is, I think, the only one that is really direct and scientific.

The critical reader will probably notice the complete absence of reference anywhere in this work to a certain term which is very familiar to students of Hydrostatics—namely, *the whole pressure of a fluid on a curved surface*. This term is one of the unfortunate misconceptions of Physics—not, indeed, so fatally charged with mischief as that other venerable illusion, *centrifugal force*, by the balancing of which with centripetal force, in the science of previous generations, the planets are enabled to revolve round the Sun. This notion of *whole pressure on a curved surface* has served as a peg from which to hang very many visionary problems of pure mathematics; but it has done more than this; for, the author of one of our elementary text-books of Hydrostatics tells his readers that, when a cylindrical vessel is filled with water, the 'whole pressure' of the liquid on its curved surface is the measure of the total strain to which the vessel is subjected! So far as I am aware, there is no reality in Physics corresponding to this notion of whole pressure on a curved surface, and certainly none is indicated by those writers who use the term. There is, indeed, in the theory of Electrostatics, and in that of Newtonian gravitation in general, the result that the surface-integral of the normal component of force-intensity taken over any closed curved surface is proportional to the quantity of attracting matter contained within the surface; but, apart from this, the arithmetical (non-vectorial) addition of a number of forces which are not parallel is a process devoid of physical meaning.

Some long controversies have recently taken place between the representatives of the practical engineers, on the one hand, and those who may be termed the scientific reformers, on the other, with regard to the way in which we should speak of the common gravitation measure of force. The former speak of 'a force of one pound,' for example, while the latter prefer 'a force of one pound weight.' As in the last edition of my *Statics*, I have everywhere throughout the present work adopted the latter mode of speaking, because I regard the distinction between the *mass* and the *weight* of a body as absolutely fundamental—the former being an essential and invariable, while the latter is a purely contingent and variable, attribute of the body. Indeed, were it not for the fact that the Earth is nearly spherically symmetrical as regards shape and density about its centre, it seems scarcely possible that men, communicating with each other over long distances, could ever have adopted the former mode of speaking.

Throughout the treatise I have made considerable use of the Principle of Virtual Work.

It will also be noticed that in several instances in which there is a conflict between the algebra and the physics of the subject, or a difficulty or misconception which is very likely to arise in the student's mind, I have been at pains to emphasise and enlarge on the difficulty, and to invite an attempt at explanation on the part of the student. This I have done because I am convinced that more than one-half of the efficiency of the teaching of any subject consists in the anticipation and removal of difficulties which are certain to occur to the mind of the student, and which, if left unnoticed, (like uncaptured fortresses in the rear of an advancing army) will greatly hinder progress and perhaps necessitate the beginning of the work all over again. I have no sympathy with the view which commonly prevails in the writing of scientific treatises, that the reader, so long as he is supplied with the bare truths and dry facts of the subject, has all that he wants and has no right to complain.

In the English language there are at present very few treatises dealing with the advanced portions of theoretical Hydrostatics; and among these the treatise of Dr. Besant has long held the foremost place. The present work is not intended to supplant that of Dr. Besant, but rather to supplement it. Some portions of the subject will, I think, be found treated at greater length in the one treatise and some in the other; so that there is, I hope, ample room for both.

As regards the second portion of this work—that devoted to elementary Hydrokinetics—some explanation is necessary.

The critical reader will, I hope, remember that it is meant to be strictly *elementary*, and to be merely the necessary complement of that portion of the Hydrostatics which can be studied by those who desire to attain a useful working knowledge of the subject without attempting the application of the higher pure mathematics.

Three treatises dealing specially with Hydrokinetics have recently appeared in English—namely, Lamb's *Treatise on the Motion of Fluids*, Lord Rayleigh's *Theory of Sound*, and Basset's *Hydrodynamics*. These treatises are practically exhaustive, and, in all probability, will continue to be so for many years to come. Hence it is certain that every student who desires to carry his study of Hydrokinetics to the limits of our existing knowledge will turn to these works for information. The attempt, therefore, to treat at length of this part of the subject is unnecessary, and it would have resulted in nothing better than a mere copy of the works of these authors. Probably a Chapter dealing with the generalities of Hydrokinematics and Hydrokinetics would have been found useful as an introduction to the higher treatment of these subjects; but this has been reserved for consideration and, perhaps, a future edition.

In the passage of the work through the press I have had the advantage of the assistance of my colleague Professor Stocker, to whose practical knowledge of the subjects of Gases and Capillarity, in particular, I am indebted for very useful criticism and information.

GEORGE M. MINCHIN.

COOPERS HILL,
   September, 1892.

TABLE OF CONTENTS

CHAPTER I.
General Properties of a Perfect Fluid . . . . PAGE 1

CHAPTER II.
Centre of Parallel Forces (Elementary Cases) 27

CHAPTER III.
Liquid Pressure on Plane Surfaces 37

CHAPTER IV.
General Equations of Pressure 77

CHAPTER V.
Pressure on Curved Surfaces 111

CHAPTER VI.
Gases 185

CHAPTER VII.
Hydraulic and Pneumatic Machines . . . 273

CHAPTER VIII.
Molecular Forces and Capillarity 298

CHAPTER IX.
Steady Motion under the action of Gravity . . . 368

CHAPTER X.
Wave Motion under Gravity (Simple Cases) . . . 394

Index 421

HYDROSTATICS

AND

ELEMENTARY HYDROKINETICS.

ERRATUM

Page 179, line 13, omit the words "i.e., if the external forces have a potential"

Minchin's Hydrostatics

---

affairs at any point, P, in the substance of the string. If at P we imagine a very small plane having the position pq (represented in the left-hand figure) perpendicular to the direction of the string, it is clear that the molecules of the body at the under side of this element-plane experience an upward pull from the portion of the body at the upper side of pq. This pull is called a *stress*, and it results from the strain of the substance at P—a strain which, in this

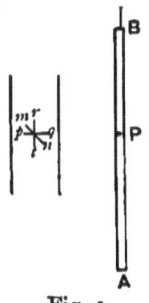

Fig. 1.

HYDROSTATICS

AND

ELEMENTARY HYDROKINETICS.

CHAPTER I.

GENERAL PROPERTIES OF A PERFECT FLUID.

1. Strain and Stress in a Body. Whenever any material body is subjected to the action of force, whether throughout its substance or merely at certain points of its surface, the natural distances between its molecules become altered—even though it be only to an infinitesimal extent—and the body is said to be in a state of strain.

For example, let AB (Fig. 1) represent an elastic string with the end A fixed while the end B is pulled by any force. Consider the state of affairs at any point, P, in the substance of the string. If at P we imagine a very small plane having the position pq (represented in the left-hand figure) perpendicular to the direction of the string, it is clear that the molecules of the body at the under side of this element-plane experience an upward pull from the portion of the body at the upper side of pq. This pull is called a *stress*, and it results from the strain of the substance at P—a strain which, in this

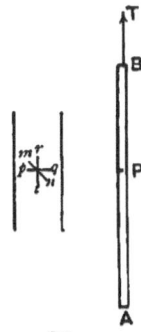

Fig. 1.

instance, consists of the increase of natural distances between molecules.

At the same time the molecules at the upper side of pq experience a downward pull, exactly equal in magnitude to the previously named pull.

The stress on the element-plane has, therefore, two aspects: it is an upward force when considered with reference to the molecules on its under face, and a downward force in reference to the molecules on its upper face.

This double aspect is a characteristic of every stress, and of every force in the Universe, however exerted, whether within the substance of what we call a single body or between two bodies influencing each other by attraction or repulsion. The double aspect is just as necessary a feature of every force as it is of every surface, which we are compelled to recognise as having two sides.

Let us now, in imagination, consider a little element-plane having the position rt, and separating molecules right and left. A moment's reflection shows that the molecules at the right experience no force (or only an infinitesimal force) from those at the left side. Practically we may say that the stress on the element-plane rt is zero.

In the same way, if we consider an element-plane at P having a position mn, intermediate to pq and rt, the force exerted on the molecules at the under side by the substance on the upper side is an upward pull whose direction is oblique to the plane.

Hence in this case of a stretched string we see that—

on the element-plane pq the stress is normal tension,
,, ,, mn ,, oblique tension,
,, ,, rt ,, zero.

If we imagine AB to represent an iron bar, we have exactly the same results when the end A is fixed and the end B is pulled; but if the end A rests on the ground, while a load is put on B, it is evident that the previous state of stress is exactly reversed in sign; that, for example, the molecules on the under side of pq experience a normal *pressure* from those on the upper side, and that the stress on mn is oblique pressure. This reversal of tension into pressure could not practically take place if the body AB were a perfectly flexible string; in other words, the body would at once collapse if we attempted to produce pressure at B.

Again, if AB is an iron column whose base A is fixed on the ground while a great horizontal pressure is exerted from right to left at the top B, the column will be slightly bent and its different horizontal sections have evidently a tendency to slip on each other: in other words, the molecules at the under side of an element-plane having the position pq, experience a force from right to left in their own plane from the substance above pq.

The stress, therefore, on an element-plane at any point inside a solid body such as iron may have any direction in reference to the plane—it may be normal pressure, normal tension, or force wholly tangential to the plane, according to the manner in which the body is strained by externally applied force. Inside a body such as a flexible string the nature of any possible stress is, as we have said, more limited, inasmuch as it cannot be normal pressure.

We shall now imagine a body in which the nature of any possible stress is still more limited—namely, a body in which the stress on every element-plane, however imagined at a point, can never be otherwise than *normal*. Such a body is a *perfect fluid*; and then the stress is, in all ordinary circumstances, *pressure*—such in the sequel we shall assume it to be—although in some recent experiments large *tensions*, or negative pressures, have been observed in some fluids. Formally, then, we define that *a perfect fluid* is a body such that, whatever forces act upon it, thus producing strain, the stress on every element-plane throughout it is normal.

2. Intensity of Stress. If we take any element-plane, mn, at a point and take the whole amount of the stress exerted on either side of the plane, and then divide the amount of the stress by the area of the element-plane, we obtain the average stress on the little plane. Thus, if the area of mn is ·001 square inches, and the stress on either side is ·02 pounds' weight, the *rate* of stress on the plane is $\frac{·02}{·001}$, or 20 pounds' weight per square inch. The stress on the plane is not uniformly distributed; but the smaller the area of the element-plane, the less the error in assuming the stress to be uniformly distributed over it. Hence, according to the usual method of the Differential Calculus, if we take an element-plane of indefinitely small area, δs, and if δf is the amount of the stress exerted on either side, the limiting value of the fraction

$$\frac{\delta f}{\delta s},$$

when δs (and therefore also δf) is indefinitely diminished, is the rate, or intensity, of stress at P on a plane in the direction mn.

It is obvious from what has been explained that such an expression as 'the intensity of stress at a point P in a strained body' is indefinite, because different element-planes at the same point P may have very different intensities of stress exerted upon them. We shall presently show, however, that in the case of a perfect fluid the above expression is perfectly definite, because all element-planes at the same point, if they have the same area, experience normal pressures of exactly the same amount; but for any other body than a perfect fluid such would not be the case, and we should be obliged to speak of 'the intensity of stress at P on a plane having such or such particular position.'

Intensity of stress, then, is measured in *pounds' weight per square inch*, or *kilogrammes' weight per square centimètre*, or *dynes per square centimètre*, or generally, in *units of force per unit area*.

3. Principle of Separate Equilibrium. The following principle is very largely employed in the consideration of the equilibrium or motion of a fluid, or, indeed, of any material system:—

We may always consider the equilibrium or motion of any limited portion of a system, apart from the remainder, provided we imagine as applied to it all the forces which are actually exerted on it by the parts imagined to be removed.

Thus, suppose Fig. 2 to represent a fluid, or other mass, at rest under the action of any forces, and let us trace out in imagination any closed surface enclosing a portion, M, of the mass. Then all the portion of the mass outside this surface may be considered as non-existent, so far as M is concerned, if we supply to each element of the surface of M, the stress which is actually exerted on it by the mass outside it. The stresses exerted on the elements of surface of M, when the body is a perfect fluid, are represented by the arrows in the figure.

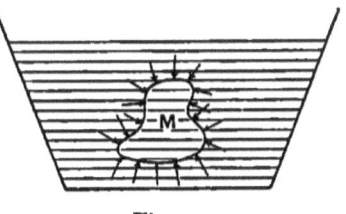

Fig. 2.

The portion M is, then, in equilibrium under the action of these pressures and whatever external forces (gravity, &c.) also act upon it.

Again, it is evident that, having traced out in imagination any surface enclosing a mass, M, of the fluid, we might, without altering anything in the state of this mass

M, replace the imagined enclosing surface by an actual material surface, and then remove all the fluid outside this surface; for the enclosing material surface will, by its rigidity, supply to M at each point the pressure which is exerted at that point by the surrounding fluid.

4. Equality of Pressure Intensity round a point. We shall now prove that the intensity of pressure is the same on all planes, pq, mn, rt (Fig. 1), at the same point, P, in a perfect fluid, to whatever system of external forces the fluid may be subject. Let any two planes, $ACca$ and $ABba$ (Fig. 3), of indefinitely small equal areas be described at P, the figure of each area being taken, for simplicity, as that of a rectangle, the side Aa being common to both, and P its middle point. These areas being indefinitely small, each may be assumed to be uniformly pressed, and the resultant pressure on it acts at its middle point.

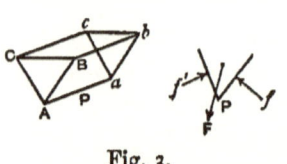

Fig. 3.

Now isolate in imagination the fluid contained within the prism $CBAabc$. This prism of fluid is kept in equilibrium under the influence of five pressures and the resultant external force; and for the equilibrium of these forces we shall take moments round the line Aa. Of the five pressures those on the faces ABC and abc are parallel to Aa, and they have, therefore, no moments about Aa. The pressure on the face $BbcC$ intersects Aa, and gives no moment. There remain the pressures on $ABba$ and $ACca$, which are represented by f, f', at the right—where the figure is a middle section of the prism through P—and the component, F, of the external force which acts in this middle section.

Let $AB = AC = m$, $Aa = n$, and let ϵ be the perpendicular from P on the line of action of F.

Now if p is the intensity of pressure on the element-plane $ABba$, the force $f = mn \cdot p$; and if p' is the intensity of pressure on the face $ACca$, the force $f' = mn \cdot p'$; while, if r is the external force per unit volume acting on the fluid at P, the component F will be a fraction of the quantity $r \times$ (vol. of prism); say $F = k \cdot m^2 n \cdot r$, where k is some finite number less than unity.

Hence the equation of moments about Aa is

$$f \times \frac{m}{2} - f' \times \frac{m}{2} + \epsilon \cdot F = 0,$$

or $\qquad m^2 n (p-p') + 2k\epsilon m^2 n \cdot r = 0,$

or $\qquad p - p' + 2k\epsilon \cdot r = 0, \quad \ldots \quad (1)$

in which p, p', k, r are all finite and ϵ alone is infinitely small. Hence diminishing the size of the prism indefinitely, or, in other words, putting $\epsilon = 0$ in the last equation, we have

$$p = p', \quad \ldots \quad \ldots \quad \ldots \quad (a)$$

so that the intensity of pressure on the plane $ABba$ is the same as that of the pressure on every other plane at P.

The reason, then, why the bodily force does not interfere with the fundamental result (a) is that the pressures on the faces of the prism are finite quantities multiplied by infinitesimal *areas*, while the bodily force is a finite quantity multiplied by an infinitesimal *volume*, and, when diminishing the size of the prism indefinitely, its volume vanishes in comparison with the areas of its faces.

The proposition of this Article is a particular case of a general result in the theory of the Stress and Strain of any material body whatever. (See *Statics*, vol. ii. p. 396.)

This theorem is so important that we reproduce it here.

At any point, P, (Fig. 4), inside any body which is subject to the action of external forces, and which, therefore is strained throughout its substance, let two very small element-planes, s, s', of equal area, be placed in any two positions, however different; let Pn, Pn' be their normals drawn at P. On the upper surface of s, as seen in the figure, let the resultant stress exerted by the substance in its neighbourhood be represented in magnitude and line of action by Pf; and similarly let the stress on s' be represented by Pf'. From f let fall fr perpendicular to Pn', and from f' let fall $f'r'$ perpendicular to Pn. Then Pr is the component of the stress on s along the normal to s', and Pr' is the component of the stress on s' along the normal to s; and the important general theorem to which we refer is that—*whatever be the nature of the body, whether solid, perfect fluid*, or *imperfect fluid*,

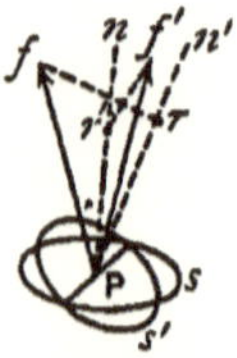

Fig. 4.

$$Pr = Pr', \quad \ldots \ldots \quad (2)$$

if, as supposed, the area of the element-plane $s =$ that of s'.

If these areas are unequal, the projections of the stress intensities, $\dfrac{Pf}{s}$ and $\dfrac{Pf'}{s'}$, along the normals Pn' and Pn are equal, i.e.,

$$\frac{Pr}{s} = \frac{Pr'}{s'}. \quad \ldots \ldots \quad (3)$$

We may designate this remarkable theorem as the *Theorem of the projections of stress-intensities*, and the following simple proof of it may be given.

At the point P in the substance (Fig. 5) let $xx'b'b$ be any element-plane whose boundary is a rectangle of area s; let $bb'c'c$ be another element-plane inclined at the angle θ to the former, its boundary being also rectangular, and such that $c'cxx'$ is a plane perpendicular to the plane $xx'b'b$. The figure therefore is that of a small rectangular prism, which we may imagine to be closed by the terminal

General Properties of a Perfect Fluid.

triangular faces bxc and $b'x'c'$. Consider the separate equilibrium of the substance enclosed within this prism. Since the areas of all the faces are very small, the stress is uniformly distributed on each of them, and the resultant stress on any face acts, therefore, at its centre of area ('centre of gravity'). Now we aim at showing that the theorem of projection holds for the two faces $xx'b'b$ and $bb'c'c$.

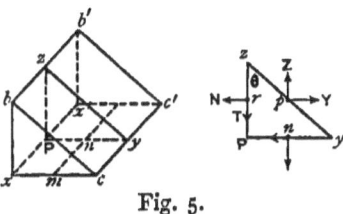

Fig. 5.

Let n be the centre of area of the face $xx'c'c$, and express the fact that the sum of the moments of all the forces acting on the prism about the line mn, parallel to xx', is zero.

To this sum of moments nothing will be contributed by the stress on the face $xx'c'c$, since this force acts at n. Resolve the stress on each face into three components parallel to Px, Py, and Pz. No force parallel to Px will give any moment, and it is easy to see that the sum of moments contributed by the two faces bxc and $b'x'c'$ will be an infinitesimal of the fourth order, the linear dimensions of the prism being infinitesimals of the first order. For, draw Py and Pz parallel to xc, and xb, and let $Px = a$, $Py = \beta$, $Pz = \gamma$; also let the components, in these directions, of the *intensity* of stress on the plane zPy be P, Q, R, each of these being a function of the co-ordinates of the point P, the co-ordinate axes being supposed to be taken at some fixed origin parallel to Px, Py, Pz. The two latter components for the face bxc are

$$Q + a\frac{dQ}{dx} \text{ and } R + a\frac{dR}{dx};$$

and those for $b'x'c'$ are

$$-\left(Q - a\frac{dQ}{dx}\right) \text{ and } -\left(R - a\frac{dR}{dx}\right).$$

10   *Hydrostatics and Elementary Hydrokinetics.*

These are stress-*intensities*, and therefore each of them must be multiplied by the area, $\frac{1}{2}\beta\gamma$, of the faces on which they act. Taking moments about mn, the first two give

$$\frac{\beta\gamma}{12}\left(R\beta + 2Q\gamma + a\beta\frac{dR}{dx} + 2a\gamma\frac{dQ}{dx}\right),$$

and last two give a moment of opposite sign in which, moreover, a is changed to $-a$; hence the whole sum of moments for these terminal faces is

$$\frac{a\beta\gamma}{6}\left(\beta\frac{dR}{dx} + 2\gamma\frac{dQ}{dx}\right) \cdot \cdot \cdot \cdot \cdot \cdot (4)$$

This, we shall see, is infinitesimal compared with the moments of the stresses on the faces $bb'x'x$ and $bb'c'c$, which act, respectively, at the middle points of Pz and zy. For clearness the mid-section, zPy, of the prism is represented in Fig. 5 with the arrows representing the forces on the faces above named, the components parallel to Px, perpendicular to the plane of the figure, not being represented. If N and T are the components of stress intensity on $bb'x'x$, the forces rN and rT are, respectively, $N.s$ and $T.s$, whose sum of moments round the point n in the previous sense is

$$-\frac{s}{2}(\gamma N + \beta T). \cdot \cdot \cdot \cdot \cdot \cdot (5)$$

If Y, Z are the components of stress-intensity on the face $bb'c'c$, the forces represented by pY and pZ in the figure are $Y.s \sec\theta$ and $Z.s \sec\theta$, whose sum of moments about n is

$$\frac{s}{2}.\gamma Y \sec\theta. \cdot \cdot \cdot \cdot \cdot \cdot (6)$$

Now since s is $a\gamma$, we see that the moment (4) is of the *fourth* order of small quantities, while (5) and (6) are each of the third, and therefore (as in p. 7) the first would disappear ultimately in any equation involving all three.

Again, it matters not whether the substance of the prism is acted upon by external force or not; for this force

General Properties of a Perfect Fluid.

would be proportional to the volume $\frac{1}{2}\alpha\beta\gamma$, and its moment about mn would involve the product of this volume and another infinitesimal of the first order; hence the moment of the external force would be of the *fourth* order, and it is, therefore, to be neglected in comparison with (5) and (6).

The equation of moments, then, is simply

$$\gamma Y \sec\theta = \gamma N + \beta T,$$
$$\therefore\ Y = N\cos\theta + T\sin\theta, \quad \ldots \quad (7)$$

which asserts the truth of the theorem, because the right-hand side of (7) is the projection of the stress-intensity on the plane $xx'b'b$ along the normal to the plane $bb'c'c$, and the left the projection of the stress-intensity on $bb'c'c$ along the normal to the first plane.

To see how simply this shows that the stress-intensities on all element-planes at a point P in a perfect fluid are equal, let us recur to our former definition of such a body (p. 3). Assume, then, the strained body to be such that the stress Pf (Fig. 4) acts in the normal nP and Pf' acts in the normal $n'P$. If ϕ is the angle between the normals, $Pr = Pf.\cos\phi$, and $Pr' = Pf'.\cos\phi$; therefore (3) gives

$$\frac{Pf}{s} = \frac{Pf'}{s'},$$

that is, $\qquad\qquad p = p',$

where p and p' are the intensities of pressure on the two element-planes. Hence the result that *if a body is such that the stresses on all element-planes at a point are normal to these planes, the* INTENSITY *of the stress is the same for all.*

This principle is sometimes loosely spoken of as the principle of 'equality of fluid pressure round a point.' The *equality* is thus shown to be a mathematical deduction from the normality.

Of course at any other point, Q, in the fluid the pressure-intensity may be very different from what it is at P, this difference always depending on forces which act *bodily* on the fluid mass; forces which act only at the *surface* of the fluid would not, as will be seen, produce different pressure-intensities at different points.

Hence, then, at each point, P, in a perfect fluid acted upon by any forces there is a certain pressure-intensity, p, which has reference simply to the point itself and not to any *direction* at the point; in other words, if (x, y, z) are the co-ordinates of P,

$$p = \phi(x, y, z), \quad \ldots \ldots \quad (\beta)$$

i.e., p is some function of the position of P, depending, of course, on the nature of the forces acting on the fluid; and no such simple result, independent of *direction*, characterises the strain and stress of a natural solid.

Hence also the difference, dp, between the pressure-intensity at P and at any very close point, P', is a *perfect differential*; i.e., if dx, dy, dz, are the excesses of the co-ordinates of P' over those of P, we must have necessarily some such result as

$$dp = L\,dx + M\,dy + N\,dz, \quad \ldots \quad (\gamma)$$

and we see from the above that L, M, N are the differential coefficients of one and the same function, ϕ, with respect to x, y, z.

Such a result, for example, as $dp = y\,dx - x\,dy$ could not hold for a perfect fluid.

5. External Bodily Forces and Surface Forces. A mass of fluid on the surface of the Earth or any planet is acted upon by the attractive force of the planet in such a way that each particle of the fluid experiences this external force, which is called the weight of the particle. Some fluids are of a magnetic nature, i.e. each particle of them feels the attractive or repulsive force of the pole of a magnet. Forces which, proceeding in this way from external bodies, are felt by the separate particles of the fluid are called *bodily forces*. If we imagine the fluid taken out into interstellar space, at a practically infinite distance from every star and magnet, its molecules would have no weight and would experience no force of any kind from external bodies; and if we imagine further that we accompany the fluid to such a region, supplied with nothing but a vessel fitted with pistons, we should be unable to influence the internal portions of the fluid in any other way than by producing pressure on various portions of its bounding surface.

External forces thus produced merely at places on the bounding surface are called *surface forces*.

6. Principle of Pascal. *If a perfect fluid is acted upon by no other than surface forces, the intensity of pressure is constant all over the surface and at all points in the interior of the mass.*

Let a perfect fluid be contained within the contour $ABCD$ (Fig. 6), and suppose pressure to be applied over its surface so that the intensity of this pressure at A is p pounds' weight per square inch. At A take a very small area, s square inches, represented by AA', and on this little area erect a right cylinder, AP, of any length. Now consider the separate equilibrium of the fluid contained within this cylinder.

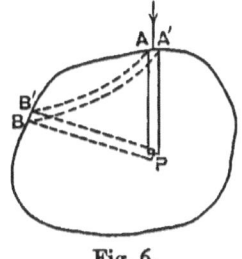

Fig. 6.

This fluid is held in equilibrium by the force $p.s$ pounds' weight acting on AA', a pressure on the base at P, and a series of pressures all over its curved surface. Resolving forces in the direction AP, we have $p.s = $ the pressure on the base at P, since the pressures on the curved surface are all at right angles to AP. But the

14   *Hydrostatics and Elementary Hydrokinetics.*

area of the base is also s square inches, therefore if $p'=$ intensity of pressure at P,

$$p.s = p'.s \quad . \quad . \quad . \quad . \quad (1)$$
$$\therefore \; p = p'.$$

Again, the pressure intensity at every point on the surface is also p. For, let the base at P be turned round through any angle, and on its new position construct a right cylinder cutting the surface obliquely at B. Let θ be the angle between the normal to the surface at B and the axis, PB, of the cylinder; let p' be the intensity of pressure exerted by the envelope at B on the fluid, and consider the separate equilibrium of the fluid in the cylinder PB. The area of the normal cross-section of the cylinder being s, the area cut off from the surface at B is $s \sec \theta$, and the total pressure on this is $p'.s \sec \theta$. Now resolving along the axis PB for the equilibrium of the enclosed fluid, we have

$$p.s = p'.s \sec\theta . \cos\theta,$$
$$\therefore \; p' = p.$$

This may also be seen by constructing on the area s at A a tube, $AA'BB'$, of any form whatever and of uniform normal cross section. The fluid inside this tube is kept in equilibrium by the terminal pressures on AA' and BB' together with the pressures of the surrounding fluid which are all normal to the sides of the tube. Hence (except that the terminal forces at A and B are *pressures* and not *tensions*) this fluid is in the same condition as a flexible string stretched over a smooth surface and acted upon by two terminal forces only, in addition to the continuously distributed normal pressure of the smooth surface; and it is obvious, by considering the equilibrium of each elementary length of the tube, that the forces per unit area at A and B are equal.

General Properties of a Perfect Fluid. 15

In the same way the intensity of pressure at any point, Q, in the mass can be proved to be p; for Q and P can be connected by a slender right cylinder having equal and parallel plane bases at Q and P.

This method of proof shows that if the stress on every element-plane in the substance were not normal, its intensity would not be the same at all points. For, if the stresses on the different elements of the curved surface of the cylinder APA' were oblique, they would furnish a component parallel to AP, and the equality (1) would cease. Hence in a viscous fluid, i.e. one in which there is friction between neighbouring molecules, the pressure-intensity is not necessarily the same at all points.

If the area AA' is an aperture, fitted accurately by a piston, in a vessel $ABCD$ containing the fluid, the pressure at A may be produced by loading this piston. Suppose the area of the base AA' of the piston to be 2 square inches, and the total load on the piston to be 40 pounds' weight; then *every* element of the surface of the containing vessel will experience pressure at the rate of 20 pounds' weight per square inch, every element-plane in the interior will also experience this intensity, and the pressure will be uniform all over every plane area imagined in the fluid, however great its area may be.

Here, however, a caution may be given. If $ABCD$ is a vessel filled with water, and if a piston at A produces an intensity of pressure of 20 pounds' weight per square inch, we shall not find the intensity at such a point as C to be 20, but something notably greater if C is at a lower level than A; and at a point of the surface higher than A the intensity would be found to be less than 20 pounds' weight per square inch. But the reason of this is sufficiently indicated in our formal enunciation of Pascal's Principle—viz., that on the surface of the Earth water is acted upon by a bodily force (gravitation). If, however, we could take the vessel of water into interstellar space and apply pressure at A by a piston, we should actually find the same intensity at all points of the containing vessel.

We may, indeed, regard the Pascal Principle as *always* holding in a perfect fluid—even when the fluid is acted upon by gravity or other bodily force—but the evidence of the Principle will be masked by a second cause of pressure, viz. bodily force. If, however, the bodily force were removed, the undiminished intensity of surface pressure produced at any point would at once evidence itself.

If the fluid were hydrogen, or any light gas, the Pascal Principle would, even on the surface of a planet, be almost accurately verified within such a moderate volume as a few cubic feet of the gas, because the bodily force (weight of an element volume of the gas) is too small to generate any appreciable pressure. In accordance with the Principle of Pascal, *we may always regard a perfect fluid, even when acted upon by gravitation, as a machine for transmitting to all points, in undiminished amount, any intensity of pressure produced at any point of its surface.* This invariable transmission of surface pressure will proceed, *pari passu*, with increase or diminution of pressure produced by gravitation; but the two causes of pressure can be kept mentally quite distinct.

Thus, for example, at a depth of 100 feet in a fresh water lake, the intensity of pressure due to the weight of the water is about $43\frac{1}{3}$ pounds' weight per square inch, as will be seen later on. But at the top of the lake there is a pressure intensity of about 15 pounds' weight per square inch produced by the weight of the atmosphere, and the water acts as a machine for transmitting this latter unaltered to *every* point below, so that the actual pressure intensity at a point 100 feet down is $58\frac{1}{3}$ pounds' weight per square inch.

General Properties of a Perfect Fluid.

7. The Hydraulic Press. A machine the action of which illustrates the Principle of Pascal is the Hydraulic Press, represented in Fig. 7.

It consists of a stout cylinder, A, in which a cast iron piston, or ram, P, works up and down. This piston has a strong iron platform fixed on the top; on this platform is placed a substance which is to be subjected to great

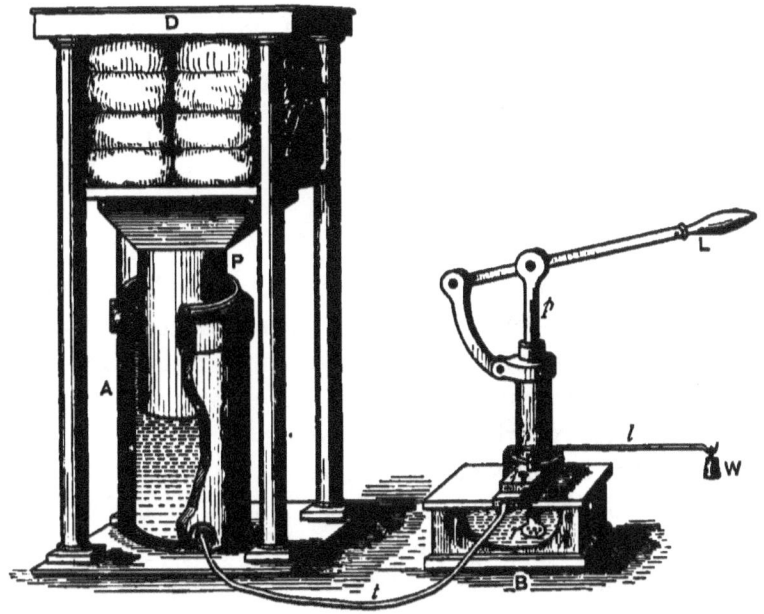

Fig. 7.

pressure between the platform and a strong plate, D, fixed to four strong vertical pillars. The pressure is applied at the bottom of the piston P by a column of water which is forced into the cylinder A through a tube, t, which communicates with a reservoir of water, B. The water is driven out of B by a force-pump whose piston, p, has a much smaller diameter than that of the piston P;

the piston p is worked up and down by means of a lever L, and the cylinder in which p works terminates inside the vessel B in a *rose*, r, the perforations in which admit water while preventing the entrance of foreign matter.

It is easy to see what an enormous multiplication of force can be produced by this machine. If F is the force applied by the hand to the lever L, n the multiplying ratio of the lever, and s the area of the cross-section of the piston p, the intensity of pressure produced on the water in the vessel B is $\frac{nF}{s}$; so that if S is the area of the cross-section of the piston P, the total force exerted on the end of this piston by the water in A is

$$nF \cdot \frac{S}{s}.$$

Thus, if $S = 100\,s$ and the ratio, n, of the long to the short arm of the lever L is 5, the upward force exerted on the piston P is $500\,F$, so that if a man exerts a force of 100 pounds' weight on the lever, a resistance of nearly 50000 pounds' weight can be overcome by the piston.

In order to prevent the intensity of pressure in the vessel B from becoming too great, a safety-valve closed by a lever loaded with a given weight, W, is employed.

The Hydraulic Press remained for a long time comparatively useless, because the great pressure to which the water was subject drove the liquid out of the cylinder A between the surface of the piston P and the inner surface of the cylinder. This defect was remedied in a very simple and ingenious manner by Bramah, an English engineer, in the year 1796. In the neck of the cylinder A is cut a circular groove all round, and into this groove is fitted a leather collar the cross-section of which is represented at c in the form ⌒. This collar is saturated with oil, in order that it may be water-tight, and it will be seen that it presses with its left-hand and upper portion against the cylinder A, while its right-hand portion is against the piston P. When, by pressure, the water is forced up between the surface of the piston and the surface of the cylinder, this water enters the lower or hollow portion of the inverted U-shaped collar and firmly presses the leather against both the piston P and the surface of the groove, thus preventing any escape of water from the cylinder.

In consequence of this great improvement in the machine, it is very commonly called *Bramah's Press*.

In order to prevent the return of the water from the cylinder A on the upward stroke of the piston p, there is a valve, represented at i in Fig. 7, and shown more clearly at

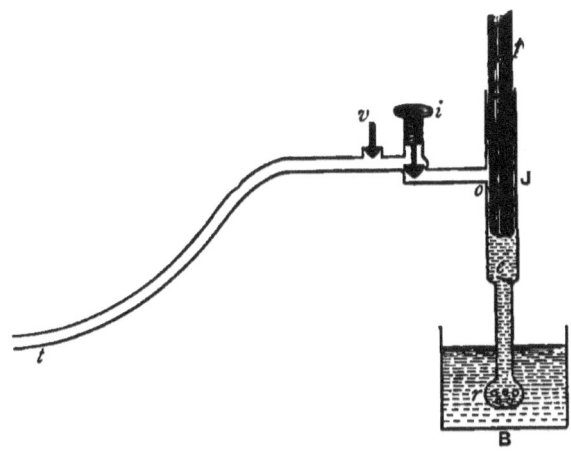

Fig. 8.

i in Fig. 8, which is a simple sketch of the essentials of the force-pump. When the piston p rises, a valve, e, in the pipe dipping into the reservoir B, opens upwards and allows the water to fill the cylinder J and to flow through o to the valve i. When p descends, the water closes the valve e and is forced to open the valve i which is pressed down by a spiral spring. When the piston p moves upwards, the water which has passed the valve i into the cylinder A cannot return into the cylinder J because it obviously assists the spring in closing the valve i. The safety-valve is represented at v in Fig. 8.

The piston p works in a stuffing-box in the upper part of the cylinder J, this stuffing-box playing the same part as the leather collar round the ram—i. e., preventing leakage. The piston must not fit the lower part of the cylinder J tightly, because when p in its downward motion passes o, the cylinder would be burst if the water above the closed valve e could not escape round the piston and out through the valve i.

Another machine depending essentially on the same principles and illustrating the Principle of Pascal is the *Hydrostatic Bellows*, which is formed by two circular boards connected, in bellows fashion, by water-tight leather, the boards being the ends of a cylinder the curved surface of which is formed by the leather. One of these boards being placed on the ground, the other lies loosely on top of it. A narrow tube communicates with the interior of this cylinder. If this tube is a long one and held vertically, when water is poured into it at its upper end, the upper board of the bellows, and any load that may be placed on it, will be raised by the pressure of the water, the intensity of which pressure depends (as will be subsequently explained) on the height to which the narrow tube is filled.

8. Liquids and Gases. An absolutely incompressible perfect fluid is called a *liquid;* but the term liquid is also applied to fluids which can be compressed, but which require very great intensity of superficial pressure to produce even a small compression.

The *cubical compressibility* of any substance (see *Statics*, vol. ii., Art. 387) is thus measured: take any volume, v, of

General Properties of a Perfect Fluid.

the substance (for simplicity a spherical, cubical, or cylindrical, volume) and let its surface be all over subject to pressure of uniform intensity, δp; measure the decrement, $-\delta v$, of volume produced; then the fractional compression is $-\dfrac{\delta v}{v}$; and if we divide the intensity of pressure which produces this by the fractional compression we obtain the measure, viz.,

$$\frac{\delta p}{-\dfrac{\delta v}{v}}, \text{ or } -v\frac{dp}{dv}, \quad \ldots \ldots (a)$$

of the modulus of cubical compressibility of the substance, this modulus being evidently a force per unit area.

Thus, if the volume and its decrement are measured in cubic centimètres, while the intensity of the pressure is measured in dynes per square centimètre, we obtain the modulus of compressibility in absolute C. G. S. units.

If k is this modulus, we have

$$-v\frac{dp}{dv} = k. \quad \ldots \ldots (\beta)$$

If k is a constant, we have the case of a homogeneous solid or a liquid—extremely large values of k characterising a body of the latter kind.

If k varies sensibly with the intensity of pressure, p, in any way, we have bodies of various physical natures, according to the mode of dependence of k on p.

If, for instance, k is equal to p, the body is a *perfect gas*. Putting $k = p$ in (β) and integrating, we have

$$pv = \text{constant},$$

the well-known equation expressing the law of Boyle and Mariotte for a perfect gas whose temperature remains unaltered while its volume and intensity of pressure vary. Hence for a gas compressed at constant temperature *the modulus of cubical compressibility is equal to its intensity of pressure.* This modulus is sometimes called the 'resilience of volume' of the strained substance. It may be expressed in terms of the density instead of the volume; for if ρ is the density of the substance inside the volume v, since the mass remains unaltered, we have

$$v\rho = v_0\rho_0 = \text{constant},$$

where v_0 and ρ_0 are the volume and density of the element considered before strain. Hence (β) becomes

$$\rho \frac{dp}{d\rho} = k. \quad \ldots \ldots \ldots (\gamma)$$

Employing the units of the C. G. S. system (forces in dynes, &c.) the following is a table of resiliences of volume for various liquids *:

	Temp. Cent.	Coefficient of Resilience of Vol.
Distilled Water	15	2.22×10^{10}
Sea Water	17.5	2.33×10^{10}
Alcohol	0	1.21×10^{10}
,, ...	15	1.11×10^{10}
Ether	0	$.93 \times 10^{10}$
,, ...	14	$.792 \times 10^{10}$
Bisulphide of Carbon . .	14	1.60×10^{10}
Mercury	15	54.20×10^{10}

9. Specific Weight. By the term *specific weight* of any homogeneous substance we shall understand its *weight per unit volume.*

If w is the weight of any homogeneous substance per unit volume, a volume V will have a weight given by the equation

$$W = V.w.$$

---

* Taken from Everett's *Units and Physical Constants.*

General Properties of a Perfect Fluid.

It will be useful to remember that

1 cubic foot of water has a mass of about 62½ lbs.
1 ,, ,, ,, ,, ,, ,, ,, 1000 ounces.
1 ,, inch of mercury ,, ,, ·491 lbs.

These numbers are, of course, only approximate, because the mass of a cubic foot of any substance depends on the temperature at which it is.

A *gramme* is defined to be the mass, or quantity of matter, in 1 cubic centimètre of water when the water is at its temperature of maximum density; this temperature is very nearly 4° C.

A term in frequent use is the *specific gravity* of a substance, which ought, apparently, to signify the same thing as its specific weight; but it does not. The specific gravity of any homogeneous solid or liquid means, in its ordinary employment, the *ratio* of the weight of any volume of the substance to the weight of an equal volume of distilled water at the temperature 0° C. Thus, for example, in the following table of specific gravities:

gold	19·3
silver	10·5
copper	8·6
platinum	22·0
sea-water	1·026
alcohol	·791
mercury	13·596

the number opposite the name of any substance does not tell us the weight of a cubic foot, or of any other volume, of the substance; it merely tells, with regard to platinum, for example, that a cubic foot of it, or a volume V of it, is 22 times as heavy as a cubic foot, or a volume V, of distilled water. A table of specific gravities is a table of *relative* weights of equal volumes. In the C. G. S. system, since the unit of weight is that of 1 cubic centimètre of water, and since water is the substance with which in a table of specific gravities all solids and liquids are compared, the number (specific gravity) opposite any substance expresses the actual mass, in grammes, of 1 cubic cm. of the substance.

If s is the specific gravity of any substance and w the actual weight of a unit volume of the standard substance (water), the weight of a volume V of the substance is given by the equation
$$W = Vsw.$$

The term *density* is also used to denote the *mass, per unit volume*, of a substance. Thus if mass is measured in grammes and volume in cubic centimètres, the density of silver is 10·5 grammes per cubic centimètre; the density of mercury is 13·596 grammes per cubic cm. If mass is measured in pounds and volume in cubic inches, the density of silver is ·3797 lbs. per cubic inch and that of mercury ·491 lbs. per cubic inch. These latter numbers are, of course, proportional to the former.

The term *density* has no reference to gravitation. If silver and mercury are taken from the Earth to a position in interstellar space in which there is felt no appreciable attraction from any Sun or Planet, it is still true that silver has a mass of 10·5 and mercury a mass of 13·596 grammes per cubic cm. Neither would, in this position, have any specific *weight*, since there is no external force of attraction acting on them; but the moment they are taken to the surface of any Sun or Planet, each acquires weight, and the *ratio* of the weights of equal volumes of them is the ratio, 10·5 : 13·596, of their densities. If, for example, they were carried to the surface of the Planet Jupiter, the weight of a cubic cm. of each would be nearly $2\frac{1}{2}$ times as great as it is on the surface of the Earth; but a table of *relative*

General Properties of a Perfect Fluid. 25

weights of substances on Jupiter would be exactly the same as a table of relative weights on the Earth.

If any given volumes of a number of homogeneous substances are mixed together in such a way as to make a homogeneous mixture whose volume is the *sum* of the volumes of the separate substances, the specific weight of the mixture is easily found. For, let v_1 and w_1 be the volume and specific weight of the first substance; v_2 and w_2 those of the second; and so on. Then if w is the required specific weight of the mixture, since the weight of the mixture is equal to the sum of the separate weights,

$$(v_1 + v_2 + v_3 + \ldots) w = v_1 w_1 + v_2 w_2 + v_3 w_3 + \ldots;$$

$$\therefore w = \frac{\Sigma vw}{\Sigma v}.$$

Such a mixture is called a *mechanical* mixture—as, for instance, a mixture of sand and clay. But when a chemical combination takes place between any of the substances, the volume of the mixture is not equal to the sum of the volumes mixed—as when sulphuric acid is mixed with water. If for any chemical mixture V (which must be specially measured) is the volume of the mixture, it is evident that we have, as above,

$$w = \frac{\Sigma vw}{V}.$$

Example.

A cask A is filled to the volume v with a liquid of specific weight w; another cask, B, is filled, also to the volume v, with another liquid of specific weight s; $\dfrac{v}{n}$ is taken out of A and $\dfrac{v}{n}$ also out of B, the first being put into B and the second into A, and the contents of each cask are shaken up so that the liquid in each becomes homogeneous. The same process is repeated again and again: find—

(a) the specific weight of the liquid in each cask after m such operations;

(b) the volume of the original liquid in each cask.

Ans. If $w-s$ is denoted by d, and if w_m, s_m are the specific weights of the liquids in A and B, respectively, after m operations,

$$w_m = w - \frac{d}{2}\left\{1 - \left(1 - \frac{2}{n}\right)^m\right\},$$

$$s_m = s + \frac{d}{2}\left\{1 - \left(1 - \frac{2}{n}\right)^m\right\},$$

and the volume of the original liquid in either cask is

$$\frac{v}{2}\left\{1 + \left(1 - \frac{2}{n}\right)^m\right\}.$$

[N.B. The liquids are assumed not to enter into chemical combination.]

CHAPTER II.

CENTRE OF PARALLEL FORCES (ELEMENTARY CASES).

10. Points and Associated Magnitudes. Let A and B, Fig. 9, be any two points, PP any plane whatever, AM and BN the perpendiculars from A and B on the plane, and G a point on the line AB dividing it in the ratio of any two magnitudes, m and n, of the same kind, so that $\frac{AG}{GB} = \frac{n}{m}$; then the per-

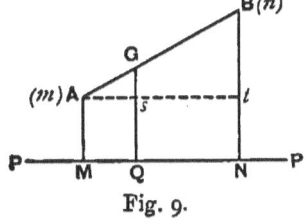

Fig. 9.

pendicular, GQ, from G on the plane is given by the equation

$$GQ = \frac{m \cdot AM + n \cdot BN}{m+n} \quad \ldots \quad (a)$$

For, draw Ast parallel to MN, meeting GQ in s. Then

$$GQ = Gs + sQ = Gs + AM.$$

But $\quad \dfrac{Gs}{Bt} = \dfrac{n}{m+n}; \quad \therefore \; Gs = \dfrac{n}{m+n}(BN - AM).$

Substituting this for Gs in the value of GQ, we have (a).

The figure represents m and n as associated with A and B, respectively, and G divides AB so that the segments next A and B are *inversely* as the magnitudes associated with these points.

If one of the magnitudes, m and n, associated with either point is negative, the dividing point G is to be taken on AB produced or on BA produced, according as n or m is *numerically* the greater (irrespective of sign). Thus if the magnitude associated with B is $-n$, the distance of the point G from the plane PP is

$$\frac{m \cdot AM - n \cdot BN}{m-n} \quad \ldots \ldots \quad (\beta)$$

The result (a) is well known in the composition of two parallel forces of like sense acting at A and B, while (β) applies to the case in which the parallel forces are of unlike sense.

If parallel forces whose magnitudes are m and n act in the same sense and in *any* common direction at A and B, equation (a) gives the distance of their 'centre' from any plane; while (β) gives the distance of the centre of parallel forces of unlike sense.

The results (a) and (β) have not, however, been restricted to the case in which m and n are *forces*. These quantities may be, as said before, any two magnitudes of the same kind—e. g., two masses, two areas, two volumes, &c. The case in which one—say n—is negative may also be represented by supposing m and $-n$ to be a positive and a negative charge of electricity.

When m and n are quantities of matter, the point G is called their *centre of mass* (see *Statics*, vol. i., Art. 90).

When m and n are positive, the magnitude $m+n$ is associated with G; and if the magnitude associated with B is $-n$, the magnitude $m-n$ is associated with G.

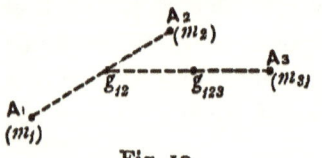

Fig. 10.

Let there be any number of given points, A_1, A_2, A_3,... (Fig. 10) with which are associated any given magnitudes m_1, m_2, m_3... respectively, and take the centre, g_{12}, of the magnitudes m_1, m_2, then take the centre, g_{123}, of the magnitude $m_1 + m_2$ at g_{12} and the magnitude m_3 at A_3, then the centre of $m_1 + m_2 + m_3$ at g_{123}

Centre of Parallel Forces (Elementary Cases). 29

and m_4 at A_4, and so on. In this way we arrive at a final point, G, which is the centre of the whole system of magnitudes. It is required to express the distance of this point from any plane in terms of the given magnitudes and the distances of their associated points from the plane.

This is easily done by (a). For, if the distances of $A_1, A_2, A_3, \ldots$ from any plane are, respectively, $z_1, z_2, z_3, \ldots$ and z_{12} is the distance of g_{12} from the plane, we have by (a),

$$z_{12} = \frac{m_1 z_1 + m_2 z_2}{m_1 + m_2}.$$

Also if z_{123} is the distance of g_{123} from the plane,

$$z_{123} = \frac{(m_1 + m_2) z_{12} + m_3 z_3}{m_1 + m_2 + m_3}$$

$$= \frac{m_1 z_1 + m_2 z_2 + m_3 z_3}{m_1 + m_2 + m_3}.$$

Hence, by repeated applications of the simple result (a), if $\bar{z}$ is the distance of G from the plane,

$$\bar{z} = \frac{m_1 z_1 + m_2 z_2 + m_3 z_3 + m_4 z_4 + \ldots}{m_1 + m_2 + m_3 + m_4 \ldots} \quad . \quad (\gamma)$$

$$= \frac{\Sigma mz}{\Sigma m}. \quad \ldots \ldots \ldots \ldots (\delta)$$

The plane of reference, PP, may be such that some of the points are at one side of it and some at the other side. In this case some of the z's are positive and some negative, the side of the plane which we take as positive being a matter of choice.

If the points $A_1, A_2, \ldots$ are not all in one plane, to determine the position of G, we shall require to find its distances from some *three* planes of reference. If the points $A_1, A_2, \ldots$ all lie in one plane, it will be sufficient to find the distances of G from any two lines in this plane. In this case PP (Fig. 9) may be supposed to be a mere line in the plane of the given points, although, strictly speaking, it represents, in this case, a plane perpendicular to the plane of the points. If the points $A_1, A_2, \ldots$ all lie on a right line, G lies on this line, and its position will be known if its distance from any other line is known.

When $m_1, m_2, m_3, \ldots$ are masses, G is their *centre of mass*, and equation (γ) expresses the *Theorem of Mass Moments*, the product mz, of any mass, m, and the distance, z, of its centre of mass from a plane being called *the moment of the mass with respect to the plane*.

Cor. The sum of the moments of any masses with respect to any plane passing through their centre of mass is zero.

Even when $m_1, m_2, m_3, \ldots$ are not masses, but any magnitudes of the same kind (forces, areas, &c.) we shall refer to (γ) as the Theorem of Mass Moments.

When $m_1, m_2, m_3, \ldots$ are the magnitudes of a system of parallel forces acting at $A_1, A_2, A_3, \ldots$ in any common direction, the point G is called the *centre of the system of parallel forces*.

It is evident that the distances $z_1, z_2, \ldots$ need not be perpendiculars; they may be oblique distances—all, of course, measured in the same direction.

The work of practical calculation is often facilitated by forming tables of masses, distances, and products, in columns, as in the following example.

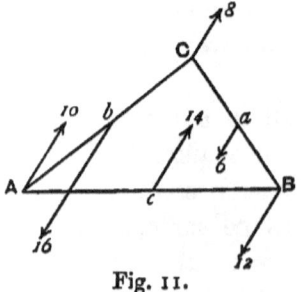

Fig. 11.

EXAMPLES.

1. At the vertices, A, B, C (Fig. 11) of a triangle and at the middle points, a, b, c, of the opposite sides act parallel forces whose magnitudes and senses are represented in the figure; find the position of the centre of the system.

The position of the centre will be known if its distances from any two sides, AB and AC, are known.

Centre of Parallel Forces (Elementary Cases). 31

To find its distance from AB, let the length of the perpendicular from C on AB be p; and form a table of forces and distances of their points of application from AB as in the following scheme, taking as positive those forces which act in the sense of that at A :—

Forces.	Distances from AB.	Products.	Distances from AC.	Products.
10	0	0	0	0
−16	$\tfrac12 p$	−8 p	0	0
8	p	8 p	0	0
−6	$\tfrac12 p$	−3 p	$\tfrac12 q$	−3 q
−12	0	0	q	−12 q
14	0	0	$\tfrac12 q$	7 q
−2		−3 p		−8 q

The sum of the first column answers to Σm, the denominator of (δ), p. 29, while the sum of the third column answers to Σmz, the numerator, so that the perpendicular distance of the centre from AB is

$$\frac{-3p}{-2}, \quad \text{or} \quad \frac{3}{2}p.$$

Drawing, then, a line parallel to AB at a distance $\tfrac{3}{2}p$ (above C), we know that G lies somewhere on this line.

Denoting the perpendicular from B on AC by q, forming a table (column 4) of distances from AC, and a column (number 5) of corresponding products, and dividing the sum of these products by the sum of the forces, we have the distance of G from AC equal to

$$\frac{-8q}{-2q}, \quad \text{or} \quad 4q.$$

Hence G lies on a line to the right of B distant $4q$ from AC. The point of intersection of this with the previous line is G.

2. From a solid homogeneous triangular prism is removed a portion by a plane parallel to the base cutting off $\dfrac{1}{n}$ of the axis measured from the vertex; find the distance of the centre of

mass of the remaining frustum from the base. Let ABC be the middle section of the prism perpendicular to its vertex edge, and let PQ represent the cutting plane.

Now the volume of the whole prism is to the volume of the removed prism as the area ACB is to the area PCQ; and since the areas of any similar plane figures are proportional to the squares of corresponding lines in them (*Euclid*, Book VI. prop. 20),

$$\text{area } PCQ = \frac{1}{n^2} (\text{area } ACB).$$

Hence if V is the volume of the whole prism, the volume of the small prism $= \dfrac{V}{n^2}$, and that of the frustum $= \left(1 - \dfrac{1}{n^2}\right)V$.

Now let h be the length of the perpendicular from C on AB, and equate the mass-moment of the whole prism with respect to the base AB to the sum of the mass-moments of the frustum and the small prism. Since the centre of mass of a prism is distant by $\frac{1}{3}$ of the height from the base, the distance of the centre of mass of PCQ from AB is $\frac{1}{3}\dfrac{h}{n} + \left(1 - \dfrac{1}{n}\right)h$, or $\dfrac{3n-2}{3n}h$. If, then, x is the distance of the centre of mass of the frustum

$$V \cdot \frac{h}{3} = \left(1 - \frac{1}{n^2}\right) V \cdot x + \frac{V}{n^2} \cdot \frac{3n-2}{3n} h,$$

$$\therefore\ x = \frac{(n-1)(n+2)}{3n(n+1)} \cdot h.$$

3. From a solid homogeneous cone is removed a portion by a plane parallel to the base cutting off $\dfrac{1}{n}$ of the axis measured from the vertex; find the distance of the centre of mass of the remaining frustum from the base.

(The volumes of similar cones are proportional to the cubes of corresponding linear dimensions. Hence if V is the vol. of the whole cone, the vol. of the small cone $= \dfrac{1}{n^3} V$.)

Ans. The distance $= \dfrac{(n-1)(n^2+2n+3)}{4n(n^2+n+1)} \cdot h.$

Centre of Parallel Forces (Elementary Cases).

4. From the middle point of one side of a triangle is drawn a perpendicular to the base; find the distance, from the base, of the centre of area of the quadrilateral thus formed.

Ans. If h is the height of the triangle, $\frac{5}{14}h$.

5. Find the position of the centre of area of a trapezium.

Ans. It is on the line joining the middle points of the two parallel sides, and if the lengths of these sides are a and b, and h the perpendicular distance between them, its distance from the side a is
$$\frac{2b+a}{3(a+b)} \cdot h.$$

6. Prove that the distance of the centre of area of a triangle from any plane is one-third of the algebraic sum of the distances of its vertices from the plane.

(The centre of area of any triangle is the same as the centre of mass of three equal particles placed at its vertices.)

7. Prove that the distance of the centre of area of a plane quadrilateral from any plane is
$$\tfrac{1}{3}(\Sigma z - \zeta),$$
where Σz is the sum of the distances of its vertices, and ζ the distance of the point of intersection of its diagonals, from the plane.

11. Continuously Distributed Forces. We shall now find the centre of a system of parallel forces distributed continuously over a plane area, in a few simple cases which do not require the application of the Integral Calculus.

(1) If normal pressure acts all over any plane area in such a way that its intensity is the same at all points, the resultant pressure acts at the centre of area ('centre of gravity,' so called) of the figure. For, if at any two points, A, B, in the given figure we take any two elements of area, the pressures on them are directly proportional to the areas themselves, and the resultant of these forces acts at a point in the line AB dividing it into segments inversely as the forces, i.e., inversely as the areas; hence

it acts at the centre of area of the two elements of area. The process, therefore, of finding the point of application of the resultant of the whole system of pressures acting on the indefinitely great number of elements of area into which the given figure can be broken up is precisely the same as that of finding the centre of area of the figure.

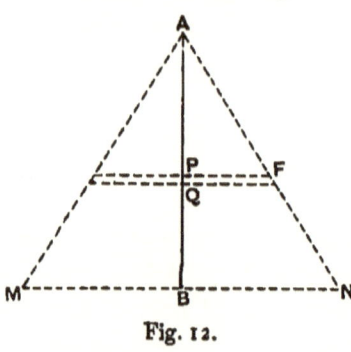

Fig. 12.

(2) If parallel forces act at all points of a right line, AB, Fig. 12, in such a way that the force at any point P is directly proportional to the distance, PA, of P from one extremity, A, of the line, the resultant force acts at the point on AB which is $\frac{2}{3}$ of the length AB from A.

For, imagine AB to be broken up into an indefinitely great number of small equal parts, PQ; describe any isosceles triangle, MAN, having AB for height, and from all the points, P, Q, ... of division of AB draw parallels to the base MN, thus dividing the area MAN into an indefinitely great number of narrow strips. The area of any strip, QF, is simply proportional to the distance, PA, of the strip from A. Hence the areas of the strips are exactly proportional to the given system of parallel forces; but the centre of area of the strips, or centre of area of the whole triangle, is $\frac{2}{3} AB$ from A. This point is, then, the centre of the given system of forces.

(3) If parallel forces act at all points of a right line, AB, Fig. 12, in such a way that the force at any point P is proportional to the *square* of the distance, PA, of P from one extremity, A, of the line, the resultant force acts at a point on AB which is $\frac{3}{4}$ of the length AB from A.

For, imagine AB to be broken up, as before, into equal

Centre of Parallel Forces (Elementary Cases). 35

elements, such as PQ; describe any solid cone having AB for its axis, and let this cone be represented by MAN. From all the points of division of AB draw planes parallel to the base MN of the cone, thus dividing the cone into an indefinitely great number of thin circular plates. The volume of the plate at P is $\pi PF^2 \times PQ$, and since the thicknesses of the plates are all equal to PQ, the volume of the plate is proportional to PF^2, i.e., to PA^2. Hence the volumes of the plates vary exactly as the forces of the given system, and therefore the centre of volume of the plates is identical with the centre of the force system; but the former (centre of volume of the cone) is $\frac{3}{4} AB$ from A; therefore, &c.

(4) If parallel forces act at all points of a right line AB, in such a way that the force at any point P is proportional to the product of the distances PA, PB of P from the extremities of the line, the resultant force acts at the middle point of AB.

For, taking a point, P', whose distance from B is equal to that of P from A, the forces at P and P' are evidently equal; their resultant therefore acts at the middle of AB. Hence the system of forces from A to this middle point is the same as the system from B to this point; the resultant, therefore, of the whole system acts at the middle point of AB.

(5) If each infinitesimal element of any plane area is acted upon by normal pressure proportional conjointly to the magnitude of the area and to the distance of the

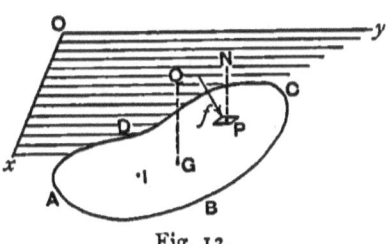

Fig. 13.

element from a given plane, the *magnitude* of the resultant pressure will be proportional to the product of the whole plane area and the distance of its centre of area from the given plane.

For, let yOx, Fig. 13, represent the given plane, and $ABCD$ the given plane area. Take any point, P, in this area, and round P describe a very small closed curve whose area is s. Let PN, the perpendicular from P on the plane yOx, be denoted by z; then, by hypothesis, the amount of force, f, on the element at P is given by the equation

$$f = k \cdot sz,$$

where k is a given constant. If s', s'',... are any other elements of area whose distances from the given plane are z', z'',... the resultant pressure, being equal to the sum of $f, f', f'', \ldots$ of the individual pressures on the elements, is equal to
$$k\left(sz + s'z' + s''z'' + \ldots\right).$$

But if $A =$ the area of the whole plane figure, and $\bar{z}$ is the distance, GQ, of the centre of area, G, from yOx, we have
$$A\bar{z} = sz + s'z' + s''z'' + \ldots.$$

Hence if P is the resultant pressure,

$$P = k \cdot A \cdot \bar{z}. \quad \ldots \ldots (a)$$

The student must be careful to observe that the resultant pressure *does not act at* G, but evidently at some such point as I, whose distance from the plane yOx is greater than the distance of G from the plane.

In this case, then, the mean intensity of pressure on the area is that which exists at G.

CHAPTER III.

LIQUID PRESSURE ON PLANE SURFACES.

Elementary Cases.

12. Intensity of Pressure produced by Gravity. Let ACB, Fig. 14, be a vessel of any shape containing water or other homogeneous liquid. Then at each point, P, of the liquid the action of gravity produces a certain intensity of pressure, the magnitude of which we proceed to find. At P draw an in-

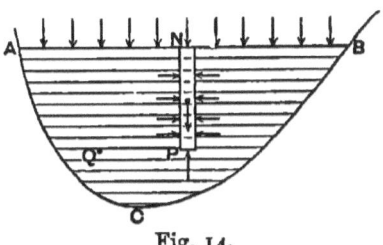

Fig. 14.

definitely small horizontal element of area—s square inches, suppose—and on the contour of this area describe a vertical cylinder, PN. Consider now the separate equilibrium (Art. 3) of the liquid in this cylinder:

If PN is z inches in length, the volume of the cylinder $= z \cdot s$ cubic inches, and if the specific weight of the liquid is w pounds' weight per cubic inch, the weight of the cylinder $= wzs$. This cylinder is acted upon by a vertically upward pressure on the base s at P and a system of horizontal pressures round its curved surface, in addition to its weight—omitting, for the present, the surface pressure at N produced by the atmosphere or any other cause.

If p pounds' weight per square inch is the intensity of pressure at P, the upward pressure on the base s is $p \cdot s$. Resolving forces vertically, we have, then,

$$p \cdot s = wz \cdot s;$$
$$\therefore p = wz, \quad \ldots \quad \ldots \quad \ldots \quad (a)$$

which gives the required intensity of pressure.

If the surface intensity of pressure is p_0 pounds' weight per square inch, this will be added to the value (a), by Pascal's principle; hence the complete value of p is given by the equation

$$p = wz + p_0. \quad \ldots \quad \ldots \quad \ldots \quad (\beta)$$

Observe that we have not assumed the bounding surface AB to be horizontal.

Without any reference to the shape of the surface AB, we can see that the intensity of pressure is the same at all points P, Q, ... which lie in the same horizontal plane.

For, draw PQ; at P and Q place two indefinitely small equal elements of area, s, perpendicularly to PQ; form a cylinder having PQ for axis and these little areas for bases, and consider the separate equilibrium of the liquid enclosed in this cylinder. The forces keeping it in equilibrium are its weight, a system of pressures all round its curved surface, and the pressures on its bases at P and Q. Resolving forces along PQ for equilibrium, neither the weight nor the system of pressures on the curved surface will enter the equation; therefore the pressure on the base s at $P =$ the pressure on the (equal) base s at Q; that is, the intensity at $P =$ the intensity at Q.

From this it follows that the bounding surface AB on which at all points there is either no pressure, or pressure of constant intensity, *must be a horizontal plane*.

For, take any two points, P, Q, in a horizontal plane, and let their vertical distances below AB be z and z'.

Then by (β), we have

$$wz' + p_0 = wz + p_0;$$
$$\therefore z' = z,$$

that is, all points in the same horizontal plane are at the same depth below the surface AB—which proves AB to be a horizontal plane.

It is usual to speak of the surface, AB, of contact of the liquid with the atmosphere as the *free surface* of the liquid. It is simply a surface at each point of which the intensity of pressure is constant, the constant being the atmospheric intensity.

The result at which we have arrived may be also stated thus—*all points in a heavy homogeneous liquid at which the intensity of pressure is the same lie in a horizontal plane;* and from this it follows that if a mass of water partly enclosed by subterranean rocks, &c., has access to the atmosphere by any number of channels, the level of the water will be the same in all these channels. It is to be carefully observed that z in (α) and (β) *is the depth of the point P* (Fig. 15)

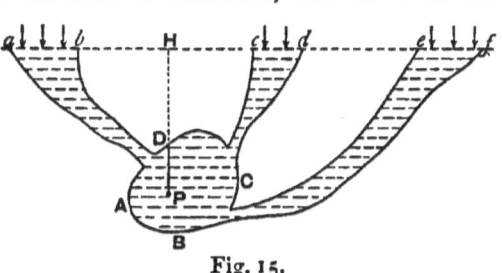

Fig. 15.

below the free surface—not the distance, PD, of the point P from the roof of the cavity in which the water is partly confined.

We may here, if we please, consider the separate equilibrium of a small vertical cylinder of the liquid terminating at D, and we shall have simply the result (β) in which, however, p_0 would now mean the vertical downward component of the pressure intensity of the roof of the cavity at D on the water. But the result (β) holds for the intensity of pressure at P if PH is put for z, where H is the foot of the perpendicular from P on the plane of the free surfaces ab, cd, ef of the water; for, nowhere in the liquid will the state of affairs be altered if we imagine the roof of the cavity to be removed, and the space bDc to be filled with water up to the level bc. In this way we shall have a vertical cylinder, PH, unobstructed by the roof, and terminating on the free surface.

It is usual to illustrate the fact that all parts of the free surface of a liquid lie in a horizontal plane by taking a vessel, ABC, of any shape and fitting into it tubes or funnels of various forms, and then pouring water in through any one of these tubes, the visible result being that the water stands at the same level in all the tubes. This is, indeed, nothing more than the principle of separate equilibrium (see end of Art. 3); for, these variously shaped funnels may be supposed to have been surfaces traced out in imagination in a large vessel of water whose free surface was af, these imagined surfaces being then replaced by material tubes, and the outside liquid removed. The level of the liquid in each tube would still be af.

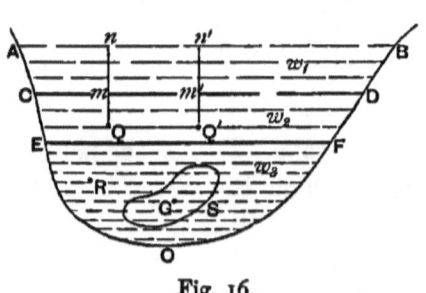

Fig. 16.

13. **Superposed Liquids.** If in a vessel, AOB, Fig. 16, several liquids be placed as layers, one on top of another, there being no chemical combination between them, the common surface of each pair of liquids is a horizontal plane. Let the specific

weights of the liquids be $w_1, w_2, w_3, \ldots$. The free surface, AB, has been already proved to be a horizontal plane (Art. 12); and the same process will prove CD, the surface of separation of w_1 and w_2, to be a horizontal plane. For, in the liquid w_2 take any two points, Q, Q' (as in Fig. 14, p. 37), in the same horizontal plane. Then by taking a slender horizontal cylinder having QQ' for axis, we prove that the intensity of pressure at $Q =$ that at Q'. Now taking a vertical cylinder Qmn, at Q, considering its separate equilibrium, we find that if p is the intensity of pressure at Q, and $Qm = x$, $mn = y$,

$$p = w_2 x + w_1 y.$$

Similarly if $Q'm'n'$ is the vertical line at Q', and $Q'm' = x', m'n' = y'$,

$$p = w_2 x' + w_1 y'.$$

Hence
$$w_2(x-x') = w_1(y'-y). \quad \ldots \quad (1)$$

But $Qn = Q'n'$, i.e., $x+y = x'+y'$, $\therefore x-x' = y'-y$; so that unless $x-x' = 0$ and $y'-y = 0$, equation (1) will give $w_1 = w_2$, which is not the case, by hypothesis.

Hence we must have

$$Qm = Q'm', \text{ and } mn = m'n',$$

and since this holds for all points Q, Q' in the same horizontal plane, all points, $m, m', \ldots$ in the surface CD are at the same height above the same horizontal plane; therefore CD is a horizontal plane. Similarly, by taking two points in the same horizontal plane in the liquid w_3, we prove that EF is a horizontal plane.

If h_1 and h_2 are the thicknesses of the layers w_1 and w_2, and if R is a point in w_3 at a depth z below the surface, EF, of w_3, the intensity of pressure, p, at R is given by the equation
$$p = w_1 h_1 + w_2 h_2 + w_3 z, \quad \ldots \quad (2)$$

to which, if atmospheric (or other) pressure acts on the uppermost surface AB, must be added p_0, the intensity of this surface pressure, so that

$$p = p_0 + w_1 h_1 + w_2 h_2 + w_3 z. \quad \ldots \quad (3)$$

Similarly for any number whatever of superposed layers.

Each layer of liquid, in fact, acts as an atmosphere, producing an intensity of pressure on the next layer below it equal to

$$wh, \quad \ldots \quad \ldots \quad \ldots \quad (4)$$

where w is the specific weight of the layer and h its thickness.

If the h's are measured in centimètres and the w's in grammes' weight per cubic centimètre, the above equations express p in grammes' weight per square cm.

The method of *regarding any layer of liquid, even when there is only one liquid in question, as an atmosphere producing an intensity of pressure given by* (4) *on the layer on which it rests, this intensity being then transmitted unaltered to all points below* (by Pascal's principle) is one which we shall frequently employ in the sequel.

From the general principle (*Statics*, vol. i., Art. 121) that, for stable equilibrium, any system of material particles acted upon by gravity only must arrange themselves into such a configuration that their centre of gravity occupies the lowest position that it can possibly occupy, it follows that in a system of superposed liquids of different densities they must arrange themselves so that the density of each liquid is greater than that of any one above it.

Again, if ABC, Fig. 17, represents a vertical section of a vessel of any shape into which are poured two different liquids, AB and BC, which do not mix, the system will settle down into a position in which the centre of gravity of the whole mass occupies the lowest position that it can occupy, and the vertical heights, h, h', of the free surfaces, A and C, above the common surface, B, of the liquids will be inversely proportional to the specific weights of the liquids.

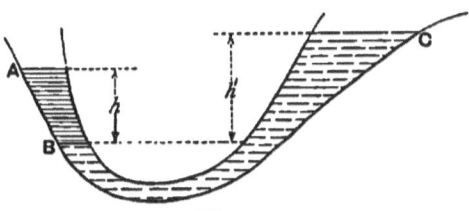

Fig. 17.

To see the latter, we may take any point (most conveniently a point in the common surface B) and equate the intensity of pressure produced there by everything at one side of the point to the intensity of pressure produced by everything at the opposite. Thus, let w and w' be the specific weights of the liquids AB and BC, respectively; select a point, P, in the common surface B. Then if h is the difference of level between P and A, the intensity of pressure produced at P by the liquid AB and the overlying atmosphere at A is

$$wh + p_0.$$

Also, h' being the difference of level of P and C, the intensity of pressure at P produced by the right-hand liquid and the atmosphere above C is

$$w'h' + p_0.$$

There is only one intensity of pressure at P; hence these must be equal;

$$\therefore\ w.h = w'.h', \quad \ldots \quad \ldots \quad (5)$$

which shows that the heights of the free surfaces above the common surface, B, of the liquids are inversely as their specific weights.

Thus, if AB is mercury and BC water, the surface C will be 13·596 times as high above B as the surface A is.

As an example, let two liquids, AB, BC, Fig. 18, be poured into a narrow circular tube held fixed in a vertical

44 *Hydrostatics and Elementary Hydrokinetics.*

plane, the lengths of the arcs occupied by the liquids being assigned; it is required to find their positions of equilibrium.

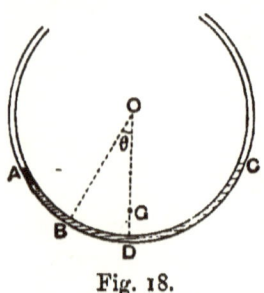

Fig. 18.

The figure of equilibrium will be defined by the angle, θ, which the radius, OB, to the common surface of the liquids makes with the vertical, OD.

Let the angles, AOB, BOC, subtended by the liquid threads at the centre of the circle be a, a'; let their specific weights be w, w', respectively; and let r be the radius of the circle.

Equate the intensity of pressure produced at B by the one liquid to that produced by the other. The difference of level between B and A is $r\{\cos\theta - \cos(\theta + a)\}$, and this multiplied by w is the pressure intensity at B due to the first liquid. The difference of level of B and C is

$$r\{\cos\theta - \cos(a'-\theta)\}.$$

Hence

$$w\{\cos\theta - \cos(\theta+a)\} = w'\{\cos\theta - \cos(a'-\theta)\}, \quad (6)$$

$$\therefore \tan\theta = 2\frac{w'\sin^2\frac{a'}{2} - w\sin^2\frac{a}{2}}{w'\sin a' + w\sin a}. \quad (7)$$

The equation (6) is easily seen to express the fact that the centre of gravity of the system of two liquid threads has, in the position of equilibrium, the greatest vertical depth below O that any geometrical displacement of the two liquid threads could give it. For, the centre of gravity of the thread AB lies on the radius bisecting AB and is at a distance $2r\dfrac{\sin\frac{a}{2}}{a}$ from O (*Statics*, Art. 166); its vertical depth below O is therefore $2r\dfrac{\sin\frac{a}{2}}{a}\cos\left(\dfrac{a}{2}+\theta\right)$, and the weight of the liquid AB is proportional to raw. Hence if $\bar{z}$ is the depth of the centre of gravity, G, of the two liquids, we have, by mass-moments,

$$(aw+a'w')\dfrac{\bar{z}}{r}$$
$$= 2w\sin\dfrac{a}{2}\cos\left(\dfrac{a}{2}+\theta\right)+2w'\sin\dfrac{a'}{2}\cos\left(\dfrac{a'}{2}-\theta\right)\ldots(8)$$

If we make θ such that $\bar{z}$ is a maximum, by equating to zero the differential coefficient of the right side of (8), we have the result (6). Of course it follows from the elements of Statics that G is in the vertical radius OD. (The reactions of the tube all pass through O, &c.)

14. Pressure on a Plane Area. Let $ABCD$, Fig. 13, p. 35, represent a plane area occupying any assigned position in a heavy homogeneous liquid whose free surface is xOy.

Then if w (pounds' weight per cubic inch, suppose) is the specific weight of the liquid, and z (inches) is the depth, PN, of any point below xOy, the intensity of pressure at P, due solely to the weight of the liquid, is wz. Hence (case 5, p. 35) the resultant pressure on one side of the area is
$$A\cdot\bar{z}\cdot w,\quad\ldots\ldots\ldots(a)$$

where A (square inches) is the magnitude of the area, and $\bar{z}$ is the depth, GQ, (inches), of its centre of area below the free surface.

If on the free surface, xOy, there is intensity of pressure (atmospheric or other) of p_0 (pounds' weight per square inch), this pressure will produce its resultant, Ap_0, at G, and the total pressure on one side of the area is

$$A\left(\bar{z}w+p_0\right).\quad\ldots\ldots\ldots(\beta)$$

46   *Hydrostatics and Elementary Hydrokinetics.*

As before pointed out (p. 36) the pressure (α) due to the liquid does not act at G, but at some point lower down.

If a plane area, S, Fig. 16, p. 40, occupies an assigned position in a liquid on the surface of which are superposed given columns of other liquids, the resultant pressure on the area is easily found. For, if $\bar{z}$ is the depth of the centre, G, of area *below the surface EF*, of the liquid w_3, the pressure of this liquid is $A\bar{z}w_3$, where A is the magnitude of the area. Also the column AD produces a resultant pressure equal to $Ah_1 w_1$, where h_1 is the thickness of the column; the second column produces $Ah_2 w_2$; so that the total pressure on S is

$$A(h_1 w_1 + h_2 w_2 + \bar{z} w_3); \quad \ldots \ldots \quad (\gamma)$$

and similarly for any number of liquids, the resultant pressure will be

$$A(h_1 w_1 + h_2 w_2 + h_3 w_3 + \ldots + \bar{z} w_n), \quad \ldots \quad (\delta)$$

where $\bar{z}$ is the depth of G *below the surface of the liquid, w_n, in which the area lies.*

Examples.

1. If a plane area, occupying any position in a liquid, is lowered into the liquid by a motion of translation unaccompanied by rotation, show that the point of application of the resultant pressure on one side of the area rises towards the centre of area, G, the more the area is lowered. (See Fig. 20.)

Draw the horizontal plane CD, touching the boundary of the area at its highest point, and consider the pressures due separately to the layer between CD and the free surface, AB, and to the mass of liquid below CD. Since there is no change in the position of the area relative to the liquid below CD, this latter pressure will always act with constant magnitude and point of application, I_0; but the pressure of the superincumbent layer, always acting at G, increases in magnitude with x, the distance between AB and CD. Hence of the two parallel forces—at I_0 and G—the first remains constant, while the second continually increases; their resultant, therefore, gets nearer and nearer to G as CD is lowered.

Liquid Pressure on Plane Surfaces. 47

2. A triangular area of 100 square feet has its vertices at depths of 5, 10, and 18 feet below the surface of water; find the resultant pressure on the area, the atmospheric intensity being 15 pounds' weight per square inch.

Ans. 127·12 tons' weight.

3. Find the depth of a point in water at which the intensity of the water pressure is equal to that due to the atmosphere.

Ans. About $34\frac{1}{2}$ feet.

4. A rectangular vessel 1 foot high, one of whose faces is 6 inches broad, is filled to a height of 4 inches with mercury, the remainder being filled with water; find the total pressure against this face, the atmospheric intensity being 15 pounds' weight per square inch.

Ans. About $1117\frac{1}{2}$ pounds' weight.

5. Into a vessel containing mercury is poured water to a height of 8 inches above the mercury. If a rectangular area 6 inches in height is immersed vertically so that part lies in the mercury and part in the water, find the length of the area immersed in the mercury when the fluid pressure on this portion is equal to that on the portion in the water.

Ans. Nearly 1·39 inches.

6. Into a vessel containing a liquid of specific gravity ρ is poured water to a height a. If a rectangular area of height h is immersed vertically, part in the water and part in the lower liquid, find the length of the area in this liquid when the fluid pressures on the two portions are equal.

Ans. $\dfrac{\sqrt{(2a-h)^2+h(2a-h)(1+\rho)}-(2a-h)}{1+\rho}.$

7. A beaker containing liquid is placed in one pan of a balance, and is counterpoised by a mass placed in the other pan. If a solid body suspended by a string held in the hand is then immersed in the liquid, what will be the effect on the balance?

If the string sustaining the solid is attached to the arm from which the pan containing the beaker is suspended, and the system counterpoised by a mass in the other pan, will the state of the balance be the same whether the body is immersed in the beaker or not?

48   *Hydrostatics and Elementary Hydrokinetics.*

8. A straight glass tube is bent so that the two portions, AB, BC, are at right angles; it is held in a vertical plane with the point B downward, and the branch BC inclined at the angle a to the horizon; into AB is poured a liquid of specific weight w, the length of the column being l; into BC is poured a liquid (w', l'); find the position of equilibrium.

Ans. The length of the branch BC occupied by the liquid of specific weight w is $\dfrac{wl - w'l' \tan a}{w(1 + \tan a)}$.

9. Columns of any number of liquids which do not mix are placed in a narrow circular tube whose plane is vertical; find the position of equilibrium.

Ans. If the specific weights are $w_1, w_2, w_3, \ldots$ the angles subtended at the centre by the threads of liquid $a_1, a_2, a_3, \ldots$, and if θ is the angle made with the vertical by the radius drawn to the free extremity of the liquid w_1,

$$\tan \theta = \frac{w_1 + (w_2 - w_1) \cos a_1 + (w_3 - w_2) \cos (a_1 + a_2) + (w_4 - w_3) \cos (a_1 + a_2 + a_3) \cdot}{(w_1 - w_2) \sin a_1 + (w_2 - w_3) \sin (a_1 + a_2) + (w_3 - w_4) \sin (a_1 + a_2 + a_3) + .}$$

10. A rectangular area, $LMRS$, Fig. 19, whose plane is vertical, has one side, LM, in the free surface of water; show how to divide the area, by horizontal lines into n strips on each of which the water pressure shall be the same.

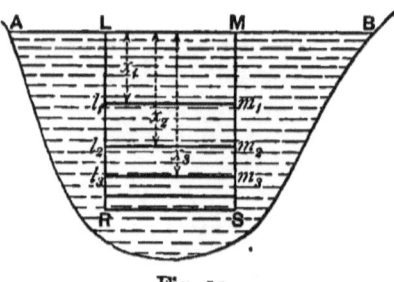

Fig. 19.

Let $LM = a$, $LR = h$, $w =$ specific weight of liquid, and measure the depths, $x_1, x_2, x_3, \ldots$ of the successive lines of division, $l_1 m_1, l_2 m_2, l_3 m_3, \ldots$ each from the surface.

Then the pressure on the rectangle $Lm_1 = \dfrac{1}{n}$ (pressure on LS);

,, ,, ,, ,, $Lm_2 = \dfrac{2}{n}$ (,, ,,);

,, ,, ,, ,, $Lm_3 = \dfrac{3}{n}$ (,, ,,);

and so on. Thus, instead of calculating the pressures on the separate strips, Lm_1, L_1m_2, L_2m_3, ... and equating them to $\dfrac{1}{n}$ of the pressure on LS, we take successive rectangles each having one side LM in the free surface. This is simpler.

Now the pressure on LS is $ah \cdot \dfrac{h}{2} \cdot w$; the pressure on Lm_1 is $ax_1 \cdot \dfrac{x_1}{2} \cdot w$; pressure on Lm_2 is $ax_2 \cdot \dfrac{x_2}{2} \cdot w$; hence

$$x_1^2 = \frac{1}{n} h^2, \quad \therefore \quad x_1 = h \sqrt{\frac{1}{n}},$$

$$x_2^2 = \frac{2}{n} h^2, \quad \therefore \quad x_2 = h \sqrt{\frac{2}{n}},$$

$$\cdots \cdots \cdots \cdots \cdots$$

$$x_r^2 = \frac{r}{n} h^2, \quad \therefore \quad x_r = h \sqrt{\frac{r}{n}}.$$

11. A triangular area, ABC, has its vertex A in the surface of water, its plane vertical, and its base BC horizontal; divide the area by horizontal lines into n strips on which the water pressures shall be equal.

Ans. The depths of the successive lines of division are

$$h\left(\frac{1}{n}\right)^{\frac{1}{3}}, \ h\left(\frac{2}{n}\right)^{\frac{1}{3}}, \ h\left(\frac{3}{n}\right)^{\frac{1}{3}}, \ \ldots \ h\left(\frac{r}{n}\right)^{\frac{1}{3}}, \ \ldots$$

12. A trapezium whose plane is vertical has one of the parallel sides in the free surface of a liquid; divide the area by a horizontal line into two parts on which the liquid pressures are equal.

Let a, b be the parallel sides, the former lying in the surface; let $h = $ height of trapezium; let $c = b - a$, and $x = $ depth of the required line; then x is given by the equation

$$2c(2x^3 - h^3) + 3ah(2x^2 - h^2) = 0 \quad \ldots \quad (1)$$

The root of this equation which is relevant lies between $\dfrac{h}{2^{\frac{1}{2}}}$ and $\dfrac{h}{2^{\frac{1}{3}}}$, which, respectively, correspond to the cases of examples 7 and 8.

By putting $\frac{a}{c} = k$ and $\frac{x}{h} = y$, and depriving (1) of its second term, we have
$$4z^3 - 3k^2 z + k^3 - 3k - 2 = 0, \quad \ldots \quad (2)$$
where $z = y + \tfrac{1}{2}k$. Now (2) can always be solved by either of the well known results,
$$4\cos^3\theta - 3\cos\theta - \cos 3\theta = 0, \quad \ldots \quad (3)$$
or
$$4\cosh^3\theta - 3\cosh\theta - \cosh 3\theta = 0. \quad \ldots \quad (4)$$

Thus, putting z in (2) equal to $k^2 \cos\theta$, we have, to determine θ,
$$\cos 3\theta = \frac{2 + 3k - k^3}{k^3}. \quad \ldots \quad (5)$$

But if the numerator of (5) is greater than the denominator, we must put $z = k^2 \cosh\theta$, and θ is to be found from the equation
$$\cosh 3\theta = \frac{2 + 3k - k^3}{k^3}. \quad \ldots \quad (6)$$

In either case 3θ is known from a table of circular or hyperbolic cosines, and thence z, &c.

If the horizontal line is to be drawn so that the pressure on the upper trapezium $= \frac{1}{n}$ of the pressure on the whole area, the equation for x is
$$2ncx^3 + 3nahx^2 = (3a + 2c)h^3,$$
which can be solved in exactly the same way.

When $b = 0$, or the area is a triangle with its base in the surface and vertex down, the values of z in (2) are 0, $\pm \frac{\sqrt{3}}{2}$, the first of which alone is relevant to the problem, since the latter two give, respectively, a value of x which is $> h$, and a negative value of x—both of which are *physically* impossible.

13. A cube is filled with a liquid, and held with a diagonal vertical; find the pressures on one of the lower and one of the upper faces.

Ans. $\frac{2}{\sqrt{3}} W$ and $\frac{1}{\sqrt{3}} W$, where $W =$ the weight of the liquid in the cube.

Liquid Pressure on Plane Surfaces.

14. A circular area is immersed in a homogeneous liquid, a tangent to the circle lying in the free surface, A being the highest point of the circle; draw a chord, BC, of the circle perpendicular to the diameter through A so that the pressure on the triangle ABC shall be a maximum.

Ans. The distance of BC from A is $\frac{5}{6}$ of the diameter.

15. A triangular area, ABC, occupies any position in a liquid; find a point, O, in its area such that the liquid pressures on the parts BOC, COA, and AOB shall be proportional to three given numbers.

Let a, β, γ be the depths of A, B, C below the free surface; let the ratios of the pressures on the above areas, respectively, to the pressure on the whole triangle ABC be ρ_1, ρ_2, ρ_3; let z be the depth of O, and put x for $z+a+\beta+\gamma$; then x is determined from the cubic

$$\frac{\rho_1}{x-a} + \frac{\rho_2}{x-\beta} + \frac{\rho_3}{x-\gamma} = \frac{1}{a+\beta+\gamma}. \quad \ldots \quad (1)$$

Assuming $a > \beta > \gamma$, the value of x in this equation which is $> a$ is the only one relevant, because the values which are between a and β and between β and γ give negative values of z.

The position of O is completely defined by its areal co-ordinates, i.e., by the ratios of the areas BOC, COA, AOB to the area ABC. If these ratios are l, m, n, respectively, the equations are

$$l(z+\beta+\gamma) = \rho_1(a+\beta+\gamma), \quad \ldots \quad (2)$$

and two similar, where z is $la + m\beta + n\gamma$. When z is known from (1), l is found from (2); &c.

16. A triangle has its base, BC, in the free surface of a liquid, and its vertex, A, down; find a point, O, in its area such that the pressures on BOC, COA, AOB shall be proportional to three given numbers.

Ans. If the pressures on these areas are to the pressure on ABC in the ratios $\rho_1 : \rho_2 : \rho_3$, and h is the depth of A, the point O is the intersection of a horizontal line at a depth $h\sqrt{\rho_1}$ with a line drawn from A to a point, P, in BC such that

$$\frac{BP}{PC} = \frac{\rho_3}{\rho_2}.$$

15. Centre of Pressure. Hitherto we have been occupied with the calculation of the *magnitude* of the resultant pressure on one side of a plane area. We have now to consider the point of the area at which this resultant pressure acts. Except in the case in which the plane of the area is horizontal, this point—which is called the *centre of pressure*—is always lower in the area than G, the centroid, or 'centre of gravity,' of the area.

The position of the centre of pressure on a given area varies with the position (depth, orientation, &c.) of the area in the fluid; and before determining its position in a few simple and frequently occurring cases, we shall lay down a general principle, founded on the remark near the end of Art. 13, which is often of great assistance in calculation. When a plane area—or, indeed, any surface whatever—occupies any position in a liquid, we may draw any horizontal plane whatever in the liquid and consider the column of liquid above this plane as playing the part of an atmosphere—i.e., as producing at all points below the plane a constant intensity of pressure, which is transmitted in virtue of Pascal's Principle. The most convenient horizontal plane for this purpose is one through the *highest point* of the given area.

Fig. 20.

Thus, for example, if *nrm*, Fig. 20, is any plane area whose plane is vertical in a liquid, and we wish to find the magnitude and point of action of the resultant pressure on one side of this area, we may draw a horizontal plane, CD, touching the contour of the area at its highest point, n, and then consider separately the pressures due to the layer of liquid between AB and CD and to the body of liquid below CD.

Liquid Pressure on Plane Surfaces. 53

With regard to the layer $ACDB$, if x is its thickness, we know that it produces at all points on CD and at all points below (Art. 13) an intensity of pressure equal to

$$w \cdot x;$$

and since this pressure is uniformly distributed over the area nrm, its resultant is (case 1, Art. 11).

$$Awx \text{ acting at } G, \quad \ldots \ldots \quad (5)$$

where G is the centre of area of nrm and A the magnitude of the area.

Hence, if we knew the magnitude and point, I_0, of application of the pressure of the liquid below CD, we should have the magnitude and point, I, of application of the pressure of the whole liquid below AB on the area by a simple composition of two parallel forces acting at G and I_0. This we shall presently illustrate by a few simple examples.

Thus we obtain the following construction for the centre of pressure, I, on a plane area (Fig. 21) occupying any position in a liquid: through the highest point, n, on the contour of the figure, draw a horizontal plane, CD, the

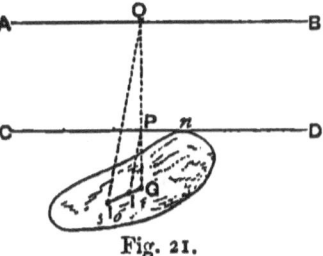

Fig. 21.

free surface of the liquid being AB; from the centroid, G, (or 'centre of gravity') of the figure draw a vertical line meeting these planes in P and Q; suppose I_0 to be the (known) position of the centre of pressure if the surface of the liquid were CD; draw QI_0, and from P draw PI parallel to QI_0, meeting GI_0 in I. This point I is the required centre of pressure on the area. We shall presently proceed to illustrate this method by some simple examples.

Special Cases of Centre of Pressure.

(1) *To find the position of the centre of pressure on a plane parallelogram, whose plane is vertical, with one side in the free surface.*

Let $ABCD$, Fig. 22, be the parallelogram. Let the area be divided into an indefinitely great number of in-

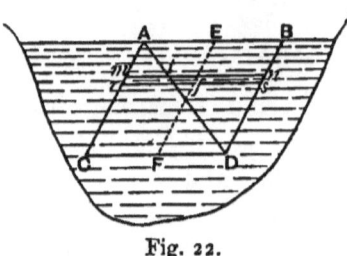

Fig. 22.

definitely narrow strips, of which $mnsr$ is the type, and let E and F be the middle points of the sides AB and CD. Then the middle point of every strip lies on the line EF. Also if x is the depth of the strip ms below AB,
and w the specific weight of the liquid, the intensity of pressure is the same at all points in the strip and (Art. 12) equal to wx, and the resultant pressure on the strip acts at its middle point, i. e., at the intersection, f, of ms with EF. Hence the resultant pressure on the whole parallelogram acts at some point on EF. Also, since the areas of the strips are all equal, the series of pressures on them are simply proportional to their distances from AB; therefore (case 2, p. 34) the point of application of the resultant pressure is $\frac{2}{3}$ of FE from E. Denote this point by T. Then

$$ET = \tfrac{2}{3} FE. \quad \ldots \ldots \quad (a)$$

If h is the height of the parallelogram, and $\bar{p}$ the perpendicular distance of the centre of pressure, T, from the surface
$$\bar{p} = \tfrac{2}{3} h. \quad \ldots \ldots \quad (a')$$

If the plane of the parallelogram is not vertical, the same point, T, will still be the centre of pressure. For, if the area is inclined to the vertical at any angle, θ, and x is the perpendicular distance of f from AB, we have

$$x = fE \cdot \cos \theta \, ;$$

and as θ is the same for all the strips, the pressures on them will still be proportional to their distances fE, &c.

(2) *To find the position of the centre of pressure on a plane triangle having one side in the free surface, and vertex down.*

Let ABD be the triangle. Divide the area, as before, into an indefinitely great number of strips, of which ts is the type. Let x be the perpendicular distance of this strip from the base AB. Now compare this with another strip, $t's'$, whose perpendicular distance from D is also x. Let h be the height of the triangle, $a = AB$, $k =$ the indefinitely small breadth of each strip. Then $tn = \dfrac{h-x}{h} \cdot a$; so that (Art. 14) the pressure on this strip is

$$\frac{ka}{h} x (h-x) w \quad . \quad . \quad . \quad . \quad . \quad (1)$$

But this is also the pressure on the second strip, $t's'$. For, $t'n' = \dfrac{x}{h} a$, and the depth of $t'n'$ is $h-x$; therefore (1) is the pressure on this strip. Since each strip is pressed at its middle point, and since all the middle points lie on ED, the resultant acts at some point on ED. Also we have just seen that the pressures along ED are equal at two points such that the distance of one from $E =$ the distance of the other from D. Hence (case 4, p. 35) the resultant pressure acts at the *middle point*, M, of ED; that is,

$$EM = \tfrac{1}{2} ED. \quad . \quad . \quad . \quad . \quad . \quad (\beta)$$

If $\bar{p}$ is the perpendicular distance of M from the surface

$$\bar{p} = \tfrac{1}{2} h. \quad . \quad . \quad . \quad . \quad . \quad (\beta')$$

Also, whatever be the angle of inclination of the plane of the triangle to the vertical, the same point, M, is the centre of pressure.

(3) *To find the position of the centre of pressure on a plane triangle having a vertex in the free surface and its base horizontal.*

Let ACD be the triangle. Then a combination of the two results just proved will enable us to find Q, the centre of pressure. For, complete the parallelogram $ABDC$. Then the pressure on the parallelogram is the resultant of the pressures on the two triangles. But since all the narrow horizontal strips into which the given triangle ACD can be divided have their middle points (centres of pressure for each of them) ranged along AF, the resultant pressure on the triangle acts somewhere on AF. Join the point, M, of application of one of the two parallel forces to the point, T, of application of the resultant, and produce MT to meet AF in Q. Then Q is the point of application of the pressure on ACD.

Now $$\frac{QF}{EM} = \frac{FT}{TE} = \tfrac{1}{2}, \quad \therefore QF = \tfrac{1}{2} EM,$$

$$\therefore AQ = \tfrac{3}{4} AF. \quad \ldots \ldots \ldots (\gamma)$$

If h is the height of the triangle, and $\bar{p}$ the perpendicular distance of Q from the *free surface*,

$$\bar{p} = \tfrac{3}{4} h; \quad \ldots \ldots \ldots (\gamma')$$

and, as before, the point, Q, of application of the resultant pressure is the same whatever be the inclination of the plane of the triangle to the vertical.

The result might have been deduced directly from case 3, p. 34. For, if the area be divided into strips, we have $mt = \frac{x}{h} a$, where $a = CD$, and x is the perpendicular from A on mt. Hence the pressure on the strip mt is $\frac{a}{h} wx^2$, so that the pressures along AF are proportional to the squares of the distances of their points of application from A. The resultant, therefore, acts at a point $\frac{3}{4}$ of the way down along AF.

These three simple cases, combined with the principle (see p. 42) of regarding any column of liquid as an atmosphere, producing its resultant pressure at the centre of area, will suffice for calculations concerning the centres of pressure of many plane polygonal and other figures occupying any positions in a liquid.

Thus, let the area be nrm, Fig. 20, p. 52; and suppose that, if all the liquid above the horizontal plane CD is removed, we know the depth, $\bar{p}_0$, of the centre of pressure, I_0, of the remaining liquid below CD. Then, if $\bar{z}_0$ is the depth of G below CD, $A =$ magnitude of the area, $w =$ specific weight of the liquid, the pressure, F_0, at I_0 is

$$A\,\bar{z}_0\,w.$$

Let $x =$ the thickness of the column AD. Then the pressure due to this column $= Axw$, and it acts at G. The resultant pressure (at I) is of course the sum of these forces; and if $\bar{p}$ is the depth of I below AB, we have, by the theorem of moments,

$$\bar{p} = (\bar{p}_0 + x)\frac{\bar{z}_0}{\bar{z}_0 + x} + x, \quad \ldots \quad (\delta)$$

the point I dividing $I_0 G$ so that

$$\frac{II_0}{IG} = \frac{x}{\bar{z}_0}. \quad \ldots \quad \ldots \quad (\epsilon)$$

(4) *To find the position of the centre of pressure on a plane triangle occupying any position in a liquid.*

Let ABC, Fig. 23, be the triangle; let A be its area, and a, β, γ the depths of its vertices below the free surface of the liquid.

We shall calculate the distance of I, the centre of pressure,

58 Hydrostatics and Elementary Hydrokinetics.

from a side BC of the triangle. Let p be the perpendicular from A on BC.

If through C we draw a horizontal plane, the column of liquid above this plane produces a pressure equal to $A\gamma w$ at the centre of area of the triangle ABC, i.e., at the point whose distance from BC is $\tfrac{1}{3}p$. Hence
$$(A\gamma w,\ \tfrac{1}{3}p)$$
represent this force and the distance of its point of application from BC.

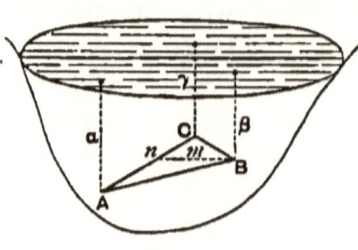

Fig. 23.

Consider now the effect of the liquid below this horizontal plane through C. From B draw the horizontal plane cutting the area ABC in the line Bn, and consider separately the pressure of this liquid on the areas BnC and BnA.

(a) If, for convenience, we let x and y be the perpendiculars from A and B on the horizontal plane through C, the area $BnC = A\dfrac{y}{x}$, and the perpendicular from n on BC is $p\dfrac{y}{x}$. Now the pressure of the liquid below C on BnC acts at a point three-fourths of the way down Cm, where m is the middle point of Bn, and the distance of this point from BC is $\tfrac{3}{8}p\dfrac{y}{x}$. Hence the pressure on BnC and the distance of its point of application from BC are represented by
$$\left(\tfrac{2}{3}A\dfrac{y^2}{x}w,\ \tfrac{3}{8}p\dfrac{y}{x}\right).$$

(b) Consider now the pressure on BnA due to the liquid below C. Treat the column above Bn as an atmosphere. It produces pressure at the centre of area of BnA; and we easily find that this force and the distance of its point of application from BC are represented by

$$\left(A\,\frac{x-y}{x}\,yw,\ \tfrac{1}{3}p\,\frac{x+y}{x}\right).$$

There is, finally, the pressure on BnA due to the liquid below Bn. This acts at the middle point of Am; so that for this force we have

$$\left(\tfrac{1}{3}A\,\frac{(x-y)^2}{x}\,w,\ p\,\frac{2x+y}{4x}\right).$$

The sum of these four forces is, of course,

$$\tfrac{1}{3}A(a+\beta+\gamma)w\,;$$

and if the perpendicular from I on BC is denoted by $\bar{p}$, we have, by moments (Art. 10) with reference to BC,

$$\tfrac{1}{3}(a+\beta+\gamma)\bar{p}$$
$$=p\left\{\tfrac{1}{3}\gamma+\frac{y^3}{4x^2}+\frac{y(x^2-y^2)}{3x^2}+\frac{(x-y)^2(2x+y)}{12x^2}\right\}$$
$$=p\left(\tfrac{1}{3}\gamma+\frac{2x+y}{12}\right)$$
$$=\tfrac{1}{12}p(2a+\beta+\gamma),$$

since $x=a-\gamma$, $y=\beta-\gamma$. Hence

$$\bar{p}=\tfrac{1}{4}p\,\frac{2a+\beta+\gamma}{a+\beta+\gamma},$$

with similar values of $\bar{q}$ and $\bar{r}$, the distances of I from AC and AB.

Now this shows that I is the centre of gravity of three particles whose masses are proportional to

$$2a+\beta+\gamma,\quad a+2\beta+\gamma,\quad a+\beta+2\gamma,$$

placed at A, B, C, respectively. Or, at A we can imagine two particles (a, s); at B two particles (β, s); at C two particles (γ, s), where $s = a + \beta + \gamma$.

Hence the following simple construction for I:—

Find the centre of gravity, G', of three particles whose masses are proportional to a, β, γ, placed at A, B, C respectively; join G' to G, the centre of area of the triangle ABC, and on GG' take
$$GI = \tfrac{1}{4} GG'.$$

It is also easily seen that I is the centre of gravity of three particles whose masses are proportional to a, β, γ, placed at the middle points of the bisectors of the sides BC, CA, AB, drawn from the opposite vertices.

And it is evident that we have also the following construction for I: at the middle points of the sides of the triangle imagine particles whose masses are proportional to the depths of these points; then their centre of gravity is the centre of pressure of the triangle.

The preceding method of finding I has been given for the purpose of illustrating the principle (Art. 15) that any column of liquid above an area may be treated like an atmosphere. For the following much more elegant investigation the author is indebted to Mr. W. S. McCay, F. T. C. D.

Take any point, O, in the area ABC; let the perpendicular from O on the surface of the liquid be ζ, the perpendiculars from O on the sides BC, CA, AB being x, y, z; also denote the areas of the triangles BOC, COA, AOB by Δ_1, Δ_2, Δ_3, respectively. Then evidently the line CO, if produced, divides the side AB into segments proportional to Δ_1 and Δ_2 inversely, while BO divides AC inversely as Δ_1 to Δ_3; hence O is the centre of mean position of the points A, B, C for the system of multiples Δ_1, Δ_2, Δ_3. If, then, Δ is the area ABC, we have

$$\Delta \cdot \zeta = \Delta_1 \cdot a + \Delta_2 \cdot \beta + \Delta_3 \cdot \gamma; \quad \cdots \quad (1)$$

or, if p, q, r are the perpendiculars from A, B, C on the opposite sides,

$$\zeta = \frac{\alpha}{p} \cdot x + \frac{\beta}{q} \cdot y + \frac{\gamma}{r} \cdot z. \quad \ldots \quad (2)$$

Now the intensity of pressure at O is $\zeta \cdot w$, so that this can be considered as a superposition of intensities

$$\frac{\alpha}{p} wx, \quad \frac{\beta}{q} wy, \quad \frac{\gamma}{r} wz,$$

i.e., of the intensities of pressure which would be produced if the sides BC, CA, and AB were placed in the surfaces of liquids whose specific weights are $\frac{\alpha}{p} w$, $\frac{\beta}{q} w$, and $\frac{\gamma}{r} w$, respectively. Now the first of these liquids would produce a resultant pressure equal to

$$\triangle \times \frac{p}{3} \times \frac{\alpha}{p} w, \quad \text{i.e., } \tfrac{1}{3} \triangle \alpha w,$$

acting at the middle point of the bisector of BC drawn from A (p. 55). Similarly for the other two liquids; so that the actual pressure on the triangle ABC is the resultant of forces

$$\tfrac{1}{3} \triangle \alpha w, \quad \tfrac{1}{3} \triangle \beta w, \quad \tfrac{1}{3} \triangle \gamma w,$$

acting at the middle points of the bisectors; and its point of application is, of course, the same as the centre of gravity of particles whose masses are proportional to α, β, γ placed at these middle points; and this is the second of the constructions which we have given above for the point I.

For simplicity and elegance this proof of our construction leaves nothing to be desired.

EXAMPLES.

1. A triangular area whose height is 12 feet has its base horizontal and vertex uppermost in water; find the depth to which its vertex must be sunk so that the difference of level between the centre of area and the centre of pressure shall be 8 inches.

Ans. Four feet.

2. Find the depth of the centre of pressure on a trapezium having one of the parallel sides in the surface of the liquid.

Ans. If the side a is in the surface and b below, h being the height of the trapezium, the depth of the centre of pressure is
$$\frac{h}{2}\frac{3b+a}{2b+a};$$
and it lies, of course, on the line joining the middle points of a and b.

3. In the last example find the position of the centre of pressure by geometrical construction.

(Break up the area into a parallelogram and a triangle, or two triangles.)

4. The plane of a trapezium being vertical, and its parallel sides horizontal, to what depth must the upper side be sunk in a liquid so that the centre of pressure shall be at the middle point of the area?

Ans. The parallel sides being a and b, of which the upper, a, must be the greater, the required depth $= \dfrac{b}{a-b}h$, where $h=$ height of the trapezium.

5. Show how to find, by geometrical construction, the position of the centre of pressure on a plane quadrilateral occupying any position.

6. Find the depth of the centre of pressure of a plane quadrilateral in terms of the depths of its vertices and the depth of the point of intersection of the diagonals.

Ans. If $a, \beta, \gamma, \delta, \zeta$ are the depth of the vertices and intersection of diagonals, and if $s = a+\beta+\gamma+\delta$, and Σa^2 denotes $a^2+\beta^2+\gamma^2+\delta^2$, the depth is

$$\frac{1}{4}\frac{s^2-2s\zeta+\Sigma a^2}{s-\zeta}.$$

7. A rectangular area of height h is immersed vertically in a liquid with one side in the surface; show how to draw a horizontal line across the area so that the centres of pressure of the parts of the area above and below this line of division shall be equally distant from it.

Ans. The line of division must be drawn at a depth

$$\frac{\sqrt{5}-1}{2}.h.$$

8. Supposing a rectangular vessel whose base is horizontal to be divided into two water-tight compartments by means of a rigid diaphragm moveable round a horizontal axis lying in the base of the vessel; if water is poured into the compartments to different heights, find the horizontal force which, applied to the middle point of the upper edge of the diaphragm, will keep this diaphragm vertical, and find the pressure on the axis.

Ans. Let a be the length of the axis, c the height of the vessel, h and h' the heights of the water in the compartments $(h > h')$; then the required force is $\dfrac{a}{6c}(h^3-h'^3)w$, and the pressure on the axis is $\frac{1}{2}a(h^2-h'^2)w - \dfrac{a}{6c}(h^3-h'^3)w$.

16. Lines of Resistance. Supposing Fig. 24 to represent a vertical transverse section, $ABCD$, of an embankment which is pressed by water on the side AB (assumed vertical), if we take any horizontal section, RQ, of the embankment and consider the equilibrium of the portion, $RQAD$, above this section, we see that it is acted upon by its weight and also by the water pressure which is a horizontal force acting at a point two-thirds of the way down AQ. Taking the resultant of these two forces, its line of action, pX, intersects the section RQ in a point X, the calculation of the position of which is of great importance in the construction of reservoirs.

As the section RQ varies in position, the point X describes a curve which is called a *line of resistance*. For the simple case in which the embankment is formed of homogeneous material and the transverse vertical section is a trapezium, we proceed to give a simple rule for tracing

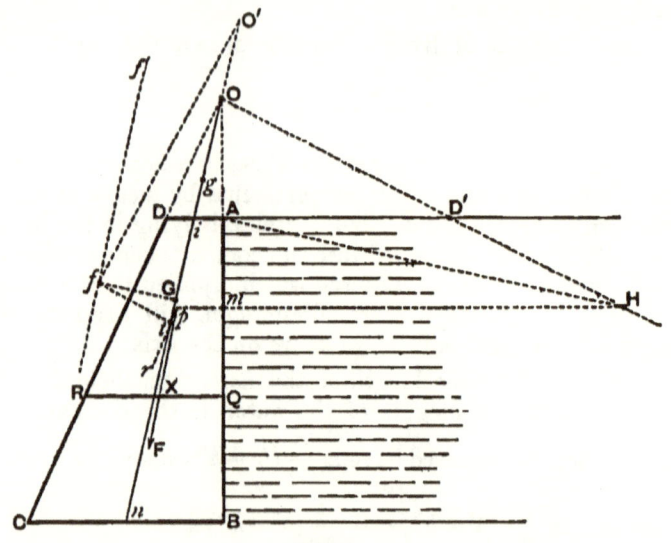

Fig. 24.

the positions of the point X for any number of horizontal sections.

To calculate the forces acting on the portion $RQAD$, produce RD and QA to meet in O, and consider its weight as the weight, W, of RQO acting through the centre of gravity, G, of this triangle accompanied by an upward vertical force, W_0, the weight of DAO acting at g, the centre of gravity of the triangle DAO.

Consider also the pressure against AQ as the pressure, P, of water against OQ, the level of the water reaching to

O—this force acting two-thirds of the way down OQ, and therefore through G—accompanied by the reversed pressure, P_0, of the water against OA, this force acting through g; and also the reversed water pressure against AQ due to a constant head OA, this force acting towards the right of the figure through m, the middle point of AQ.

Let $OQ = y$, $OA = y_0$, $w =$ specific weight of the water, $w' =$ specific weight of embankment, $m = \tan DOA$, $l =$ length (perpendicular to the plane of the paper) of the embankment.

Then

$W = \frac{1}{2} m l w' y^2$, $P = \frac{1}{2} l w y^2$, $W_0 = \frac{1}{2} m l w' y_0^2$, $P_0 = \frac{1}{2} l w y_0^2$.

If R is the resultant of W and P, R_0 the resultant of W_0 and P_0, we have

$$R = \frac{l}{2} \sqrt{w^2 + m^2 w'^2} \cdot y^2, \text{ acting at } G,$$

$$R_0 = \frac{l}{2} \sqrt{w^2 + m^2 w'^2} \cdot y_0^2, \text{ acting at } g,$$

and R is parallel to R_0, each making with the horizon an angle whose tangent is $\dfrac{w'}{w} \tan DOA$. To construct this direction, taking DA to represent w, produce DA to D' so that AD' represents w'; draw OD'; then R is perpendicular to OD', and is therefore of constant direction for all sections RQ.

The resultant of the two unlike parallel forces R at G and R_0 at g is part of the force acting on the portion $RQAD$. Suppose this resultant to act at the point t on the line gG (which, of course, passes through O and through the middle points, i, n, of AD and BC).

Then
$$\frac{tG}{tg} = \left(\frac{y_0}{y}\right)^2 = \left(\frac{gO}{GO}\right)^2,$$
$$\therefore \frac{tG}{Gg} = \frac{gO^2}{Gg \times GO},$$

gO being produced to O' so that $gO = OO'$. Hence
$$tG \cdot GO' = gO^2, \quad \ldots \ldots \quad (a)$$
and this shows that t can be found by drawing a line ff' parallel to On at a distance equal to gO, erecting a perpendicular at G to On meeting ff' in f, and then drawing ft perpendicular to fO'. Hence $R - R_0$ acts in the line tr which is perpendicular to OD'.

With this force must finally be coupled the horizontal force $lw \cdot OA \cdot AQ$ acting at m in the sense mH. Denote this force by H. Now $R - R_0 : H = OH : AO$; for
$$R - R_0 = \tfrac{1}{2} l \sqrt{w^2 + m^2 w'^2}\,(y^2 - y_0^2)$$
$$= \tfrac{1}{2} lw \sec \beta \cdot (y^2 - y_0^2), \text{ where } \beta = \angle AOD',$$
$$= lw \cdot AQ \cdot Om \sec \beta$$
$$= lw \cdot AQ \cdot OH.$$

Hence in the triangle OAH the sides AO and OH are perpendicular and proportional to the forces H and $R - R_0$; therefore the resultant, F, of these forces is perpendicular and proportional to the side AH; and therefore if we produce the lines rt and Hm to meet in p, the line of action of F, the resultant force to which the portion $RQAD$ of the embankment is subject, is the perpendicular from p on AH.

Moreover $F = lw \cdot AQ \cdot AH$, and therefore *this force is equivalent to the weight of a column of water having AQ for a horizontal base and having a height AH.*

As the section RQ varies, the locus of X—i. e. the line of resistance—is a hyperbola, as is easily proved thus. Take the horizontal line through O as axis of x, and OB as axis of y, and let $QX = x$. Now express the fact that the sum of the moments of the water pressure against AQ (acting two-thirds of the way down AQ) and the weight of the portion $RQAD$ of the embankment about X is zero.

The latter force may be resolved into the weight of RQO acting downwards at G and the (negative) weight of DAO acting upwards at g. Our equation is then

$$\tfrac{1}{6} lw (y-y_0)^3 - \tfrac{1}{2} lw'my^2 (x - \tfrac{1}{3} my) + \tfrac{1}{2} lw'my_0^2 (x - \tfrac{1}{3} my_0) = 0.$$

The left-hand side contains $y - y_0$ as a factor. Expelling this, and denoting $\tan \beta$ by t,

$$(y-y_0)^2 - 3tx(y+y_0) + mt(y^2 + y_0 y + y_0^2) = 0 \quad . \quad (\beta)$$

is the equation of the locus, which is therefore a hyperbola.

This hyperbola passes through i, the middle point of AD, and it has for one asymptote the line $y + y_0 = 0$, i.e. a horizontal line at a distance OA above O. The equation of the other asymptote is

$$3tx - (1+mt)y + 3y_0 = 0, \quad . \quad . \quad . \quad (1)$$

and this line can be easily drawn by means of its intercepts on the axes of x and y.

The intercept on the axis of x is $-\dfrac{y_0}{t}$, or $-\dfrac{AO^2}{AD}$, so that if (Fig. 25) OL is drawn perpendicular to OD', this intercept is $-AL$; therefore if AJ is drawn parallel to OL meeting Ox in J, the asymptote passes through J. The intercept on the axis of y is $\dfrac{3y_0}{1+mt}$; but

$$\dfrac{1}{1+mt} = \dfrac{OA^2}{OA^2 + D'A \cdot AD};$$

now if a semicircle is described on DD' as diameter, it will

68   *Hydrostatics and Elementary Hydrokinetics.*

cut AO at h so that $L'A \cdot AD = Ah^2$, and if AK is taken $= Ah$,

$$\frac{1}{1+mt} = \frac{OA^2}{OK^2} = \cos^2 AOK;$$

hence by drawing As perpendicular to OK and sk perpen-

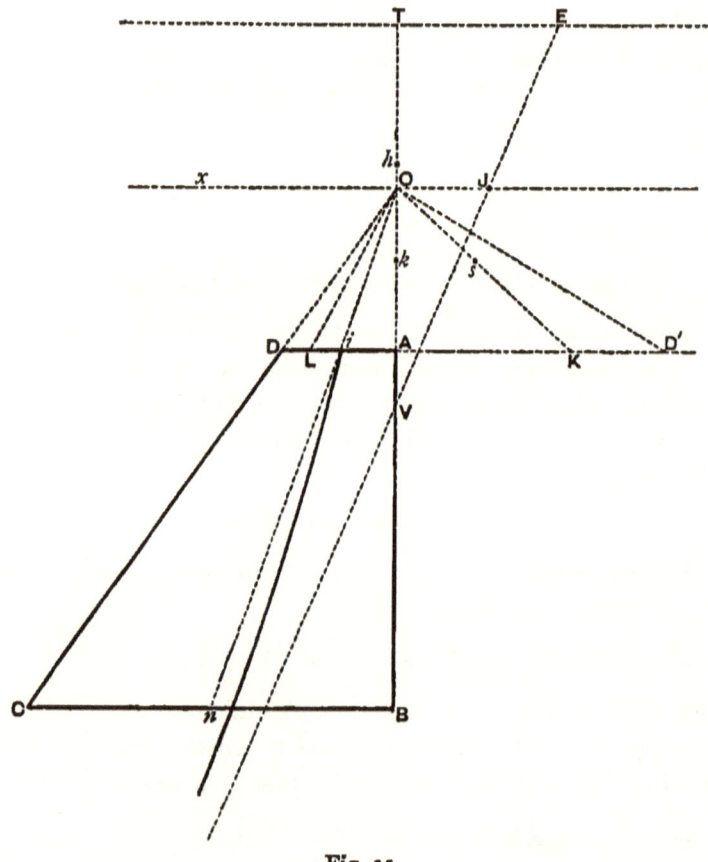

Fig. 25.

dicular to OA, the intercept on OB is $3\,Ok$, or OV; the asymptote is therefore the line VJ. The point V may also be found by producing OK through K to S, so that $OS =$

Liquid Pressure on Plane Surfaces. 69

3 Os, and drawing SV horizontal. Taking the vertical length $OT = OA$, the other asymptote is the horizontal line TE, meeting VJ in E, which is the centre of the hyperbola.

Hence we have both asymptotes and one point, i, of the curve, from which data the curve is easily constructed

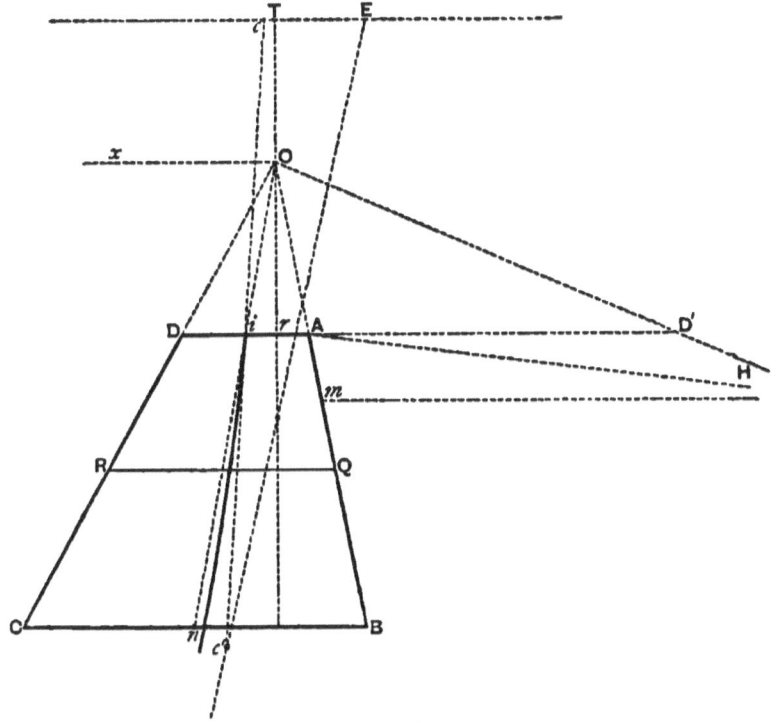

Fig. 26.

by means of the property that if any line cuts the hyperbola and the asymptotes the intercepts between the curve and the asymptotes are equal. The relevant portion of the curve is represented by the full line passing through i.

If the face AB of the embankment is not vertical, the line of resistance is still a hyperbola.

70　Hydrostatics and Elementary Hydrokinetics.

As before, take AD to represent w, and AD' to represent w'; from O draw Or perpendicular to AD; take Or as axis of y, and the horizontal line Ox as axis of x; let $Or = y_0$, $m = \tan DOr$, $n = \tan AOr$, $t = \tan D'Or$; then, by the method of moments before used, the equation of the locus (deprived of the factor $y - y_0$) is

$$3x\{ty + (t-2n)y_0\} - \{(m-n)t - mn - n^2 + 1\} \times (y^2 + y_0 y + y_0^2) - 3(n^2 - 1)y_0 y - 3n^2 y_0^2 = 0. \quad (\gamma)$$

The curve passes, as before, through the point i; the equation of one asymptote, TE, found by equating to zero the coefficient of the highest power of x, is

$$y = -\frac{t-2n}{t} \cdot y_0, \quad \ldots \quad (2)$$

so that the asymptote is found by producing rA through A to A' so that $rA = AA'$, then drawing a parallel $A'O'$ to $D'O$, and measuring $OT = OO'$.

The equation of a line through O parallel to the tangent to the curve at i is

$$4(t-n)x = \{(m-n)t - 2mn\}y. \quad \ldots \quad (3)$$

In any actual case all the constants, m, n, t, y_0, will be given, and therefore the lines (2) and (3) can be accurately drawn. A construction, independent of the numerical values of the constants, can be given for the line (2), and therefore for the tangent at i. Let a circle round the triangle $D'Oi$ cut Or in k', and let a circle round AOD cut Or in k; then, Ok' being supposed $> Ok$, measure off a length kk' along Ox towards the left-hand, and along Or measure off a length $2AD'$; the diagonal (through O) of the rectangle determined by these lengths is parallel to the tangent at i.

A point on the other asymptote can now be easily found; for, let the tangent at i cut the asymptote TE in c; then take, along this tangent, $ic' = ic$, and we have the point c' on the second asymptote.

The direction of this asymptote is that of the line
$$3tx - \{(m-n)t - mn - n^2 + 1\}y = 0, \quad . \quad . \quad . \quad (4)$$
as is obvious from the terms of the second degree in (γ). Hence when the constants are numerically assigned, the direction of this asymptote is easily constructed, and therefore (since it passes through c') the asymptote itself can be drawn.

The circles above described may be utilised for drawing this asymptote. For, if a perpendicular to OA at A meets Or in l, we see that (4) is equivalent to
$$\frac{x}{y} = \frac{y_0 + 2rk' - rk - rl}{3 \cdot rD'},$$
so that if we measure the length $\frac{1}{3}(y_0 + 2rk' - rk - rl)$ from O along Or and the length rD' along xO produced through O, and draw the diagonal, through O, of the rectangle determined by these lines, the required asymptote is perpendicular to this diagonal.

The construction of the curve then proceeds exactly as in the first case.

The resultant force to which any portion, $RQAD$, of the embankment, cut off by a horizontal plane RQ is subject is found by exactly the same construction as before—viz. bisect AQ in m; draw mH horizontal, to meet OD' in H; then the resultant force is perpendicular to AH and is equal to the weight of a column of water having AH for height and for base the vertical projection of AQ.

If we adopt the method of fictitiously completing the embankment and raising the level of the water to O, as in the first case, we shall have the forces R and R_0 acting at points G' and g' which both lie on a fixed line passing through O; g' is obtained by taking the intersection of the vertical through g with a perpendicular to OA through a point two-thirds of OA from O. The resultant, $R-R_0$, of R and R_0 acts at a point, t', on the line $Og'G'$ which is found exactly as t was before found. The forces R and R_0 are each perpendicular to OD'; and, omitting l for simplicity,

$$R = \tfrac{1}{2} wy^2 \sec\beta, \quad R_0 = \tfrac{1}{2} wy_0^2 \sec\beta, \text{ where } \beta = \angle rOD'.$$

Hence the first method of finding the line of resistance, by tracing out the locus of X (Fig. 24), applies here also.

Let us now consider the case in which a section of the embankment consists of two trapeziums, $ABCD$ and $BEFC$, Fig. 27, the level of the water being DA.

We may suppose FE to represent any horizontal section across the second trapezium, the distance between EF and BC being y; and we shall calculate the resultant force, F_2, acting on $BEFC$ by first supposing CB to be the level of the water, and then taking account of the weight of $ABCD$ and the effect of the column of water between A and B.

If CB is the water level and everything above is neglected, the line of resistance through $BEFC$ is a hyperbola starting from the middle point of BC, and the point b in which the resultant force, f_2, on $BEFC$ cuts EF is easily found. This force f_2 is found in line of action as before explained—that is, by producing EB and FC to meet in O', taking $BC':BC = w':w$, and drawing a horizontal line, nH', through the middle point, n, of BE to meet $O'C'$ in H'; then
$$f_2 = wy \cdot BH'. \quad \ldots \ldots \quad (5)$$

Now the effect of the portion $ABCD$ and the water column between A and B is twofold: firstly, the water column produces on BE a pressure, p, acting normally at n, and
$$p = wh_1 \cdot BE, \quad \ldots \ldots \quad (6)$$
or
$$= wy \cdot BI, \quad \ldots \ldots \quad (7)$$
where h_1 is the height of the upper trapezium and I is the

Liquid Pressure on Plane Surfaces. 73

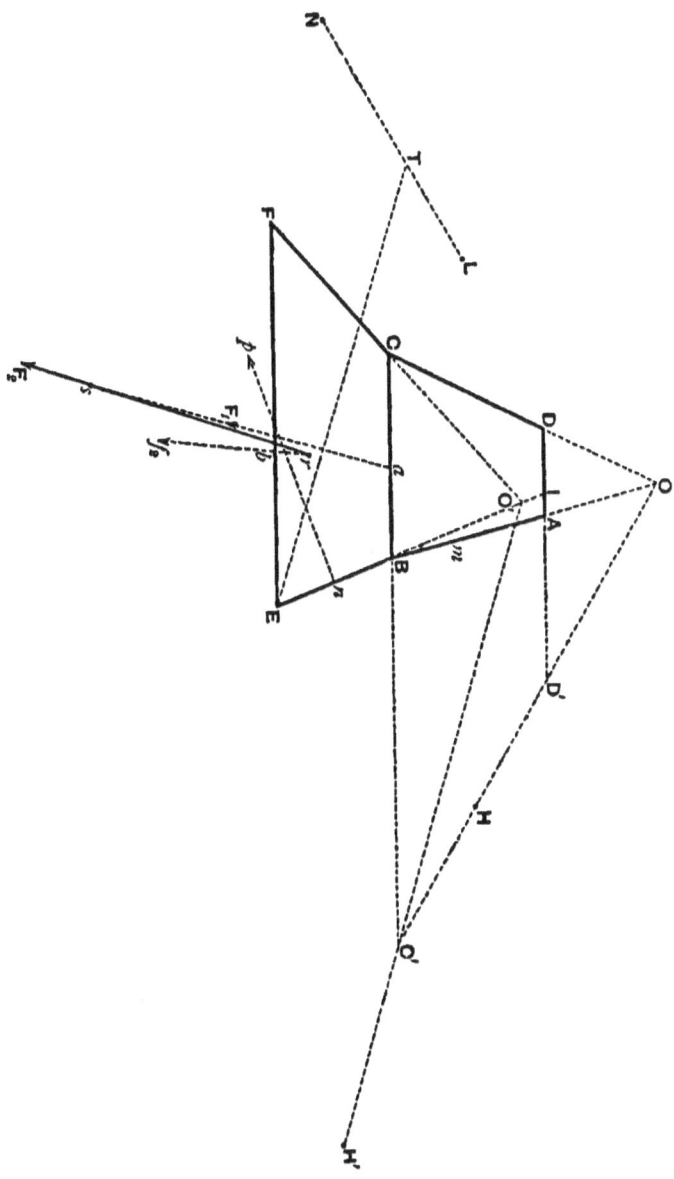

Fig. 27.

point in which EB cuts AD; secondly, the upper structure and water column will produce a force, F_1, acting at the point a in BC, this point being that in which BC is intersected by the hyperbola before described. If m is the middle point of AB, and if, as before, $AD:AD' = w:w'$, and the horizontal line mH cuts OD' in H, we have

$$F_1 = wh_1 . AH. \quad \ldots \quad \ldots \quad (8)$$

Hence we have only to find the resultant of the forces f_2, p, and F_1. This resultant will pass through the point of intersection of F_1 and the resultant of f_2, p; and also through the point of intersection of f_2 and the resultant of F_1, p.

Now taking the value (7) of p, since

$p = wy . BI$, and p is perpendicular to BI,

$f_2 = wy . BH'$, and f_2 is perpendicular to BH',

it follows that if the resultant of p and f_2 is denoted by (p, f_2),

$$(p, f_2) = wy . HH', \text{ and is perpendicular to } HH', \quad (9)$$

and this resultant intersects F_1 in s.

Again, $\quad p = wh_1 . BE$,

$\quad F_1 = wh_1 . AH$,

so that if BL is drawn equal and parallel to HA, we have for the resultant (p, F_1)

$$(p, F_1) = wh_1 . EL \text{ and is perpendicular to } EL, \quad (10)$$

and this resultant intersects f_2 in r.

Hence the line of action, rs, of F_2 is known.

If EN is drawn parallel and equal to $H'B$, we have

$$(p, F_1) = wh_1 . EL, \quad \ldots \quad \ldots \quad (11)$$

$$f_2 = wy . EN, \quad \ldots \quad \ldots \quad (12)$$

hence if T is the point dividing NL so that
$$NT : TL = h_1 : y,$$
the resultant of the forces (11), (12) is given by the equation
$$F_2 = w(h_1 + y) \cdot ET \quad \ldots \quad (13)$$
(see *Statics*, Vol. I, Art. 23). The line ET is, of course, perpendicular to rs, so that the point T may be found by drawing this perpendicular.

Observe that $h_1 + y$ is the depth of the line EF below the surface, DA, of the water, so that we have the same rule for the resultant force on the section $BEFC$ as that for the force on the upper section—viz. it is the weight of a water column having for base the vertical projection of IE and height ET.

If below EF there is another section of the embankment in the form of a trapezium, the force F_2 and the depth of E below AD play the same part in the calculation of the resultant force on this lower section as that which was played by F_1 and the depth of B in the calculation just given; so that this process can be employed for the complete construction of the line of resistance through any embankment the section of which can be broken up into successive trapeziums.

In Masonry Dams for reservoirs the vertical section of the upper portion, $ABCD$, Fig. 24, has often the simple form of a rectangle. If in this case the level of the water reaches to AD, the top of the dam, the line of resistance is a parabola whose equation referred to the vertical line in as axis of y and the horizontal iD as axis of x is
$$y^2 = 6 AD' \cdot x,$$
so that in is the tangent at the vertex, and the directrix is a vertical line to the right of i at a distance from i equal to $\frac{3}{2} AD'$. The line $D'H$ (Fig. 24) being the vertical through

D', the resultant force on any section RQ is perpendicular and proportional to AH, where H is the point (as in the figure) in which the horizontal line mH through m meets the vertical line through D'; the resultant force

$$= w \cdot AQ \cdot AH.$$

If the level of the water does not reach to the top of the dam, let the top line of the dam be $A_0 D_0$; then the force on any section RQ is found by tracing the parabola and combining the force $w \cdot AQ \cdot AH$ perpendicular to AH with the weight of the portion $ADD_0 A_0$. The result is this: from A_0 draw $A_0 Z$ parallel to QD' and meeting AD in Z; then the resultant force is perpendicular to ZH and is equal to $w \cdot AQ \cdot ZH$.

(For simplicity l has been omitted in the calculations from p. 72. This omission amounts to considering the forces R, R_0, p, &c. as those exerted per unit length of the embankment.)

CHAPTER IV.

GENERAL EQUATIONS OF PRESSURE.

(This Chapter may be omitted on first reading.)

17. Equation of Equilibrium of a Fluid under Gravity. If in the case of a fluid acted upon solely by gravity we imagine the density not to be the same at all points, the expression (a), p. 38, for the intensity of pressure will no longer hold. For in Fig. 14, p. 37, the weight of the cylindrical column PN will not be wzs, since w varies from point to point of its depth. But if w is the specific weight at any depth z, the weight of this cylinder is $s \int w \, dz$, the limits of z being 0 and NP; and, as before, this weight must be equal to the upward pressure on the base at P, viz. $p \cdot s$. Hence

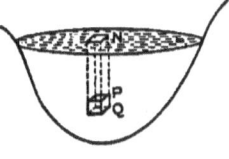

Fig. 28.

$$p = \int w \, dz$$

$$\therefore \frac{dp}{dz} = w. \quad \ldots \ldots \quad (1)$$

If, for example, the density varies directly as the depth, we have $w = kz$, and (1) gives

$$p = \tfrac{1}{2} kz^2.$$

Equation (1) could have been obtained by considering the equilibrium of a very small rectangular parallelopiped,

78 *Hydrostatics and Elementary Hydrokinetics.*

PQ, Fig. 28, described with vertical and horizontal sides at P. For, if $s =$ area of each horizontal face, and if Q is a point vertically below P so that $PQ = dz$, the weight of the element is $w \cdot s dz$, where w is the weight per unit volume of the fluid at P. Also the downward pressure on the horizontal face at P is $p \cdot s$, where p is the pressure intensity at P; and since the pressure intensity at Q is $p + \dfrac{dp}{dz} dz$, the upward pressure on the horizontal face at Q is $\left(p + \dfrac{dp}{dz} dz\right) s$. Considering the separate equilibrium of this elementary parallelopiped, and resolving forces vertically, we have

$$\left(p + \frac{dp}{dz} dz\right) s = p \cdot s + w \cdot s dz,$$

$$\therefore \quad \frac{dp}{dz} = w,$$

as before.

If we are measuring force in pounds' weight and length in inches, p will be in pounds' weight per square inch, z being the depth of the point in inches, and w the weight per cubic inch of the liquid at P in pounds' weight—in other words, w is the number of pounds mass of the liquid per cubic inch at P.

If force is measured in poundals, the weight per cubic inch of the liquid at P is about $32 \cdot 2\, w$, where w is still the number of pounds mass per cubic inch at P.

It is usual to denote the number of units of mass per unit volume by ρ. If then force is measured as a multiple of the weight of the unit mass, the equation for p is

$$\frac{dp}{dz} = \rho. \qquad \ldots \ldots \quad (2)$$

General Equations of Pressure.

But if force is measured in absolute units, the number of these in the weight of a unit mass being g (i. e. 32·2 poundals or 981 dynes, according as the "British Absolute" or the C. G. S. system is used), w in (1) is ρg, and the equation for p is

$$\frac{dp}{dz} = \rho g. \quad \ldots \ldots \ldots (3)$$

With this form of the equation, and the C. G. S. system of units, the student must observe that—

p is in dynes per square centimètre,
z ,, linear centimètres,
ρ ,, grammes per cubic centimètre,
g ,, centimètres per second per second (about 981).

If the fluid is of constant density, (1) gives $p = wz$, the result which in the previous chapters we have employed in the case of water.

In the case of a gas ρ, or w, is proportional to p: in a subsequent chapter we shall prove that

$$p = 2926 \cdot 9 \frac{T}{s} \rho, \quad \ldots \ldots (4)$$

where p is the intensity of pressure in grammes' weight per square centimètre, T is the absolute temperature of the gas on the Centigrade scale, s is the specific gravity of the gas referred to air, and ρ is the mass of the gas in grammes per cubic centimètre.

Using equation (2), and denoting $2926 \cdot 9 \frac{T}{s}$ by k, we have

$$\frac{dp}{p} = \frac{dz}{k}, \quad \ldots \ldots \ldots (5)$$

z being measured vertically downwards. If z is measured vertically upwards, we have

$$\frac{dp}{p} = -\frac{dz}{k}. \quad \ldots \ldots \ldots (6)$$

Integrate this, assuming T constant throughout the gas, and suppose that when $z = 0$ the value of p is p_0; then
$$p = p_0 e^{-\frac{z}{k}}. \quad \ldots \quad \ldots \quad (7)$$

This gives the intensity of pressure at a height of z centimètres in the atmosphere, on the assumption of constant temperature, s being unity, and p_0 being the intensity of pressure at the ground.

Suppose any gas contained in a pipe, vertical or not, closed at the upper end. Let O be the point at which $z = 0$ and p is p_0; let P be any point above O; let the pipe at O be open to the air, so that p_0 is produced by the atmosphere in contact with the gas at O. At the point P in the pipe the intensity of pressure of the enclosed gas is given by (7), and at P just outside the pipe the intensity of the pressure, p_1, of the air is given by the equation

$$p_1 = p_0 e^{-\frac{z}{k_1}}, \quad \ldots \quad \ldots \quad (8)$$

where $k_1 = 2926 \cdot 9 \, T$, the gas and the air being assumed to be at the same temperature.

Now if the gas is lighter than air—suppose hydrogen or coal gas—k is $> k_1$, and therefore

$$p > p_1,$$

i.e. the gas would escape into the surrounding air if an aperture were opened in the pipe at P. At O the gas does not rush out of the pipe, although a communication is established with the air; the gas at O would *diffuse* into the air, but we may suppose that the pipe at O contains a piston which restrains the gas and on the top of which the atmosphere presses.

The higher the point P in the pipe, the greater the ratio of p to p_1, and therefore the more rapid the escape of the gas when a communication is opened. Hence the gas

General Equations of Pressure.

lights at the top of a house are, if the taps are opened to the same extent, brighter than those at the bottom of the house; and, in consequence of this, it is commonly said that 'the pressure of the gas at the top of the house is greater than that at the bottom'—a thing which could not possibly be true, since gravitation must diminish the pressure as the height increases. It is not the pressure of the gas that is greater at the top but the velocity of its escape.

When a balloon ascends, the neck is, for safety, left open to the air, so that the intensity of pressure of the gas at the neck is that of the atmosphere at this point; the gas does not rush out at the neck; but if a valve is opened at the top of the balloon, the gas will escape for the reason already given—viz. that the intensity of pressure of the enclosed gas at this point is greater than that of the adjacent air.

If the gas in the pipes were heavier than air, p would be $< p_1$, and the reverse of the above would be true.

When density is measured in pounds per cubic foot, intensity of pressure in pounds' weight per square foot, and T is $460 + t$, the absolute temperature on the Fahrenheit scale,

$$p = 53 \cdot 30222 \frac{T}{s} \rho, \quad \ldots \quad (9)$$

and at a height of z feet in the column of gas we have (7), in which k has the value given in (9).

Since T will usually be a large number, if z does not exceed one or two hundred feet, we may take $e^{-\frac{z}{k}} = 1 - \frac{z}{k}$, and we have

$$p - p_1 = p_0 \frac{z(1-s)}{53 \cdot 30222 \times T}, \quad \ldots \quad (10)$$

for the excess of gas pressure in the pipes over that of the

G

82 *Hydrostatics and Elementary Hydrokinetics.*

outside air at a height of z feet; and in this equation the pressures may be estimated in any units whatever.

18. General Equations of Equilibrium. If the forces acting on the fluid are any assigned system, let the force per unit mass at P have for components parallel to any three rectangular axes the values X, Y, Z, so that on an element of mass dm these forces will be Xdm, &c. At P draw a small rectangular parallelopiped, with edges Pa, Pb, Pc, or dx, dy, dz, parallel to the co-ordinate axes. Then, if ρ is the density of the fluid at P, the mass contained in this parallelopiped is $\rho\, dx\, dy\, dz$.

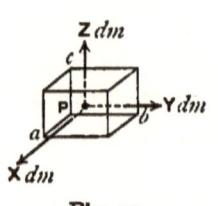

Fig. 29.

Consider the separate equilibrium of this fluid. If p is the pressure intensity at P, the pressure on the face bPc is $p \cdot dy\, dz$, and since the pressure intensity on the opposite face is $p + \dfrac{dp}{dx} \cdot dx$, the pressure on the face is

$$\left(p + \frac{dp}{dx} dx\right) dy\, dz.$$

For the equilibrium of the element, equate to zero the component of force acting on it parallel to the axis of x, and we have

$$X dm + p \cdot dy\, dz - \left(p + \frac{dp}{dx} dx\right) dy\, dz = 0,$$

or $\dfrac{dp}{dx} = \rho X$, ... (1)

Similarly $\dfrac{dp}{dy} = \rho Y$, ... (2)

$\dfrac{dp}{dz} = \rho Z$, ... (3)

by resolving forces parallel to the other axes.

Each of these equations is a particular expression of the general result that *at every point in the fluid the line-rate of* variation of the intensity of pressure in any direction is equal to the component in this direction of the external force per unit volume of the fluid—a result which is easily seen thus.

Let PQ, Fig. 30, be an element, ds, of length of any curve through P; round this as an axis describe a cylinder of small uniform cross-section, σ; consider the separate equilibrium of the fluid contained within this cylindrical element of volume. If F is the external force per unit mass exerted on the fluid in the neighbourhood of P, the force on the enclosed fluid is $F.\rho\sigma ds$; if p is the pressure intensity at p and p' $\left(\text{which is } p + \dfrac{dp}{ds} ds\right)$, the pressure intensity at Q, the forces on the ends of the cylinder at P and Q are $p\sigma$ and $p'\sigma$. In addition to these there are side pressures which are all at right angles to the axis PQ.

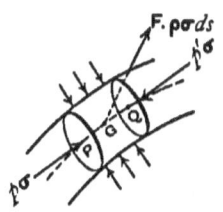

Fig. 30.

Resolving forces along PQ, we have
$$p\sigma - p'\sigma + F.\rho\sigma ds . \cos\theta = 0,$$
where θ is the angle between F and PQ; and this is the same as
$$\dfrac{dp}{ds} = \rho F \cos\theta. \quad \ldots \ldots \quad (4).$$

The equations (1), (2), (3) lead to a certain condition which the components X, Y, Z, of external force intensity at each point must satisfy in order that the equilibrium of the fluid may be possible.

For, we have
$$\left. \begin{array}{c} \dfrac{d}{dy}(\rho X) = \dfrac{d}{dx}(\rho Y); \quad \dfrac{d}{dz}(\rho Y) = \dfrac{d}{dy}(\rho Z); \\ \dfrac{d}{dx}(\rho Z) = \dfrac{d}{dz}(\rho X). \end{array} \right\} \quad . \quad (5)$$

84 *Hydrostatics and Elementary Hydrokinetics.*

Multiplying both sides of the first by Z, of the second by X, of the third by Y, and adding, we have

$$X\left(\frac{dY}{dz}-\frac{dZ}{dy}\right) + Y\left(\frac{dZ}{dx}-\frac{dX}{dz}\right) + Z\left(\frac{dX}{dy}-\frac{dY}{dx}\right) = 0, \quad (6)$$

which is the necessary condition of equilibrium.

This condition may be thus expressed: *for the equilibrium of any fluid under external bodily force it is necessary that at all points the resultant force and its curl should be at right angles.* (This term *curl* is due to Clerk Maxwell.) Regarded analytically, (6) expresses the fact that the expression $Xdx + Ydy + Zdz$, if not already a perfect differential, must be capable of being rendered so by means of a factor, the factor being, as we see, the density of the fluid at each point at which the expression $Xdx + \ldots$ is taken.

Since at each point p is some function of x, y, z, we have

$$dp = \frac{dp}{dx}dx + \frac{dp}{dy}dy + \frac{dp}{dz}dz.$$

Hence from (1), (2), (3),

$$dp = \rho(Xdx + Ydy + Zdz), \quad \ldots \quad (a)$$

from which p is known by integration.

So far, we have assumed only that the body is a perfect fluid, so that our equations all hold for compressible fluids as well as for liquids.

If the fluid is a gas, $p = k\rho$, and (a) becomes

$$\frac{dp}{p} = \frac{1}{k}(Xdx + Ydy + Zdz). \quad \ldots \quad (7)$$

Now in all cases in which equilibrium is possible the expression

$$\rho(Xdx + Ydy + Zdz)$$

must be a perfect differential—since it $= dp$ (see p. 12); and (6) expresses the condition that this should be the case; but in many cases $Xdx + Ydy + Zdz$ is a perfect differential, i.e. the external forces have a Potential.

Assuming that the forces have a potential, V, (a) becomes

$$dp = \rho dV. \qquad (8)$$

If the fluid is a homogeneous liquid, (1), (2), (3), give

$$\nabla^2 p = \rho \nabla^2 V, \qquad (9)$$

where $\nabla^2 \equiv \dfrac{d^2}{dx^2} + \dfrac{d^2}{dy^2} + \dfrac{d^2}{dz^2}$.

In all cases in which the external forces have a potential, their *level surfaces*, or equipotential surfaces (*Statics*, vol. ii, Art. 327) are also surfaces of equal pressure of the fluid; for (8) shows that in passing from a point P to another close point such that $dV = 0$, we have $dp = 0$. If the density of the fluid is variable, it will be constant all over a level surface of the external forces; for, since in (8) the left side is a perfect differential of a function of x, y, z, the right side must be so, and this requires that ρ is some function of V, i.e.

$$\rho = f(V), \qquad (10)$$

so that at all points for which V is constant ρ is also constant.

For a slightly compressible fluid, whose resilience of volume is k (Art. 8), equation (8) becomes

$$k \frac{d\rho}{\rho^2} = dV, \qquad (11)$$

and for a gas, since $p = \lambda \rho$, where λ is a constant,

$$\lambda \frac{d\rho}{\rho} = dV. \qquad (12)$$

If the temperature of the gas varies, $p = c\rho(1 + at)$, where c is a constant, t is the temperature at any point in the gas, and a is a constant. (This will be explained more fully in a subsequent chapter.) Hence, in general,

$$\frac{dp}{p} = \frac{X\,dx + Y\,dy + Z\,dz}{c\,(1+at)}, \quad \ldots \quad (13)$$

and if the applied forces have a potential, V,

$$\frac{dp}{p} = \frac{dV}{c\,(1+at)}, \quad \ldots \quad \ldots \quad (14)$$

so that, since the left-hand side is a perfect differential and the right side must also be one, it is necessary that t should be some function of V; in other words, t is constant along each equipotential surface of the external forces. Hence for a gas subject to any conservative system of forces (i. e. forces having a potential) each level surface of the forces is at once a surface of constant pressure intensity and a surface of constant temperature.

19. Non-conservative Forces. If $X\,dx + Y\,dy + Z\,dz$ is not a perfect differential, and if at any point, P, in the fluid we describe the surface of constant pressure, whose equation is

$$p = \text{const.}, \quad \ldots \quad \ldots \quad \ldots \quad (1)$$

this surface will not coincide with the surface drawn through P along which the density is constant, i. e. the surface whose equation is

$$\rho = \text{const.} \quad \ldots \quad \ldots \quad \ldots \quad (2)$$

These two surfaces will intersect in some curve, which is called the curve of constant pressure and constant density at the point P.

We propose to find the direction of this curve at P.

Let l, m, n be the direction-cosines of the tangent to the curve at P; then if ds is the indefinitely small element of its length between P and a neighbouring point Q; and if $p = f(x, y, z)$ at P, the value of p at Q is

$$f(x + l\,ds,\ y + m\,ds,\ z + n\,ds),$$

i.e. $p + \left(l\dfrac{dp}{dx} + m\dfrac{dp}{dy} + n\dfrac{dp}{dz}\right)ds$.

Hence, since there is no change in the value of p,

$$l\frac{dp}{dx} + m\frac{dp}{dy} + n\frac{dp}{dz} = 0, \quad \ldots \quad (3)$$

or, by the general equations of equilibrium,

$$lX + mY + nZ = 0, \quad \ldots \quad (4)$$

so that PQ is at right angles to the direction of the resultant force at P.

Similarly, since there is no change in ρ from P to Q, we have

$$l\frac{d\rho}{dx} + m\frac{d\rho}{dy} + n\frac{d\rho}{dz} = 0; \quad \ldots \quad (5)$$

and therefore from (4) and (5) we have

$$l:m:n = Y\frac{d\rho}{dz} - Z\frac{d\rho}{dy} : Z\frac{d\rho}{dx} - X\frac{d\rho}{dz} : X\frac{d\rho}{dy} - Y\frac{d\rho}{dx}. \quad (6)$$

Now, denoting by λ, μ, ν the components of the curl of the force, i.e.

$$\left.\begin{aligned}\lambda &= \frac{dZ}{dy} - \frac{dY}{dz}, \\ \mu &= \frac{dX}{dz} - \frac{dZ}{dx}, \\ \nu &= \frac{dY}{dx} - \frac{dX}{dy},\end{aligned}\right\} \quad \ldots \quad (a)$$

equations (5) of Art. 18 are

$$Z\frac{d\rho}{dy} - Y\frac{d\rho}{dz} + \rho\lambda = 0, \quad \ldots \quad (7)$$

$$X\frac{d\rho}{dz} - Z\frac{d\rho}{dx} + \rho\mu = 0, \quad \ldots \quad (8)$$

$$Y\frac{d\rho}{dx} - X\frac{d\rho}{dy} + \rho\nu = 0. \quad \ldots \quad (9)$$

88 Hydrostatics and Elementary Hydrokinetics.

These last show that (6) become
$$l:m:n = \lambda:\mu:\nu, \quad \ldots \ldots \quad (10)$$
and the differential equations of the curve are
$$\frac{dx}{\lambda} = \frac{dy}{\mu} = \frac{dz}{\nu}, \quad \ldots \ldots \quad (11)$$
from which, by integration, the equations of these curves are found.

Hence the direction of any such curve at any point coincides with the direction of the curl of the external force.

If the fluid is a gas whose temperature varies from point to point, we have $p = c\rho(1+at)$, where t is the temperature at P, and c and a are constants. Now the previous result is absolutely general, whatever be the connection between p and ρ; and if p and ρ are both constant along any curve, t must also be constant along the curve.

When in any case the components of the external force per unit mass are assigned—of course satisfying the necessary condition (6), Art. 18—there will be several laws of density which permit the fluid to be in equilibrium. In fact, ρ may be any of the integrating factors of the expression $Xdx + Ydy + Zdz$. We shall illustrate this in some of the following examples.

EXAMPLES.

1. If a mass of fluid is acted upon by attractive forces directed towards any number of fixed centres, find the equations of the surfaces of equal pressure.

Let the fixed centres be $A_1, A_2, \ldots$; let the distances of any point, P, in the fluid from them be $r_1, r_2, \ldots$; let the forces per unit mass at unit distances from them be $\mu_1, \mu_2, \ldots$. Then (*Statics*, Vol. II, Art. 332) the forces have a potential, V, given by the equation
$$V = \frac{\mu_1}{r_1} + \frac{\mu_2}{r_2} + \ldots,$$

and, the equipotential surfaces being also surfaces of equal pressure, the equation of any surface of constant pressure is

$$\frac{\mu_1}{r_1} + \frac{\mu_2}{r_2} + \ldots = C.$$

In the case in which there are only two centres, if one of the forces is repulsive, one of the series of surfaces is a sphere, viz. that for which $C = 0$, since if $\frac{\mu_1}{r_1} - \frac{\mu_2}{r_2} = 0$, each point, P, on the surface is such that the ratio $PA_1 : PA_2$ is constant, and the locus is well known to be a sphere having for diameter the line joining the points which divide the line $A_1 A_2$ internally and externally in the ratio $\mu_1 : \mu_2$.

2. If a fluid is acted upon by force whose components, per unit mass, are at any point proportional to

$$y^2 + yz + z^2, \quad z^2 + zx + x^2, \quad x^2 + xy + y^2,$$

determine a law of variation of the density of the fluid, the corresponding surfaces of constant pressure, &c.

If f denotes a constant force intensity and c a constant length, the components of force per unit mass will be of the forms $\frac{f}{c^2}(y^2 + yz + z^2)$, &c.

Hence

$$\frac{c^2}{f} dp = \rho(y^2 + yz + z^2) dx + \rho(z^2 + zx + x^2) dy + \rho(x^2 + xy + y^2) dz. \quad \ldots (1)$$

It is to be observed at the outset that the force and its curl are at right angles, so that equilibrium is possible, by (6), Art. 18.

The problem is to determine ρ, as an integrating factor, so that the right-hand side of (1) shall be a perfect differential. The analytical mode of procedure is to consider z at first as constant, and to find an integral of the equation

$$(y^2 + yz + z^2) dx + (z^2 + zx + x^2) dy = 0.$$

This is at once found to be

$$\frac{2}{z\sqrt{3}} \tan^{-1} \frac{2x+z}{z\sqrt{3}} + \frac{2}{z\sqrt{3}} \tan^{-1} \frac{2y+z}{z\sqrt{3}} = C,$$

or (z being merely a constant)
$$\frac{x+y+z}{z^2-2xy-zx-zy}=C.$$

Now take the function U, where
$$U=\frac{x+y+z}{z^2-2xy-zx-zy}, \quad \ldots \ldots (2)$$

and differentiate both sides with respect to x, y, and z, considering all of them variable. Then, using D for the denominator of the right-hand side of (2),

$$dU = \frac{2}{D^2}\{(y^2+yz+z^2)\,dx+(z^2+zx+x^2)\,dy$$
$$+\tfrac{1}{2}(x^2+y^2-z^2-2xz-2yz)\,dz\}\ldots (3)$$

Using (1) to simplify this, we have
$$dU = \frac{2}{D^2}\left\{\frac{c^2}{f}\frac{dp}{\rho} - \tfrac{1}{2}(x+y+z)^2\,dz\right\} \quad \ldots (4)$$

$$= \frac{2c^2}{f\rho D^2}\,dp - U^2\,dz, \text{ by (2)}. \quad \ldots \ldots (5)$$

Hence
$$\frac{dU}{U^2}+dz = \frac{2c^2}{f\rho D^2 U^2}\cdot dp, \quad \ldots \ldots (6)$$

and since the right side must now be a perfect differential, we must have $\rho D^2 U^2$ a constant or any function of p. Since by (2) $DU = x+y+z$, we have, then,
$$\rho(x+y+z)^2 = k, \quad \ldots \ldots (7)$$

where k is a constant. Hence the density at any point varies inversely as the square of the distance of the point from the plane $x+y+z=0$.

From (6) we have, then,
$$p = \frac{fk}{2c^2}\left(z-\frac{1}{U}\right)$$
$$= \frac{fk}{c^2}\cdot\frac{xy+yz+zx}{x+y+z}, \quad \ldots \ldots (8)$$

which shows that the surfaces of constant pressure are hyperboloids.

The direction-cosines of the line of constant density and constant pressure are proportional to the components of the curl of the force, and are, therefore, proportional to $y-z$, $z-x$, $x-y$; and the differential equations of these lines are

$$\frac{dx}{y-z} = \frac{dy}{z-x} = \frac{dz}{x-y} \ldots \ldots (9)$$

To integrate these, put each fraction equal to θ, where θ is unknown.
Then
$$dx = \theta(y-z), \ldots \ldots (10)$$
$$dy = \theta(z-x), \ldots \ldots (11)$$
$$dz = \theta(x-y), \ldots \ldots (12)$$

and from these, by addition,
$$dx + dy + dz = 0,$$
whose integral is
$$x + y + z = a, \ldots \ldots (13)$$
where a is any constant. Hence the curves in question are plane curves lying in planes obtained by varying a in (13).

Also multiplying (10), (11), (12) by x, y, z and adding,
$$x\,dx + y\,dy + z\,dz = 0,$$
whose integral is
$$x^2 + y^2 + z^2 = b^2, \ldots \ldots (14)$$
where b is any constant. Hence the curves lie on spheres obtained by varying b; and as they are plane curves, they are circles.

Of course they lie on the surfaces of constant pressure.

But, as previously pointed out, this is only one special law of density which we have investigated. Many other laws may be found thus.

From (5) we have
$$dU + U^2 dz = \frac{2c^2}{f\rho D^2} dp. \ldots \ldots (15)$$

Now if ϕ is any integrating factor of the left-hand side, $\dfrac{\phi}{\rho D^2}$ = constant will express a possible value of ρ. But any function of the form
$$\frac{1}{U^2} \phi\left(z - \frac{1}{U}\right)$$

will render the left-hand side a perfect differential; hence

$$\rho = \frac{1}{(x+y+z)^2} \cdot \phi\left(\frac{xy+yz+zx}{x+y+z}\right), \quad \ldots \quad (16)$$

where ϕ is any function, will express a possible value of ρ. In particular, choosing $\phi(v) = \frac{k}{v^2}$, we have

$$\rho = \frac{k}{(xy+yz+zx)^2}, \quad \ldots \quad (17)$$

which gives another simple law of density. The surfaces of constant pressure remain the same as before; for, when (15) is multiplied across by $\frac{1}{U^2} \phi\left(z - \frac{1}{U}\right)$, if $dp = 0$, we have still $z - \frac{1}{U} = C$. The *actual value* of p is, of course, different; for p will now be proportional to $\int \phi\left(z - \frac{1}{U}\right) \cdot d\left(z - \frac{1}{U}\right)$, or $\psi\left(z - \frac{1}{U}\right)$, suppose.

3. If the components of force per unit mass are proportional to
$$(y+a)^2, \quad cz, \quad -c(y+a),$$
find a law of variation of density, the corresponding surfaces of constant pressure, and the curves of constant density and constant pressure.

Ans. $\rho = \frac{k}{(y+a)^2}$, $p = k\left(x - \frac{cz}{y+a}\right)$, showing the surfaces of constant pressure to be hyperbolic paraboloids; the curves of constant density and constant pressure are right lines.

Also $\rho = \frac{1}{(y+a)^2} \cdot \phi\left(x - \frac{cz}{y+a}\right)$; so that

$$\rho = \frac{k}{\{x(y+a) - cz\}^2}$$

will also be possible.

4. If the components of force per unit mass are proportional to
$$x-a, \quad -\sqrt{h^2 - z^2 - (x-a)^2}, \quad z,$$
find the same things as in the last problem.

Ans. $\rho = \dfrac{k}{\sqrt{h^2-z^2-(x-a)^2}}$; $p = k\{\sqrt{h^2-z^2-(x-a)^2}-y\}$;
the curves are circles determined by the intersections of planes $y = C$ with cylinders $(x-a)^2+z^2 = C'$.

Generally, $\rho = \dfrac{1}{\sqrt{h^2-z^2-(x-a)^2}} \cdot \phi(\sqrt{h^2-z^2-(x-a)^2}-y)$.

5. The components of force being proportional to

$$cy-bz, \quad az-cx, \quad bx-ay,$$

show that the surfaces of constant pressure are planes passing through the line $\dfrac{x}{a} = \dfrac{y}{b} = \dfrac{z}{c}$, and the curves of constant pressure and constant density are right lines parallel to this.

6. In a spherical mass of homogeneous liquid, self-attracting according to the law of nature, find the pressure intensity at any point.

Ans. If γ is the constant of gravitation, at any point, P, distant r from the centre, $V = 2\pi\gamma\rho(a^2-\tfrac{1}{3}r^2)$, where a is the radius of the sphere (see *Statics*, Vol. II, p. 299, 4th ed.). Hence

$$p = 2\pi\gamma\rho^2(a^2-\tfrac{1}{3}r^2).$$

[The homogeneity of this is thus verified: if m denotes mass, f denotes force, and l denotes length, we know that $\gamma \cdot \dfrac{m^2}{l^2} = f$; also $\rho = \dfrac{m}{l^3}$; hence p is of the nature $\dfrac{f}{l^2}$, as it ought to be. Since the pressure intensity in a homogeneous sphere is thus proportional to the square of the density, we see the nature of the assumption made by Laplace (*Statics*, vol. i, Art. 174) in the case of the Earth—that in passing from stratum to stratum the change in the pressure intensity is proportional to the change in the square of the density.]

20. Equations of Equilibrium in Polar and Cylindrical Co-ordinates. Let P, Fig. 31, be any point in a fluid, at which the components of force acting on the fluid, per unit mass, are X, Y, Z parallel to the rectangular axes,

Ox, Oy, Oz; and let the position of P be defined by the usual polar co-ordinates, viz. the radius vector $OP (= r)$, the colatitude, POz ($= \theta$), and the longitude, xOn ($= \phi$), this last being the angle between the plane xz and the plane containing P and the axis of z. The arcs in the figure are those determined on a sphere whose centre is O and radius OP by the axes and the line OP.

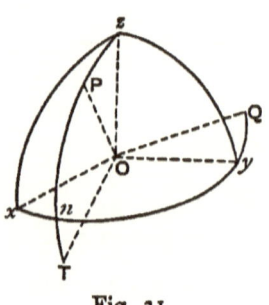

Fig. 31.

Sometimes it is convenient to consider the resultant force per unit mass at P as resolved into three rectangular components corresponding to radius vector, latitude, and longitude, i. e. components along OP, along the line at P perpendicular to OP in the plane POz, and along the tangent at P to the parallel of latitude.

Producing the great circle zPn to T so that $nT = zP = \theta$, the second of these directions is parallel to OT; and since the third is at right angles to the plane POz at P, if we produce the arc xy to Q so that $yQ = xn = \phi$, the line OQ is parallel to the tangent at P to the parallel of latitude.

Let R be the component of the force-intensity (i. e. force per unit mass) at P in the direction OP; let Θ be its component in the second and Φ its component in the third of these directions.

Now, the axis of x being in any direction, we have proved the equation

$$\frac{dp}{dx} = \rho X,$$

so that if ds is the element of length of a curve drawn in any direction at P, and S the force-intensity along the

General Equations of Pressure.

tangent to this curve at P in the sense in which ds is measured, it follows that

$$\frac{dp}{ds} = \rho S.$$

Taking ds along OP, we have $ds = dr$; taking ds along the meridian zP at P, we have $ds = rd\theta$; and taking ds along the parallel of latitude at P in the sense OQ, we have $ds = r\sin\theta\, d\phi$, since the radius of the parallel of latitude is $r\sin\theta$. Hence the equations of equilibrium are

$$\left.\begin{aligned}\frac{dp}{dr} &= \rho \cdot R, \\ \frac{dp}{d\theta} &= \rho\, r \cdot \Theta, \\ \frac{dp}{d\phi} &= \rho r \sin\theta \cdot \Phi.\end{aligned}\right\} \quad \ldots \quad (1)$$

Equations of equilibrium in Cylindrical Co-ordinates. By the cylindrical co-ordinates of P are meant the distance, z, of P from the plane xy, the perpendicular distance, ζ, of P from the axis of z, and the longitude, ϕ, i.e. the angle between the plane xz and the meridian plane zOP. Hence if Z, Z_1, Φ denote the components of force-intensity at P parallel to Oz, perpendicular to Oz, and along the tangent to the parallel of latitude,

$$\left.\begin{aligned}\frac{dp}{dz} &= \rho Z, \\ \frac{dp}{d\zeta} &= \rho Z_1, \\ \frac{dp}{d\phi} &= \rho r \sin\theta \cdot \Phi.\end{aligned}\right\} \quad \ldots \quad (2)$$

EXAMPLES.

1. If a homogeneous liquid is acted upon by gravity, and each particle is, in addition, acted upon by a force emanating from a vertical axis proportional to the distance of the particle from that axis, find the intensity of pressure at any point.

Adopting the C. G. S. system, measuring forces in dynes and masses in grammes, $Z = -g$ (the axis of Z being drawn vertically upwards), $Z_1 = f \cdot \dfrac{\zeta}{a}$, where f is a constant force in dynes and a a constant length in centimètres, and $\Phi = 0$. Hence

$$\frac{dp}{dz} = -\rho g; \quad \frac{dp}{d\zeta} = \rho f \cdot \frac{\zeta}{a}; \quad \frac{dp}{d\phi} = 0; \quad \ldots \quad (3)$$

and since $dp = \dfrac{dp}{dz} dz + \dfrac{dp}{d\zeta} d\zeta + \dfrac{dp}{d\phi} d\phi$, we have

$$dp = -\rho g \, dz + \rho f \cdot \frac{\zeta}{a} d\zeta; \quad \ldots \quad (4)$$

$$\therefore \quad p = -\rho g z + \frac{\rho f}{2a} \cdot \zeta^2 + C, \quad \ldots \quad (5)$$

where C is a constant. If on the free surface $p = p_0$ at all points, and the origin is taken at the intersection of the axis of z with this surface, we have

$$p = p_0 - \rho g z + \frac{\rho f}{2a} \cdot \zeta^2, \quad \ldots \quad (6)$$

which shows that all the surfaces of constant pressure are paraboloids of revolution round the axis of z, their parameters being $\dfrac{2g}{f} \cdot a$.

2. If the fluid is compressible and follows the law that ρ is proportional to p at each point; then, supposing that if at each point the intensity of pressure were p_0, the density would be ρ_0, we have $\rho = \rho_0 \cdot \dfrac{p}{p_0}$, and (4) becomes

$$\frac{dp}{p} = \frac{\rho_0}{p_0} \left(-g \, dz + \frac{f}{a} \zeta \, d\zeta \right), \quad \ldots \quad (7)$$

General Equations of Pressure.   97

the integral of which is

$$p = C e^{\frac{\rho_0}{p_0}\left(-gz + \frac{J}{2a}\zeta^2\right)}, \quad \ldots \ldots \quad (8)$$

where C is a constant. If the origin is taken as before, C is the constant intensity of pressure on the free surface, and the surfaces of constant pressure are still paraboloids of revolution.

3. If a mass of homogeneous liquid surrounds a sphere of uniform density, and is subject to the attraction of this sphere as well as to a force emanating from a fixed axis through the centre of the sphere and proportional to the distance, as above, we may employ either the general equations of Art. 18 or cylindrical co-ordinates, or we may take as co-ordinates r and ζ simply. Thus, if the attraction of the sphere (not represented in the figure) per unit mass at P is represented by $k\dfrac{c^2}{r^2}$, where c is a constant length in centimètres and k a constant force in dynes, we have

$$\frac{dp}{dr} = -\rho k \frac{c^2}{r^2}; \quad \frac{dp}{d\zeta} = \rho f \frac{\zeta}{a}; \quad \ldots \ldots \quad (9)$$

and since $dp = \dfrac{dp}{dr}dr + \dfrac{dp}{d\zeta}d\zeta$, we have

$$dp = -\rho k \frac{c^2}{r^2} dr + \rho f \frac{\zeta}{a} d\zeta, \quad \ldots \ldots \quad (10)$$

$$\therefore \; p = \;\; \rho k \frac{c^2}{r} + \frac{\rho f}{2a}\zeta^2 + C, \quad \ldots \ldots \quad (11)$$

where C is a constant. If Oz cuts the free surface where $r = R$, and if p_0 is the pressure intensity on the free surface,

$$C = p_0 - \rho k \frac{c^2}{R}.$$

4. Suppose a homogeneous sphere at rest surrounded by an atmosphere whose particles attract each other and are attracted by the sphere according to the law of nature; it is required to find an equation for the intensity of pressure at any point of the atmosphere.

The surfaces of constant density in the atmosphere will be spheres concentric with the nucleus. Let P be a point in the

H

atmosphere distant r from the centre, O, and $M =$ mass of nucleus. Then, if γ is the constant of gravitation, with the units of the C. G. S. system the attraction at P per unit mass is $\dfrac{\gamma M}{r^2}$ due to the nucleus, and in addition to this there is the attraction of that portion of the atmosphere contained within the sphere of radius OP: the portion outside this sphere, since its layers of constant density are spherical shells with centre O, exerts no attraction at P (*Statics*, vol. II, Art. 319).

To find the attraction due to the portion of the atmosphere within the sphere of radius OP, describe a sphere of radius x, less than OP; let ρ' be the density (grammes per cubic cm.) on its surface, and describe another sphere of radius $x + dx$; then the mass of the shell contained between these spheres is $4\pi\rho' x^2 dx$, and its attraction per unit mass at P is $4\pi\gamma\rho'\dfrac{x^2 dx}{r^2}$ (*Statics*, ibid.).

Hence the equation for p at P is, since the resultant force is directed from P to O,

$$-\frac{1}{\rho}\frac{dp}{dr} = \frac{\gamma M}{r^2} + \frac{4\pi\gamma}{r^2}\int_a^r \rho' x^2 dx, \quad \ldots \quad (1)$$

where a is the radius of the nucleus.

To form a differential equation for p, multiply both sides by r^2, and then differentiate both sides with respect to r. Now observe that ρ' is some function of x, and that we are differentiating the integral at the right-hand side with regard to its upper limit, so that the result of this differentiation will be the function under the integral sign with the upper limit, r, substituted for x (see Williamson's *Integral Calculus*, Art. 114, or Greenhill's *Diff. and Int. Cal.*, Art. 207).

Hence

$$\frac{d}{dr}\left(\frac{r^2}{\rho}\frac{dp}{dr}\right) = -4\pi\gamma\rho r^2, \quad \ldots \quad (2)$$

in which $p = k\rho$, where $k = 2926 \cdot 9\,\dfrac{gT}{s}$, as will be seen in a subsequent Chapter; therefore

$$\frac{d}{dr}\left(\frac{r^2}{p}\frac{dp}{dr}\right) + \frac{4\pi\gamma}{k^2} p r^2 = 0. \quad \ldots \quad (3)$$

The integral of this equation can be presented in a series. If

General Equations of Pressure.

$\frac{4\pi\gamma}{k^2}$ is denoted by μ, a particular value of p is given by the equation $p = \frac{2}{\mu r^2}$. Now assume

$$p = \frac{2\phi}{\mu r^2}, \quad \dots \quad \dots \quad (4)$$

where ϕ is an unknown quantity, and change the independent variable from r to $\frac{1}{r}$.

Denoting $\frac{1}{r}$ by x, and putting $\log \phi = \psi$, equation (3) becomes

$$x^2 \frac{d^2\psi}{dx^2} + 2(e^\psi - 1) = 0. \quad \dots \quad (5)$$

Let $\frac{1}{a}$ be denoted by c, and expand ψ in powers of $x-c$ according to the formula

$$\psi = \psi_0 + \frac{x-c}{\lfloor 1} \left(\frac{d\psi}{dx}\right)_0 + \frac{(x-c)^2}{\lfloor 2} \left(\frac{d^2\psi}{dx^2}\right)_0 + \dots, \dots (6)$$

where ψ_0, $\left(\frac{d\psi}{dx}\right)_0$, ... are the values of ψ and its differential coefficients at the surface of the solid nucleus, i.e. where $x = c$. In calculating the successive differential coefficients

$$\left(\frac{d^2\psi}{dx^2}\right)_0, \left(\frac{d^3\psi}{dx^3}\right)_0, \dots$$

by successive differentiations of (5) in terms of the unknown and arbitrary constants ψ_0 and $\left(\frac{d\psi}{dx}\right)_0$, it will be found convenient to take $e^{\psi_0} = A$ and $\left(\frac{d\psi}{dx}\right)_0 = \frac{2}{c} B$, so that A and B are the two arbitrary constants which belong to the integral of (5).

It will be seen that all the coefficients in (6) at the right-hand side vanish for the particular values $A = 1$, $B = 0$, and the integral then obtained is $\psi = 0$, i.e. $\phi = 1$, or $p = \frac{2}{\mu r^2}$.

Hydrostatics and Elementary Hydrokinetics.

21. Centre of Pressure. Hitherto in finding the position of the centre of pressure on a plane area we have confined our attention to areas of simple forms, such as triangles, quadrilaterals, &c. We shall now consider a plane area of any form occupying an assigned position in water, or other liquid, subject to the action of gravity only. Let the area be rnm, Fig. 20, p. 52. If we draw at a depth z below the surface AB of the liquid a horizontal line, and another line parallel to this at the infinitesimal distance dz below it, denoting the length of the line intercepted by the area rnm by y, the pressure on the strip contained by these close lines is $wzy\,dz$, where w = specific weight of the liquid; and this pressure acts at the middle point of the strip, therefore its moment about AB is $wz^2y\,dz$. Now the whole pressure on the area rnm is $A\bar{z}w$, where A is the magnitude of the area and $\bar{z}$ the depth of G, its centre of area; hence if $\bar{p}$ is the depth of the centre of pressure,

$$\bar{p} = \frac{\int z^2 y\,dz}{A\bar{z}}, \quad \ldots \quad \ldots \quad (1)$$

in which y is a known function of z when the form and position of the area are assigned.

The position of the centre of pressure may be otherwise expressed thus. At any point in the surface of the liquid draw two rectangular axes, Ox, Oy, and draw the axis of Oz vertically downwards; break up the area rnm into infinitesimal elements; let x, y, z be the co-ordinates of any point, P, in the area at which the element of area is dS. Then the pressure on this element is $wz\,dS$. Hence the co-ordinates of the centre of pressure, I, are

$$\frac{\int xz\,dS}{\int z\,dS}, \quad \frac{\int yz\,dS}{\int z\,dS}, \quad \frac{\int z^2\,dS}{\int z\,dS}.$$

If at all points on the contour, rnm, of the area we draw vertical lines terminated by the surface of the liquid, the

centre of gravity of the cylinder of fluid enclosed by these lines and the area rnm lies on the vertical through I, midway between I and the surface. For, if $d\sigma$ is the projection of dS on the free surface of the liquid, the elementary cylinder standing on dS has for volume $zd\sigma$, and the co-ordinates of its centre of gravity are x, y, $\dfrac{z}{2}$; hence the co-ordinates of the centre of gravity of the whole cylinder are
$$\frac{\int x z\, d\sigma}{\int z\, d\sigma},\ \frac{\int y z\, d\sigma}{\int z\, d\sigma},\ \tfrac{1}{2}\frac{\int z^2\, d\sigma}{\int z\, d\sigma},$$
and if θ is the angle between the plane of the area rnm and the horizon, it is evident that $d\sigma = \cos\theta \cdot dS$, so that in the numerator and denominator of each of these last expressions we may replace $d\sigma$ by dS, and the result is then obvious.

The position of the centre of pressure on a plane area can be very easily expressed with reference to the principal axes of the area at its centre of gravity, G. Thus, let CDE (Fig. 32) be the plane area, its plane being inclined to the vertical at any angle, θ; let GA and GB be its principal

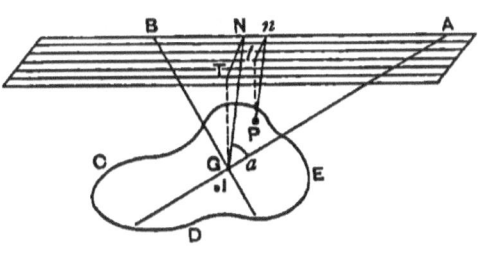

Fig. 32.

axes at G, intersecting the surface of the liquid in A and B; let $GN(=h)$ be the perpendicular from G (in the plane of the area) on the line AB, and let GN make the angle a with GA.

The equation of AB with reference to GA and GB as axes of x and y, respectively, is
$$x \cos a + y \sin a - h = 0;$$

102 *Hydrostatics and Elementary Hydrokinetics.*

and if P is any point in the area at which the element of area dS is taken, the perpendicular Pn from P on AB is $h - x\cos a - y\sin a$, if x, y are the co-ordinates of P with reference to GA and GB. Hence the perpendicular, Pt, from P on the surface of the liquid is

$$(h - x\cos a - y\sin a)\cos\theta,$$

and the pressure on dS is

$$w(h - x\cos a - y\sin a)\cos\theta . dS, \quad \ldots \quad (a)$$

where w is the weight of the liquid per unit volume. Now take the sum of the moments of the elementary pressures of which (a) is the type about GA and equate it to the moment of the resultant pressure, $A \times GT . w$, where A is the area and GT the perpendicular from G on the surface. If (ξ, η) are the co-ordinates of I, the centre of pressure, we have

$$w A h \cos\theta . \eta = \quad w\cos\theta \int (h - x\cos a - y\sin a) y\, dS . \quad (2)$$
$$= - w\cos\theta \sin a \int y^2 dS,$$

the other integrals vanishing since the principal axes at the centre of area are those of co-ordinates. Now $\int y^2 dS$ is the moment of inertia of the area about GA, which we shall denote by $A . k_1^2$, k_1 being the radius of gyration of the area about GA. Hence, finally,

$$\eta = -\frac{k_1^2}{h}\cos a\,; \quad \ldots \quad \ldots \quad (3)$$

and in the same way, equating the moment of the whole pressure about GB to the sum of the moments of the elementary pressures of the type (a), we have

$$\xi = -\frac{k_2^2}{h}\sin a. \quad \ldots \quad \ldots \quad (4)$$

Thus the co-ordinates are independent of the inclination of the given plane area to the vertical, as we have previously pointed out, so that if the area were turned round the line AB in which its plane intersects the surface of the liquid through any angle, the centre of pressure, I, would continue to be absolutely the same point in the area.

The expressions (3), (4) lead at once to an obvious geometrical interpretation, viz.—construct the ellipse whose equation with reference to GA and GB is

$$\frac{x^2}{k_1^2} + \frac{y^2}{k_2^2} = 1;$$

take the pole, Q, of the line AB with reference to this ellipse; the co-ordinates of Q are $(-\xi, -\eta)$, so that if the line QG is produced through G to I so that $GI = QG$, we arrive at I, the centre of pressure.

These expressions (3), (4) give us at once some simple results concerning the motion of the centre of pressure produced by various displacements of the given area.

Thus, if the area is rotated in its own plane about G, while G is fixed, the only variable in the values of ξ, η is a; and if this is eliminated from (3), (4), we have

$$\frac{\xi^2}{k_2^4} + \frac{\eta^2}{k_1^4} = \frac{1}{h^2}, \quad \ldots \ldots (5)$$

which is the locus described in the area by the centre of pressure—viz. an ellipse.

To find the locus described in this case by the centre of pressure with reference to fixed space, refer its position to the line GN and the horizontal line through G in the area as axes of x' and y', respectively. If (x', y') are the co-ordinates of I with reference to these axes, we have

$$\xi = x' \cos a - y' \sin a,$$
$$\eta = x' \sin a + y' \cos a.$$

Substituting the above values of ξ, η, and eliminating a, we have

$$\left(x' + \frac{k_1^2 + k_2^2}{2h}\right)^2 + y'^2 = \left(\frac{k_1^2 - k_2^2}{2h}\right)^2, \quad \ldots (6)$$

104 *Hydrostatics and Elementary Hydrokinetics.*

which shows that I describes a circle in fixed space, the centre of the circle being on the vertical through G.

Again, if the area is lowered into the liquid without rotation, h is the only variable in (3) and (4), by eliminating which we have

$$\frac{\eta}{\xi} = \frac{k_1^2}{k_2^2} \tan a, \quad \ldots \ldots \quad (7)$$

which shows that I describes a right line in the area; and it describes also a right line in space, for (7) gives a linear relation between x', y', and constants.

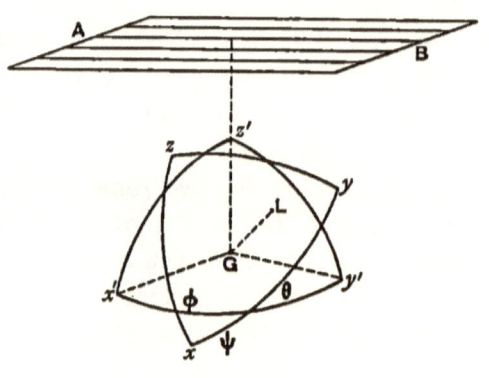

Fig. 33.

Let us next suppose the plane of the area to have any position whatever, which we shall define in the usual way by the Precession and Nutation angles θ, ϕ, ψ.

Take the vertical Gz' as axis of z', and any two rectangular horizontal lines, Gx', Gy', as axes of x' and y'. Let Gx, Gy be the principal axes at G in the plane of the given area, while Gz is the axis perpendicular to this plane.

Then, since the direction-cosines of Gz' with reference to the axes of x, y, z are

$$-\sin\psi \sin\theta, \quad \cos\psi \sin\theta, \quad \cos\theta,$$

the length of the perpendicular from any point $(x, y, 0)$ in the given area on the free surface AB is

$$h + x \sin\psi \sin\theta - y \cos\psi \sin\theta.$$

This multiplied by $w\,ds$ gives the pressure on the element of area, and the total moment of pressure about Gx is $-Ak_1^2 w \cos\psi \sin\theta$. Hence, as before,

$$\xi = \frac{k_2^2}{h} \sin\psi \sin\theta, \quad \ldots \ldots \quad (8)$$

$$\eta = -\frac{k_1^2}{h} \cos\psi \sin\theta. \quad \ldots \ldots \quad (9)$$

Suppose the area to be rotated, like a rigid body, round any line, GL, fixed in space, the direction-cosines of this line being l, m, n with reference to the space axes Gx', Gy', Gz'. Then the angles between GL and the principal axes Gx, Gy, Gz are all constant, and if we denote their cosines by λ, μ, ν, respectively, we find, by eliminating θ and ψ from (8) and (9) by means of the constants λ, μ, ν, the equation

$$\frac{\xi^2}{k_2^4} + \frac{\eta^2}{k_1^4} + \frac{1}{\nu^2}\left(\frac{\lambda\xi}{k_2^2} + \frac{\mu\eta}{k_1^2} + \frac{n}{h}\right)^2 = \frac{1}{h^2}, \quad \ldots \quad (10)$$

which gives the curve described in the area by the centre of pressure

This agrees with (5) for the case in which the plane of the area is always kept vertical; for in this case

$$\theta = \frac{\pi}{2}, \; \psi = -a, \; \nu = 1, \; n = 0, \; \lambda = \mu = 0.$$

If the line GL is any one in the plane of the area $\nu = 0$, and the locus described by I in the area is a right line,

$$\frac{\lambda\xi}{k_2^2} + \frac{\mu\eta}{k_1^2} + \frac{n}{h} = 0. \quad \ldots \ldots \quad (11)$$

EXAMPLES.

1. Find the position of the centre of water pressure on a circular area whose plane is vertical and whose centre is at a given depth.

Take as the principal axes of reference at G the vertical and horizontal diameters, and let h be the depth of G. Then in (3) and (4) we have $a = 0$ and $k_1^2 = k_2^2 = \tfrac{1}{4}r^2$, where r is the radius of the circle.

Hence
$$\xi = -\frac{r^2}{4h}, \quad \eta = 0,$$

so that I is on the vertical diameter at a depth $h + \dfrac{r^2}{4h}$ from the free surface.

If the area is just immersed, $h = r$, and the depth of I is
$$\tfrac{5}{4}r.$$

In the case of an elliptic area whose centre is at a depth h, and whose major axis makes an angle a with the vertical
$$\xi = -\frac{a^2}{4h}\cos a, \quad \eta = -\frac{b^2}{4h}\sin a,$$

where a and b are the semi-axes.

2. Find the pressure on a plane vertical area whose position is given in a slightly compressible liquid.

Supposing, definitely, the units of the C. G. S. system adopted, we have for the intensity of pressure at any point P,
$$\frac{dp}{dz} = \rho g, \quad \ldots \ldots \ldots (1)$$

where z is the depth of P in centimètres below the free surface.

Also from (γ), p. 22, $\dfrac{dp}{dz} = \dfrac{k}{\rho}\dfrac{d\rho}{dz}$, where k is the resilience of volume.

These give by integration
$$\rho = \frac{k\rho_0}{k - g\rho_0 z}, \quad \ldots \ldots \ldots (2)$$

where ρ_0 is the density at the surface of the liquid. Now since k is very great, we may neglect $\dfrac{1}{k^2}$, and we have
$$\rho = \rho_0\left(1 + \frac{g\rho_0}{k}z\right),$$

which indicates a uniform density superposed on a density varying directly as the depth.

General Equations of Pressure.

Substituting this in (1), we have

$$p = g\rho_0\left(z + \frac{g\rho_0}{2k}z^2\right). \quad \ldots \quad (3)$$

Let A be the magnitude of the given area, and $\bar{z}$ the depth of its centre of area.

If dS is the element of area at P, the whole pressure is $\int p\, dS$; and if $\int z^2 dS$, which is the moment of inertia of the area about the line AB (Fig. 32) in which its plane intersects the surface, is denoted by $A\lambda^2$, while $\int z\, dS = A\bar{z}$, the resultant pressure is

$$Ag\rho_0\left(\bar{z} + \frac{g\rho_0}{2k}\lambda^2\right)$$

in dynes. Dividing this expression by g, we have the pressure in grammes' weight.

3. Find the position of the centre of pressure on any vertical area which is symmetrical with respect to its principal axes at its centre of area, when immersed in a slightly compressible fluid.

With the notation of p. 103, and denoting the radius of gyration of the area about the line AB (Fig. 32) by λ, we have

$$\xi = -\frac{k_2^2}{h}\cos\alpha\left\{1 + \frac{g\rho_0}{2kh}(2h^2 - \lambda^2)\right\},$$

$$\eta = -\frac{k_1^2}{h}\sin\alpha\left\{1 + \frac{g\rho_0}{2kh}(2h^2 - \lambda^2)\right\}.$$

4. Assuming the resilience of volume of sea water to be, in C. G. S. units, $2\cdot 33 \times 10^{10}$ (see p. 22), and that 1 mile $= 160933$ centimètres, find the fractional increase in density at a depth of 1 mile in the ocean.

Ans. From the equations

$$\rho\frac{dp}{d\rho} = 2\cdot 33 \times 10^{10}$$

$$\frac{dp}{dz} = 981\rho,$$

we have

$$2\cdot 33 \times 10^{10}\frac{d\rho}{\rho^2} = 981\, dz,$$

$$\therefore \frac{\rho - \rho_0}{\rho_0} = \frac{981\, \rho_0 z}{2\cdot 33 \times 10^{10} - 981\, \rho_0 z},$$

where ρ_0 is the density at the surface. Taking $\rho_0 = 1.026$, we find at the depth of a mile

$$\frac{\rho - \rho_0}{\rho_0} = \frac{1}{142.8}, \text{ nearly.}$$

5. Assuming the resilience of volume of sea water to be constant at all depths, find what the depth of the ocean should be at a point where the density of the water is double the surface density.

Ans. Nearly 71·92 miles.

6. Represent graphically the densities of sea water at points on a vertical line drawn downwards from the surface.

Let O be the point on the surface, OA the vertical line drawn downwards to the point, A, at which the density would be doubled; produce OA to C so that $OA = AC$; draw a horizontal line, CH, through C. Then if the densities at various points on OC are represented by ordinates drawn at these points perpendicularly to OC, their extremities trace out a hyperbola whose centre is C and asymptotes CO and CH.

7. If the density of a fluid varies as any given function of the depth, find the depth of the centre of pressure on a plane vertical area.

Ans. If $\rho = f'(z)$, the depth of I is $\dfrac{\int zf(z)\,dS}{\int f(z)\,dS}$.

8. A rectangular area, $ABCD$, has one side AB in the surface of water and its plane vertical. If it is rotated about a horizontal axis at A, find the curve described in the area by the centre of pressure so long as the whole area continues immersed.

Ans. If $AB = 2a$, $AD = 2b$, the centre of pressure traces out a right line in the area. This line is thus constructed: let G be the centre of area of the rectangle, m and n the middle points of DC and CB respectively; take a point p on Gm such that $Gp = \tfrac{1}{3}Gm$, and a point q on Gn such that $Gq = \tfrac{1}{3}Gn$; then the line joining p to q is that described in the area by the centre of pressure.

9. A plane area of any form occupies a vertical position in water. If it is rotated in its plane about any point in the area, find the curve traced out in the area by the centre of pressure.

General Equations of Pressure.

Ans. Let O be the point about which the area turns, G the centre of area, a, β the co-ordinates of O with respect to the principal axes at G, $h =$ the perpendicular from O on the surface of the liquid, the radii of gyration being as in p. 103; then the equation of the locus referred to the principal axes at G is

$$(a k_1^2 x + \beta k_2^2 y + k_1^2 k_2^2)^2 = h^2 (k_1^4 x^2 + k_2^4 y^2),$$

which will be a hyperbola, an ellipse, or a parabola according as

$$GO > h, \quad GO < h, \quad \text{or} \quad GO = h.$$

If O is in the surface of the liquid, so long as none of the area is raised out of the liquid by rotation about O, the locus is a right line. Thus if a plane polygon of any shape has a corner in the surface of the liquid round which the polygon is turned, the centre of pressure describes a right line so long as the whole area remains immersed.

10. Find the position of the centre of pressure of a semicircular area whose diameter is in the surface of water.

Ans. If r is the radius of the circle, the centre of pressure is at a distance $\dfrac{3\pi}{16} r$ from the horizontal diameter. (The centre of gravity of a semicircle is $\dfrac{4r}{3\pi}$ from the centre.)

11. If the diameter is horizontal and at a depth h, find the depth of the centre of pressure.

Ans. $\dfrac{r}{4} \cdot \dfrac{16h + 3\pi r}{4r + 3\pi h}$ below the horizontal diameter.

12. Find the position of the centre of pressure on a semicircular area whose bounding diameter is vertical with one extremity in the surface of water.

Ans. Its distance from the vertical diameter is $\dfrac{4r}{3\pi}$, and its depth is $\frac{3}{4}r$.

(The point is on the vertical through the centre of gravity, G, of the area, since this is one of the principal axes at G.)

13. Find the position of the centre of pressure on a semicircular area completely immersed in water, the bounding diameter being inclined at an angle a to the horizon and having one extremity in the surface of the water.

110 *Hydrostatics and Elementary Hydrokinetics.*

Ans. The distances of the centre of pressure from the bounding diameter and the diameter perpendicular to it are, respectively,

$$\frac{r}{4} \cdot \frac{3\pi \cos a + 16 \sin a}{4 \cos a + 3\pi \sin a} \quad \text{and} \quad \frac{r}{4} \cdot \frac{3\pi \sin a}{4 \cos a + 3\pi \sin a}.$$

14. An elliptic area is immersed vertically in water; if it is displaced by rolling along the surface of the water, find the locus described in the area by the centre of pressure.

Ans. A similar, similarly placed, and concentric ellipse whose axes are each ¼ of the corresponding axes of the given ellipse.

CHAPTER V.

PRESSURE ON CURVED SURFACES.

22. Principle of Buoyancy. If any curved closed surface, M (Fig. 2, p. 5), be traced out in imagination in a heavy fluid the pressures exerted on all the elements of this surface by the surrounding fluid have a single resultant, which is equal and opposite to the weight of the fluid enclosed by M.

This is evident, because the fluid inside M is in equilibrium under its own weight and the pressure exerted on its surface by the surrounding fluid; hence this pressure reduces to a vertical upward force equal to the weight of the fluid inside M and acting through the centre of gravity of this fluid.

This is obviously true whatever be the nature of the fluid—liquid or gaseous, homogeneous or heterogeneous.

If the curved surface M is not one merely traced out in imagination in the fluid, but the surface of a solid body displacing fluid, the result is the same—

the resultant pressure of a heavy fluid on the surface of any solid body M is a vertical upward force equal to the weight of the fluid which could statically replace M, and this force acts through the centre of gravity of this replacing fluid.

112 *Hydrostatics and Elementary Hydrokinetics.*

Let Fig. 34 represent the solid body, which we may imagine to be a mass of iron, the surrounding fluid being water, air, or any fluid acted upon by gravity. The body is represented as held in its position by cords attached to fixed points, $C, D, \ldots$, and the arrows represent pressures exerted on its surface by the fluid at various points.

Now it is quite clear that if the iron body were replaced by one having exactly the same surface and occupying exactly the same position, the pressure on each element of its surface would be identically the same as before, of whatever substance the new body may be. If the new body

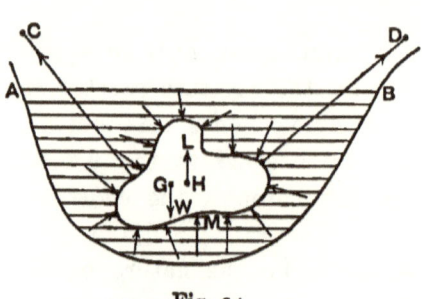

Fig. 34.

were of wood, or instead of being solid were a thin hollow shell, it might be necessary to keep it in the position represented by means which prevent its rising up out of the fluid; but we are not at all concerned with the forces which keep the body M in equilibrium; our object is merely to ascertain the resultant, if any, of the *fluid pressures* exerted in the given position on its surface.

In general, a number of forces acting in various lines which do not lie in one plane have no single resultant: their simplest reduction is to *two* forces whose lines of action do not meet (*Statics*, vol. ii., chap. xiii.). But it is remarkable that the pressures exerted on the various elements of any closed surface by a heavy fluid *have* a single resultant; and the truth of this we see by imagining the place occupied by M to be occupied by a portion of the fluid itself, placed in the vacancy without disturbing any of the surrounding fluid.

Pressure on Curved Surfaces. 113

With regard to this replacing fluid, observe two things: firstly, it is in equilibrium; secondly, it is kept so by its own weight and the very same system of pressures as that which acted on the body M, since this body and the replacing fluid present identically the same surface to the surrounding fluid. Hence, then—

the system of pressures has a single resultant which is a vertical upward force equal to the weight of the statically replacing fluid and acting through the centre of gravity of this fluid.

The centre of gravity, H, of the replacing fluid is called the *centre of buoyancy*; and, so far as the general principle of buoyancy is concerned, there is no relation between H and the centre of gravity, G, of the body; nor is there any relation between the weight, W, of this body and the weight, L, of the displaced fluid.

If the fluid is water, or any homogeneous liquid, the resultant pressure is the weight of the liquid which would flow into the vacant space if M were removed; but if the density of the fluid is different in different layers, we must not imagine the replacing fluid to be that which would flow in when M is removed, but rather to be a continuation of the surrounding fluid placed in the vacancy without any disturbance of the external fluid, and having the same surfaces of equal density as this fluid. The distribution of this replacing fluid is unique and determinate, as will be subsequently proved.

COR. I. Pressure of uniform intensity exerted over any closed surface produces no resultant.

For, imagine the closed surface to be one traced out in a perfectly weightless fluid—or a very light gas—whose surface is subject to any pressure.

The intensity of pressure will be uniform throughout the whole fluid, and therefore over the given surface; and by what has just been said, the resultant pressure over this surface is equal to the weight of the enclosed fluid—that is to say, zero.

This result can, of course, be proved mathematically; for, let dS be an element of the surface at any point, P, (x, y, z), let p be the constant intensity of pressure, and let the origin of co-ordinates be any point inside the surface. Then there is no resultant force parallel to the axis of x; for, if the element dS be projected orthogonally on the plane yz by a slender cylinder, and the sides of this cylinder be produced through the plane yz so as to meet the surface again in an element of area dS' at a point P', the projections of dS and dS' on the plane yz are numerically equal but of opposite signs; hence the pressures pdS and pdS' at P and P' neutralize each other in the direction of the axis of x; and in the same way the pressures at the ends of all other cylinders parallel to the axis of x neutralize each other in this direction, so that there is no force parallel to this axis; and, similarly, no force in any direction.

Thus the area of the projection of any closed surface on any plane must be considered as zero. In symbols, $\iint dy\,dz$ taken over any closed surface is zero.

In fact, in several investigations of mathematical physics, each element, dS, of a surface may usefully be represented by a *vector*, i.e. *a directed magnitude*, by drawing at the mean point, P, of the element dS a normal to the surface from the surface into the surrounding space—always from the same side, or aspect, of the surface into this surrounding space—taking a length on this normal proportional to the area of the corresponding element, dS, and marking the end of this normal length by an arrow head. The orthogonal projection, or component, of this marked line along any line, L, will then represent the projection of dS on any plane perpendicular to L. Some of these marked lines, or vectors, will have components along L in one sense and others will have components in the opposite sense; and thus we understand more clearly the mathematical result that the sum of the projections of the same aspects of the elements, dS, ... of any closed surface on any plane is zero.

In the same way there is no resultant moment of the pressures round any line. For, if σ_1, σ_2, σ_3 are the projections of dS at P on the planes yz, zx, xy, respectively, the components

Pressure on Curved Surfaces.

of the force pdS parallel to the axes of x, y, z are $p\sigma_1, p\sigma_2, p\sigma_3$; and the moment of this force about the axis of x is

$$p(y\sigma_3 - z\sigma_2) \quad \ldots \quad \ldots \quad \ldots \quad (a)$$

Now if the perpendicular from P on the plane yz is produced through this plane to meet the surface again in P''', the element of area, dS''', cut off at P''' by the slender cylinder described on the contour of dS at P parallel to the axis of z, will have for projection on the plane xy the value $-\sigma_3$, and the co-ordinate y being the same for P''' as for P, the moment round the axis of x of the pressure at P''' will supply the term $-y\sigma_3$, which destroys the first part of (a). The second part is similarly destroyed by the pressure at P'', the point in which the parallel from P to the axis of y meets the surface again.

Hence there is no moment of the pressure about any line. Analytically this result is expressed thus:

$$\iint (y\,dx\,dy - z\,dz\,dx) = 0$$

for any closed surface.

The result of this Corollary may also be thus stated: given any closed curve, plane or tortuous, in space; if a surface of any size and shape be described having this curve for a bounding edge, and if pressure of uniform intensity be distributed over one side of this surface, the resultant of this pressure is the same whatever the size and shape of the surface.

Hence if the curve is plane, the resultant pressure on any surface having it for a bounding edge is the same as the resultant pressure on the plane area of the curve.

COR. 2. *The principle of Archimedes.*

The particular case in which the solid body M which displaces fluid *is in equilibrium solely under the action of its own weight and the fluid pressure over its surface* furnishes the Principle of Archimedes.

The resultant of the system of fluid pressures must then be exactly equal and opposite to the weight of the displacing body.

Thus, let Fig. 35 represent a heavy body whose centre of gravity is G, floating in equilibrium in a heavy fluid.

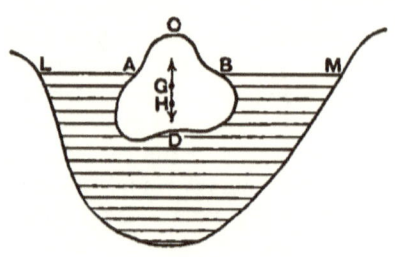

Fig. 35.

The surface over which the fluid pressure is exerted is ADB which is not a closed surface; but, as there is no pressure due to the fluid exerted over the free surface, LM, of the fluid, we can suppose the immersed surface ADB to be closed by the section of the body made by the horizontal plane AB. Hence the resultant of the pressures is the weight of the fluid that would fill the space ADB; and if H is the centre of gravity of this fluid, the resultant pressure acts up through H, so that G and H must be in the same vertical line. Hence there are two distinct conditions of equilibrium of a body floating freely in a heavy fluid, viz.

1°. *the weight of the body must be equal to the weight of the fluid which it displaces*; and

2°. *the centre of gravity of the body and the centre of gravity of the fluid that would statically fill its place (centre of buoyancy) must be in the same vertical line.*

We have hitherto supposed that the only fluid displaced by the body is that represented in the vessel below the surface LM; but if above this there is air, whose weight is considered, there is also displaced a volume of air represented by ACB, and the resultant effect of the air is to produce an *upward* vertical force, even though (as in the figure) the air pressure exerted by the air actually in contact with the displacing body should be a *downward* force; for, we must remember that the surface LM of the lower fluid is all subject to air pressure which is (by Pascal's Principle) transmitted undiminished all through this fluid, so that the lower part, ADB, of the surface of the body is really acted upon all over by air pressure of constant intensity. Now by Cor. 1, the resultant of this system of air pressures on the curved surface ADB, is the same as if the pressure was applied over the lower side of the plane area AB in which the surface LM cuts the body. The resultant air pressure is, therefore, an upward force equal to the weight of the air that would statically fill the space ACB, and it acts through the centre of gravity of this air.

The case of a balloon floating in the air is also an instance of the principle of Archimedes; the force of buoyancy is the weight of the air that could statically replace all the solid portions of the balloon and the gas which it contains. It must not be supposed that, since the balloon is a comparatively small body, the intensity of the air pressure is constant all over its surface—a not unnatural error; for, if this air pressure were of constant intensity all over the surface, its resultant would be absolutely zero, as we have already seen, and there would be no force of buoyancy. If the medium surrounding a body is ever so slightly acted upon by gravitation, its intensity of pressure cannot be constant, and hence the densities of the air at the top and at the bottom of the balloon are not the same.

23. Introduction of Fictitious Forces. In the case in which a body is partially immersed in a fluid, or a part of the body is in one fluid and the remainder in another, it is often very convenient to introduce fictitious forces of buoyancy in one part of the calculation and to take them away in another.

Thus, suppose Fig. 35 to represent a body of which the portion ADB is immersed in water while the portion ACB is in vacuo. Then the actual force of buoyancy is due to the volume ADB of water; but we can complete the volume of the displaced water by supposing the portion ACB to be also surrounded by water, and then supposing that there is a *downward* force, in addition, due to the action of this portion ACB of water taken negatively. Thus the actual force of buoyancy—viz. an upward force at H equal to the weight of the volume ADB of water—can be replaced by an *upward* force equal to the weight of the whole volume $ADBC$ of water acting at the centre of gravity of the homogeneously filled volume $ADBC$ (not G, the c. g. of the body, unless the body is itself a homogeneous solid), together with a *downward* force equal to the weight of the volume ACB of water acting at the centre of gravity of the homogeneously filled volume ACB.

In the same way, if the portion ACB is in a liquid of specific weight w_1, and ADB in one of specific weight w_2, we may regard the force of buoyancy as consisting of an upward force equal to the weight of the whole volume $ADBC$ of the liquid w_2 together with a downward force equal to the weight of a fictitious liquid of specific weight $w_2 - w_1$ acting at the centre of gravity of the homogeneously filled volume ACB.

24. Resultant Pressure on an unclosed curved surface.

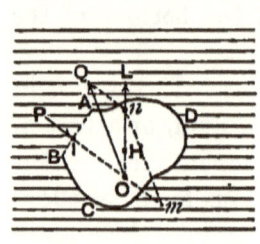

Fig. 36.

Suppose $BCDA$, Fig. 36, to represent any unclosed surface in a heavy fluid, and suppose its bounding edge to be a plane curve so that the surface can be closed by a plane base, represented by AB. It is required to find the resultant of the fluid pressures exerted on one side of the unclosed surface.

Closing the surface by means of the plane base AB, the resultant of the pressures all over the outside of the completely closed surface is the vertical upward force, L represented by the line HL drawn through the centre of gravity, H, of the fluid which would fill the volume. But if P is the resultant fluid pressure on the plane base AB, acting at the centre of pressure, I, the force L is the resultant of P and the resultant pressure over the unclosed part. This latter force, Q, is therefore found by producing the lines of action of L and P to meet—at O, suppose,—and drawing On and Om to represent L and P, respectively; then the required force Q is represented by the line OQ which is equal and parallel to mn.

If the fluid is a homogeneous liquid of specific weight w, if A is the area of the plane base AB, $\bar{z}$ the depth of the centre of area of AB below the free surface, and V is the volume of the closed surface,

$$P = A\bar{z}w, \text{ and } L = Vw.$$

Hence if θ is the inclination of the plane base AB to the horizon,
$$Q = w\sqrt{V^2 + 2VA\bar{z}\cos\theta + A^2\bar{z}^2};$$

horizontal component of $Q = A\bar{z}w\sin\theta$,

vertical component of $\quad Q = (A\bar{z}\cos\theta - V)w$.

EXAMPLES.

1. Suppose a right cone whose axis is vertical and vertex downwards to be filled with a liquid; find the resultant pressure on one-half of the curved surface determined by any plane containing the axis.

Let ACB, Fig. 37, be the vertical plane of section, and $ACDB$ the half of the curved surface on which we desire to find the resultant liquid pressure.

Consider the separate equilibrium of the fluid contained between this curved surface and the triangle ACB. It is kept at rest by its weight, the pressure of the remaining fluid on the area ACB acting at I, the centre of pressure on this triangle,

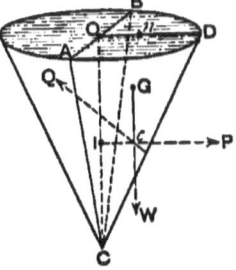

Fig. 37.

and by the pressure of the curved surface $ACDB$. The weight acts through G, the centre of gravity of the semi-cone; and if on the diameter, OD, which is perpendicular to AB, we take the point n such that $On = \dfrac{4r}{3\pi}$, where r is the radius of the base, this point n is the centre of area of the semicircle ADB, so that G lies on nC and $Gn = \frac{1}{4}Cn$ (*Statics*, Vol. I, Art. 163). The point I is half way down OC (Art. 15). If P is the pressure on the triangle ACB, $h =$ height of cone,

$$P = \tfrac{1}{3} r h^2 w; \quad \text{and} \quad W = \tfrac{1}{6}\pi r^2 h w,$$

where $W =$ weight of liquid. The lines of action of P and W meet in a point c, whose position is thus completely known; and by drawing cP and cW to represent P and W on any scale, the diagonal through c of the rectangle thus determined will represent Q, the resultant pressure of the curved surface on the fluid in the semi-cone. The line cQ is drawn to represent this pressure, and this force reversed is the pressure of the fluid on the surface.

2. If the cone is closed by a base, and the axis is held horizontal, find the resultant pressure on the lower half of the curved surface.

Ans. If X and Y are the horizontal and vertical components of the resultant pressure,

$$X = \left(\dfrac{\pi}{2} + \dfrac{2}{3}\right) r^3 w,$$

$$Y = \left(1 + \dfrac{\pi}{6}\right) r^2 h w,$$

and the line of action of the pressure passes through a point whose distances from the base and the axis of the cone are

$$\dfrac{h}{4} \cdot \dfrac{\pi + 8}{\pi + 6} \quad \text{and} \quad \dfrac{r}{4} \cdot \dfrac{3\pi + 16}{3\pi + 4}$$

(see example 10, p. 109).

3. If a hollow cylinder is filled with liquid and held with its axis vertical, determine the magnitude and line of action of the resultant pressure on one half of the curved surface cut off by a vertical plane through the axis.

Pressure on Curved Surfaces.

Ans. It is a horizontal force equal to $r^2 hw$ acting in a line $\dfrac{h}{3}$ from the base.

4. If the cylinder is closed at both ends and held with its axis horizontal, find the resultant pressure on the lower half of the curved surface.

Ans. A vertical force $= \left(2 + \dfrac{\pi}{2}\right) r^2 hw$.

5. In example 2 find the magnitude and line of action of the resultant pressure on the upper half of the curved surface.

Ans. If X and Y are the horizontal and vertical components of the pressure,

$$X = \left(\dfrac{\pi}{2} - \dfrac{2}{3}\right) r^3 w,$$

$$Y = \left(1 - \dfrac{\pi}{6}\right) r^2 hw,$$

and the line of action of the resultant passes through a point whose distances from the base and the axis of the cone are

$$\dfrac{h}{4} \cdot \dfrac{8-\pi}{6-\pi} \quad \text{and} \quad \dfrac{r}{4} \cdot \dfrac{16-3\pi}{3\pi-4}.$$

6. A spherical shell is filled with liquid; find the magnitude and line of action of the resultant pressure on the curved surface of either hemisphere cut off by any vertical central plane.

Ans. The line of action passes through the centre of the sphere; the horizontal component is $\pi r^3 w$, and the vertical $\tfrac{2}{3} \pi r^3 w$.

7. A spherical shell is filled with liquid; find the magnitude and line of action of the resultant pressure on each of the hemispheres into which the sphere is divided by any diametral plane.

Ans. If θ is the inclination of the plane section to the horizon, the pressure on one hemisphere is the resultant of two forces $\pi r^3 w$ and $\tfrac{2}{3} \pi r^3 w$, respectively perpendicular to the plane section and vertical, the lines of action of these forces including an angle θ, while the pressure on the other curved surface is the resultant of the same forces including an angle $\pi - \theta$; and both pass through the centre.

122 *Hydrostatics and Elementary Hydrokinetics.*

8. If a hole is made in the top of the shell and fitted with a funnel, find the height to which the funnel must be filled with the liquid in order that the resultant pressure on one of the hemispheres shall be a horizontal force.

Ans. The height $= r(\frac{2}{3}\sec\theta - 1)$.

25. General Principle of Buoyancy. In Art. 22 we have enunciated the principle of buoyancy in the case in which the buoyant medium is acted upon by the attraction of the earth—that is, the principle has been applied in the case of a heavy fluid of uniform or variable density, compressible or incompressible. It is evident that the same principle holds in general for any medium the particles of which are acted upon by any system of external forces, electric, magnetic, or other.

Thus, in Fig. 38, let AB be a closed surface traced out in imagination in a medium of any kind the particles of which

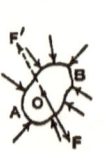

Fig. 38.

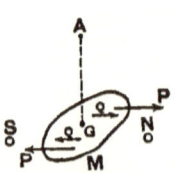

Fig. 39.

are acted upon by any system of forces, and let the resultant of these forces on the particles contained within the surface AB—if they have a single resultant at all—be a force represented by OF. Then, considering the separate equilibrium of the portion of the medium within AB, we see that this portion is kept in equilibrium by the force OF and by the pressures (represented as normal, though not necessarily so) exerted by the surrounding part of the medium on the various elements of the surface AB. It follows that these pressures have a single resultant, OF', exactly equal and opposite to OF.

Hence if AB were the surface of a foreign body the particles of which are subject to the same external forces as those acting on the medium, though the resultant, R, of these on the body AB may now be different both in magnitude and in direction from OF, the resultant of the pressures exerted on this body by the surrounding medium will still be OF'; and in order that this body should be in equilibrium without the aid of further forces (tensions of cords, &c.) it must take such a position in the medium that R is exactly equal and opposite to OF'—a condition which it may or may not be possible to fulfil.

Again, in Fig. 39, let M be any foreign body immersed in a medium—air, suppose,—and let its molecules be subject to the attractive and repulsive forces of two magnetic poles, N and S; suppose the resultant action of these poles on the body to reduce to two equal and opposite forces, P, P, forming a couple; and, to eliminate the effect of gravitation, suppose the body supported by a cord attached to its centre of gravity, G. Then the body is also acted upon by the pressures of the surrounding medium on its various elements of surface. What is the resultant action of these pressures? To answer this question, we must imagine the place of the body occupied statically by a portion of the medium itself. If the magnetic forces do not produce any effect on the particles of the medium, the pressures on the elements of surface of the replacing medium are simply equivalent to a vertical upward force acting through the centre of gravity of this portion of the medium and equal to its weight—which would usually be a very small force; but if the medium is affected by magnetic forces, and if the resultant action of these forces consists of a couple, Q, Q, the resultant action of the surrounding medium must consist—in addition to the previously mentioned small gravitation force of buoyancy—of a couple equal and opposite to the couple Q, Q. Hence, neglecting gravitation forces, if the foreign body is in equilibrium, it must place itself in such a position that the magnetic couple, P, P, produced on it;

124 *Hydrostatics and Elementary Hydrokinetics.*

together with the magnetic couple of buoyancy, $-Q, -Q$, are in *stable* equilibrium with the couple produced by the torsion of the suspending cord.

If the moment of the couple P, P, is greater than that of the couple Q, Q of buoyancy, the position of stable equilibrium of the body M will be different from that assumed on the contrary supposition. Thus, if M is a bar of iron, the couple P, P is greater than the couple Q, Q, and the bar will set axially, i.e. in the line NS joining the two magnetic poles; but if M is a bar of bismuth, Q, Q is greater than P, P, and the bar will set equatorially, i.e. at right angles (or inclined) to the line NS.

Such is, in a general way, the explanation of the behaviour of diamagnetic bodies in a magnetic field, or of magnetic bodies placed in media more strongly acted upon by magnetic forces than are the bodies themselves.

Examples.

1. A solid homogeneous right cone floats in a given homogeneous liquid; find the position of equilibrium, firstly, when the vertex is down and base up; and, secondly, when the base is down and the vertex up.

Let w', V, h be the specific weight, volume, and height of the cone; let w be the specific weight of the liquid, and x the length of the axis immersed when the vertex is down. Then since the volumes of similar solids are proportional to the cubes of their corresponding linear dimensions, the volume of the displaced liquid $= \dfrac{x^3}{h^3} V$. Hence, equating the force of buoyancy to the weight of the cone,

$$\frac{x^3}{h^3} Vw = Vw',$$

$$\therefore x = h \left(\frac{w'}{w}\right)^{\frac{1}{3}}.$$

In the second case, if x is the length of the axis above the

liquid, the volume of the displaced liquid $= \left(1 - \dfrac{x^3}{h^3}\right) V$, and we have
$$x = h\left(1 - \dfrac{w'}{w}\right)^{\frac{1}{3}}.$$

2. A solid homogeneous isosceles triangular prism floats in a given homogeneous liquid; find the position of equilibrium in each of the two previous cases.

If x is the depth of its edge below the surface, h the height of the isosceles triangle which is the section of the prism by a plane perpendicular to the edge, and A the area of this section, since the areas of similar figures are as the squares of their corresponding linear dimensions, $\dfrac{x^2}{h^2} A$ is the area of the face of the immersed prism in the first case, and if $l = $ length of edge, the volume of the prism is lA, so that the volume of the immersed prism is $\dfrac{x^2}{h^2} V$. Hence
$$\dfrac{x^2}{h^2} Vw = Vw',$$
$$\therefore \ x = h\left(\dfrac{w'}{w}\right)^{\frac{1}{2}}.$$

In the second case,
$$\left(1 - \dfrac{x^2}{h^2}\right) Vw = Vw',$$
$$\therefore \ x = h\left(1 - \dfrac{w'}{w}\right)^{\frac{1}{2}}.$$

3. A uniform rod, AB, of small normal section and weight W has a mass of metal of small volume and weight $\dfrac{1}{n} W$ attached to one extremity, B; find the condition that the rod shall float at all inclinations in a given homogeneous liquid.

Let $AB = 2a$, let m be the middle point of AB, Fig. 40, G the centre of gravity of the rod and the metal, w' the specific weight of the rod, w that of the liquid, and s the area of the normal section of the rod.

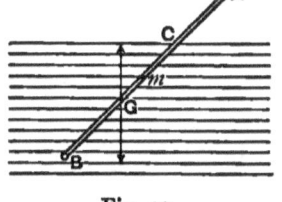

Fig. 40.

126 *Hydrostatics and Elementary Hydrokinetics.*

Then $W = 2asw'$, and $BG = \dfrac{n}{n+1}a$. Also G must be the centre of buoyancy if the rod floats in the oblique position represented, and the length, BC, of the displaced column of liquid $= \dfrac{2n}{n+1}a$, so that if the weight of this column $= \left(1 + \dfrac{1}{n}\right)W$, both conditions of equilibrium will be satisfied, whatever be the inclination of the rod. Equating the weight of the body to the force of buoyancy,

$$\left(1 + \frac{1}{n}\right) 2asw' = \frac{2n}{n+1} asw;$$

$$\therefore \quad (n+1)^2 w' = n^2 w,$$

which is the relation required between the specific weights.

4. A solid homogeneous cylinder floats, with its axis vertical, partly in a homogeneous liquid of specific weight w_1 and partly in one of specific weight w_2, the former resting on the latter; find the position of equilibrium.

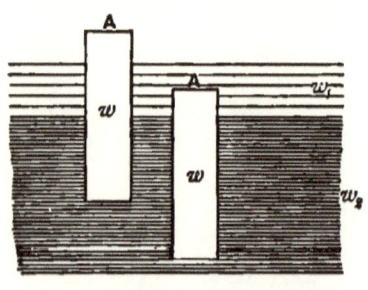

Figs. 41, 42.

Let h be the height of the cylinder, A the area of its base, w its specific weight, and c the thickness of the upper liquid column.

Then if we assume the top, A, of the cylinder to project a distance x above the upper surface of the upper liquid, as in Fig. 41, and equate the weight of the cylinder to the sum of the forces of buoyancy due to the displacements of the liquids, we have

$$hw = cw_1 + (h-c-x)w_2, \quad \dots \dots \quad (1)$$

$$\therefore \quad x = h\left(1 - \frac{w}{w_2}\right) - c\left(1 - \frac{w_1}{w_2}\right). \quad \dots \quad (2)$$

If $c\left(1 - \dfrac{w_1}{w_2}\right) > h\left(1 - \dfrac{w}{w_2}\right)$, we must write

$$x = -\left\{c\left(1 - \frac{w_1}{w_2}\right) - h\left(1 - \frac{w}{w_2}\right)\right\}, \quad \dots \quad (3)$$

and it would appear that the position of equilibrium is one in which, as in Fig. 42, A is below the upper surface of the upper liquid by the distance

$$c\left(1-\frac{w_1}{w_2}\right)-h\left(1-\frac{w}{w_2}\right), \quad \ldots \quad (4)$$

in virtue of the usual interpretation of a negative co-ordinate in algebra.

To take a numerical case, suppose $c = \frac{1}{2}h$, and

$$w : w_1 : w_2 = 5 : 2 : 6 ;$$

then $x = -\frac{1}{6}h$, and it would appear that A is $\frac{1}{6}h$ below the upper surface of the upper fluid.

Now if we had originally assumed A to be, as in Fig. 42, at an unknown distance, x, below the surface, our equation would have been
$$hw = (c-x)w_1 + (h-c+x)w_2, \quad \ldots \quad (5)$$

$$\therefore \quad x = c - h\frac{w_2 - w}{w_2 - w_1}, \quad \ldots \ldots \ldots \quad (6)$$

which disagrees with (4), and which in the particular numerical case gives $x = \frac{1}{4}h$, instead of $x = \frac{1}{6}h$, which we had been led to expect by interpretation of the negative value (3).

Why the disagreement? Because the continuity of the values of variables in algebra and algebraic geometry finds no corresponding characteristic in the hydrostatical conditions. In fact, the supposition that the negative value (3) harmonises with the physical assumptions leading to the first solution is untrue; for, in this solution we assume that, whatever be the unknown position of equilibrium of the body, the *whole column of the upper liquid is operative in producing its force of buoyancy*, as is evident from the first term, cw_1, at the right-hand side of (1); whereas the supposition that A is below the upper surface of this liquid is an explicit assumption that the whole column of the liquid may *not* be so operative. Hence we ought not to expect the two solutions to agree.

In the case, therefore, in which the value of x in (2) is negative, the correct result is (6) and not (3).

5. A heavy uniform bar, AB, of small cross-section is freely moveable round a horizontal axis fixed at one extremity, A, at a given height above the surface of a homogeneous liquid in which the rod partly rests; find the position of equilibrium and the pressure on the axis.

Let $AB = 2a$; let h be the height of A above the liquid; let s = area of cross-section of the rod; let w' and w be the specific weights of the rod and the liquid; and let θ = the angle between AB and the vertical.

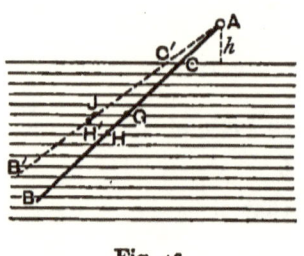

Fig. 43.

Then if BC is the part immersed, the centre of buoyancy, H, is the middle point of BC. If W=weight of rod, $W = 2asw'$; also

$$BC = 2a - h\sec\theta,$$

∴ the force, L, of buoyancy

$$= (2a - h\sec\theta)\,sw.$$

The rod is in equilibrium under the action of L, W, and the pressure at A, which last must be vertical and $= W - L$.

Taking moments about A for equilibrium,

$$W \cdot AG \sin\theta = L \cdot AH \sin\theta, \quad \ldots \ldots \quad (1)$$

and if we reject the factor $\sin\theta$, i.e. omit the consideration that $\sin\theta = 0$ gives one position of equilibrium (the vertical one), we have

$$4a^2 w' = (4a^2 - h^2 \sec^2\theta) w, \quad \ldots \ldots \quad (2)$$

$$\therefore \cos\theta = \frac{h}{2a}\left(\frac{w}{w-w'}\right)^{\frac{1}{2}}. \quad \ldots \ldots \quad (3)$$

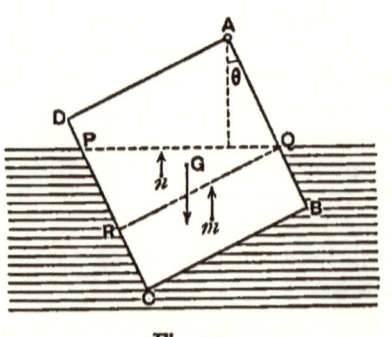

Fig. 44.

The oblique position requires w to be greater than w' and also

$$w > \frac{4a^2}{4a^2 - h^2} \cdot w';$$

so that, for example, if the bar were of metal and the liquid water, the only position of equilibrium would be the vertical one.

6. A uniform square board, $ABCD$, is moveable in a vertical plane about a smooth horizontal axis fixed at the corner A, at a given height above the surface of a liquid; find an equation for its position of

equilibrium, assuming the figure of the immersed portion to be a trapezium.

Let θ be the inclination of AB to the vertical, $AB = 2a$, height of A above liquid $= 2h$, $w' =$ specific weight of board, $w =$ specific weight of liquid; let PQ be the line of floatation, and draw QR parallel to BC.

Then we must equate the moment of the weight of the board about A to the moment of the force of buoyancy about A. But we may consider the force of buoyancy as consisting of two forces, viz. that due to the weight of the portion $QRCB$ acting upwards through m, the centre of gravity of this parallelogram, and that due to the weight of the triangular portion PQR acting upwards through n, the centre of gravity of this triangle.

Now the area of $QRCB = 4a(a - h\sec\theta)$, and the distance of m from the vertical through $A =$ half difference of distances of C and $Q = a(\cos\theta - \sin\theta) - h\tan\theta$. Also the area

$$PQR = 2a^2 \tan\theta,$$

and distance of n from vertical through A

$$= \tfrac{1}{3}(\text{dist. of } P + \text{dist. of } R - \text{dist. of } Q),$$

by ex. 6, p. 33, and therefore

dist. of $n = \tfrac{2}{3}\{a(\cos\theta + \sec\theta) - 3h\tan\theta\}$.

Also distance of G from vertical through A is $a(\cos\theta - \sin\theta)$. Hence the equation of moments about A is

$$4a(a - h\sec\theta)\{a(\cos\theta - \sin\theta) - h\tan\theta\}w$$
$$+ \tfrac{4}{3}a^2\tan\theta\{a(\cos\theta + \sec\theta) - 3h\tan\theta\}w$$
$$= 4a^3 w'(\cos\theta - \sin\theta),$$

or $(3\cos^3\theta - 2\sin\theta\cos^2\theta + \sin\theta)a^2 - 3ah + 3h^2\sin\theta$

$$= 3a^2 \frac{w'}{w} \cos^2\theta(\cos\theta - \sin\theta).$$

7. Solve the previous problem on the assumption that the unimmersed portion is triangular.

Let P, Q be now the points in which AD and AB, respectively, cut the surface of the liquid. Then we may consider the force of buoyancy, i.e. the weight of $PQBCD$, as consisting of the weight of the whole $ABCD$ (of specific weight w) acting upwards at G, together with a *downward* force equal to the weight

of a triangular portion PAQ acting at the centre of gravity of this triangle. Hence the forces which give equal and opposite moments about A are $4a^2(w-w')$ acting upwards at G and $4h^2 w \cosec 2\theta$ acting downwards at the centre of gravity of the triangle PAQ. Now (ex. 6, p. 33) the distance of this point from the vertical through $A = \frac{2}{3}h(\cot\theta - \tan\theta)$. Hence, on expelling the factor $\cos\theta - \sin\theta$, the equation of moments becomes
$$3a^3(w-w')\sin^2 2\theta - 4h^3 w(\cos\theta + \sin\theta) = 0,$$
or if we take ϕ as the inclination of the diagonal AC to the vertical, since $\theta = \dfrac{\pi}{4} - \phi$, we have
$$3a^3(w-w')\cos^2 2\phi - 4h^3 w \sqrt{2}\cos\phi = 0. \quad . \quad . \quad (a)$$

The factor expelled indicates the vertical position of AC—a position which is *à priori* evident.

26. Equilibrium of a Body in Heterogeneous Fluid. If the fluid in which the body $CADB$, Fig. 35, is floating under the influence of gravity alone is one of variable density, the positions of equilibrium are still found by the principle of buoyancy—viz. that the weight of the body is equal to the weight of the fluid displaced, and that the two centres of gravity are in the same vertical line.

Thus, supposing the specific weight, w, of the fluid at any depth, z, to be any given function of z, draw a horizontal section of the body at depth z, and let A be the area of this section. Drawing another horizontal section at depth $z + dz$, the volume between these sections is $A dz$, so that, if $W = $ weight of body,

$$W = \int_0^c w A dz,$$

where c is the depth of the lowest horizontal tangent plane to the surface of the body. This equation will determine c.

We shall confine our attention to a few simple examples, as the question is one of small importance.

Examples.

1. If a uniform solid cone floats, vertex down, in a fluid in which the density is proportional to the depth, find the length of the axis immersed.

Let w' be the specific weight of the cone, and let the specific weight of the fluid at depth z be represented by $w'\dfrac{z}{\beta}$, where β is some constant length. Taking a section of the cone at a distance x from the vertex, $A = \dfrac{\pi r^2}{h^2} \cdot x^2$; therefore

$$\tfrac{1}{3}\pi r^2 h w' = \dfrac{\pi r^2}{\beta h^2} w' \int_0^{h'} x^2 (h' - x)\, dx,$$

h' being the length of the axis, h, which is immersed. From this we have
$$h' = (4\beta h^3)^{\frac{1}{4}}.$$

2. If the cone in the last is replaced by a solid cylinder, find the length immersed.

Ans. $h' = (2\beta h)^{\frac{1}{2}}$.

3. Find the height at which a spherical balloon of given weight and radius will rest in the atmosphere, assumed of uniform temperature, and the variation of gravity being neglected.

Ans. If r feet = radius, h = height of centre of balloon, w_0 the mass in pounds of a cubic foot of air at the ground, B = mass of balloon, and k has the value given in (9), p. 81, the equation which determines h is

$$\pi k^2 w_0 \left[r \cosh \dfrac{r}{k} - k \sinh \dfrac{r}{k} \right] e^{-\frac{h}{k}} = B.$$

4. Find the position of the centre of buoyancy of the balloon on the same suppositions.

Ans. Its distance below the centre is

$$\dfrac{(r^2 + 3k^2) \sinh \dfrac{r}{k} - 3kr \cosh \dfrac{r}{k}}{r \cosh \dfrac{r}{k} - k \sinh \dfrac{r}{k}}.$$

132   *Hydrostatics and Elementary Hydrokinetics.*

27. Stability or Instability of Equilibrium. In determining the *positions* of equilibrium of a system in Hydrostatics, and in Statics generally, we are not, so far, concerned with the *nature* of the equilibrium. A particular position which we determine by the ordinary statical equations (those of resolution and moments of forces) may be one which exists only theoretically and could not exist in practice, because it may be one from which if the system is displaced, ever so slightly, and then abandoned to the forces in action, the displacement becomes greater and greater, instead of being corrected and destroyed. If a position (A) is one in which the conditions of equilibrium are all satisfied, and if the system is displaced by any means, or imagined to be displaced, into a position (A') very slightly differing from the first, and in this new position abandoned to the forces in action, then, unless the forces are such as to drive the system back from (A') to (A), the equilibrium is unstable and is practically useless. Thus, a pin resting a horizontal plane of glass will theoretically be in equilibrium if it stands vertically on its point; but this position does not practically exist, because it is unstable for the slightest displacement. A uniform rod, AB, freely moveable round a smooth horizontal axis at the end A is theoretically in equilibrium if it is placed vertical with the end B uppermost, but, again, this position does not practically exist.

Before discussing particularly the criteria of stability or instability, we may take a few simple examples in which the principle to be employed is obvious.

Thus, to find whether the oblique position of the rod AB in example 5, p. 128, is stable or unstable, imagine the rod displaced into the position $A'B'$, such that the small angle $CAC' = \delta\theta$; then if the sum of the moments about A of the forces acting on the rod in this position is in the sense opposed to that of the angular displacement, these forces will drive the rod back into the position AB, and the equilibrium will be stable.

Generally, without assuming that AB is a position of equilibrium, let M be the sum of the moments of the force of buoyancy and the weight about A in a clockwise sense. Then we have

$$M = \frac{s}{2} \sin\theta \, \{4a^2(w-w') - h^2 w \sec^2\theta\},$$

and the positions of equilibrium are determined by putting $M = 0$. These have been determined in the example referred to. Now if when θ is changed to $\theta + \delta\theta$, the increment of M is δM, and if δM and $\delta\theta$ have the same sign, the newly introduced forces produce a moment which is in the sense of the displacement and the equilibrium is unstable. In other words, if in any position of equilibrium

$$\frac{dM}{d\theta} \text{ is } +, \text{ the equilibrium is unstable,}$$

$$\frac{dM}{d\theta} \text{ is } -, \text{ the equilibrium is stable.}$$

Examining first the vertical position, $\theta = 0$, we have

$$\frac{dM}{d\theta} = \frac{s}{2}\{4a^2(w-w') - h^2 w\},$$

so that this position is stable when $h^2 w > 4a^2(w-w')$, i.e. when the oblique position does not exist.

Examining the oblique position (when it exists), we have

$$\frac{dM}{d\theta} = -sh^2 w \sin^2\theta \sec^3\theta,$$

which value is necessarily negative; therefore the oblique position, when it exists, is stable.

EXAMPLES.

1. Find whether the equilibrium of the body in example 6, p. 128, is stable or unstable in the positions determined.

Calculate the sum, M, of the moments of the forces about A, in a clockwise sense, in any position, θ; and supposing θ to be increased by $\delta\theta$, let δM be the change in the value of M. Since the sense of M is opposed to that in which θ increases, if δM and $\delta\theta$ have the same sign, the restoring moment will be increased by the displacement, and the equilibrium will be stable.

In other words, the equilibrium will be stable if $\dfrac{dM}{d\theta}$ is positive in the position of equilibrium.

Now we find, as in the example quoted, if l is the thickness of the board,

$$M = \frac{4l}{3}(\cos\theta - \sin\theta)[3a^3(w-w') - 4h^3 w(\cos\theta + \sin\theta)\operatorname{cosec}^2 2\theta].$$

We may, for simplicity, omit l, or consider it to be the unit of length, since its actual value is immaterial to the discussion.

It will be more simple to take ϕ, the inclination of AC to the vertical as the variable; then

$$M = \frac{4\sqrt{2}}{3}\sin\phi\,[3a^3(w-w') - 4\sqrt{2}\,h^3 w\cos\phi\,\sec^2 2\phi]; \quad . \;(1)$$

and in any equilibrium position the equilibrium will be stable or unstable according as $\dfrac{dM}{d\phi}$ is negative or positive.

Examining the symmetrical position, $\phi = 0$, we have

$$\frac{dM}{d\phi} = \frac{4\sqrt{2}}{3}[3a^3(w-w') - 4\sqrt{2}\,h^3 w],$$

which shows that in all cases in which $w' > w$ the equilibrium is stable; and, in fact, the equilibrium is stable if

$$\frac{w'}{w} > 1 - \frac{4\sqrt{2}}{3}\frac{h^3}{a^3},$$

and unstable if $\dfrac{w'}{w}$ is less than this.

When the symmetrical position, $\phi = 0$, is stable, the inclined position does not exist, as is seen thus: let

$$\frac{w'}{w} = 1 - \frac{4\sqrt{2}}{3}\frac{h^3}{a^3}(1-k),$$

where k is any number.

If k is positive, the symmetrical position is stable, and if k is negative, this position is unstable. Then (1) becomes

$$M = \tfrac{3\,2}{3} h^3 w \sin\phi \left[1 - k - \cos\phi \sec^2 2\phi\right], \quad . \quad . \quad . \quad (2)$$

and the inclined position is given by the equation

$$\frac{\cos\phi}{\cos^2 2\phi} = 1 - k. \quad . \quad . \quad . \quad . \quad . \quad . \quad (3)$$

Now this equation cannot be satisfied by any admissible value of ϕ if k is positive, because the least value of the left-hand side is 1, which it has when $\phi = 0$. Hence, unless k is negative, (3) cannot be satisfied by any admissible value of ϕ.

When the symmetrical position is unstable, the inclined is, of course, stable.

2. Find the conditions of equilibrium of a solid body floating in a liquid of greater density than the solid; and show that when it is in stable equilibrium the height of the common centre of gravity of the solid and liquid is a minimum.

If a body be floating in a liquid contained in a cylindrical vessel and be pressed down through a small distance, Δz, show that the common centre of gravity of the body and the liquid will be raised through a height

$$\frac{B}{A-B} \cdot \frac{(\Delta z)^2}{2h},$$

where A and B are the areas of the cross-sections of the cylinder and the body in the plane of floatation, and h is the height of this plane above the base of the cylinder. (*Mathematical Tripos*, 1878.)

The first part is evident from the general principle that the centre of gravity of any material system (consisting, in this case, partly of a solid and partly of a liquid) under the action of gravity and of forces which do no work in displacement (pressures exerted by the sides of the containing vessel) assumes the lowest position possible.

136 *Hydrostatics and Elementary Hydrokinetics.*

Let z be the height of the centre of gravity of the body and ζ that of the centre of gravity of the liquid that would occupy the place of the immersed portion above the base of the cylinder; W = weight of body, V = volume of liquid displaced, w = specific weight of liquid. Let CD be the original surface of the liquid, and let the line of floatation be marked on the body and denoted by FN; let $C'D'$ be the new surface of the liquid, and let $C'D'$ be at a height Δx above CD.

Then, since the volume of the liquid is unaltered,

$$B \Delta z = (A-B) \Delta x. \qquad (1)$$

Now take the sum of the mass-moments of the body and the liquid about the base in the new position, and we have

$$W(z-\Delta z) + \tfrac{1}{2} A w (h+\Delta x)^2 - Vw(\zeta - \Delta z)$$
$$- \tfrac{1}{2} Bw (\Delta x + \Delta z)(2h + \Delta x - \Delta z).$$

Subtracting from this the mass-moment before displacement, and observing that $W = Vw$, there remains

$$\tfrac{1}{2} A w (2h\Delta x + \Delta x^2) - \tfrac{1}{2} Bw (\Delta x + \Delta z)(2h + \Delta x - \Delta z), \quad (2)$$

of which, by (1), the terms of the first order in the small quantities cancel—as they must by the principle of virtual work; so that (2) becomes

$$\tfrac{1}{2} A w . \Delta x^2 + \tfrac{1}{2} Bw (\Delta z^2 - \Delta x^2),$$

which by (1) is
$$\tfrac{1}{2} \frac{ABw}{A-B} \cdot \Delta z^2.$$

Dividing this by the sum of the weights, Ahw, of the body and liquid, we obtain
$$\frac{B}{A-B} \cdot \frac{\Delta z^2}{2h},$$

for the amount by which the centre of gravity of the compound system is raised.

28. Principle of Virtual Work. The general principle of Virtual Work may be thus stated: *if any material system is in equilibrium under the action of forces applied at given points in the system, and if we imagine these points to receive any small displacements whatever in space but not in the material system itself, the sum of the quantities of work done* *for such displacements by all the acting forces, external and internal to the system, is zero, or, at most, an infinitesimal of the second order if the greatest displacement in the system is regarded as an infinitesimal of the first order.*

It is fully explained in *Statics*, Vol. I, Chap. VII, and Vol. II, Chap. XV, that the internal forces may or may not enter into the equation of Virtual Work, according to the nature of the displacements imagined. Thus, if an internal force* consists of the tension of a string connecting two points, A, B, of the system, this force will not enter into the equation if in the displacements of A and B the distance between them is unaltered. Also if the internal force is the reaction at a smooth joint, A, connecting two bodies in the system, its virtual work will be zero if in the displacements the bodies are still represented as connected by the joint. If the connection were represented as severed, the stress would do work which must be included in the equation.

Now in Hydrostatics we have often to deal with forces whose points of application in the material system would be displaced not merely in space but in the material system itself (and whose magnitudes would also be slightly altered —though this is of no real consequence) if the system were *actually* displaced into any close position; and in such cases a danger of error arises in applying the equation of Virtual Work.

We shall illustrate this by solving the problem of example 5, p. 128, by the principle of Virtual Work and determining whether the equilibrium of the rod is stable or unstable.

---

* More properly *stress*—with a view to the double aspect of every internal force in a system. What we call a force *external* to a given system would also be a stress (with equal and opposite values) if *both* the bodies between which it is exerted were included in the system.

138 *Hydrostatics and Elementary Hydrokinetics.*

For any small angular displacement of the rod, from an equilibrium position, about the axis A, the sum of the works of the force of buoyancy and the weight W vanishes.

In applying this condition it is not impossible that the student would proceed thus: let ζ be the depth of H below the surface and $\bar{z}$ the depth of G; then the equation of virtual work is

$$-L \cdot \delta\zeta + W \cdot \delta\bar{z} = 0.$$

But $L = (2a - h \sec\theta) sw$ and $\zeta = a\cos\theta - \tfrac{1}{2} h$,

∴ $\delta\zeta = -a \sin\theta \cdot \delta\theta$, while $\bar{z} = a\cos\theta - h$,

∴ $\delta\bar{z} = -a \sin\theta \cdot \delta\theta$,

so that the above equation gives $L = W$, a result which is known to be false.

The fallacy involved in this solution is the following.

In applying the equation of virtual work to characterise a position of equilibrium, we must imagine the points of application of all forces to remain the same in the body; we contemplate simply displacements of these points in space—not in the body as well; and, strictly speaking, we do not contemplate new forces introduced by the displacement, although if such newly caused forces are of infinitesimal magnitude and their points of application receive, in the displacement, only infinitesimal motions, the introduction of such forces will not influence the validity of the equation.

The above expression for $\delta\zeta$ is not the one proper to the equation of virtual work.

For, if the imagined new position of the rod is AB', the new centre of buoyancy, H', is the middle point of the immersed portion $C'B'$, and the $\delta\zeta$ above is the difference of level of H and H'; whereas if we take $AJ = AH$, J is the contemplated displacement of the original point, H, of application of the force L, and the value of $\delta\zeta$ proper to the equation is the difference of level of H and J.

Now it is easily seen that the ζ of J is

$$\zeta - (a + \tfrac{1}{2} h \sec \theta) \sin \theta \, \delta\theta,$$

so that the proper value of $\delta\zeta$ is $-(a + \tfrac{1}{2} h \sec \theta) \sin \theta \cdot \delta\theta$, and the equation of virtual work is

$$L(a + \tfrac{1}{2} h \sec \theta) \sin \theta \, \delta\theta - Wa \sin \theta \, \delta\theta = 0,$$

which gives the correct value of θ.

To find the work done in forcing the rod down into the fluid from one inclination, a, to any other, ϕ, take it in any position, AB, and let us find the amount of work done by the force of buoyancy and the weight in reaching the position AB', the angle BAB' being $\delta\theta$. The work of the force of buoyancy is the work of L in the displacement of H to J plus the work of a small force of buoyancy corresponding to the submergence of a small element of length above C, and this latter work is an infinitesimal of the second order. Hence if δV is the work done by the forces acting on the body,

$$\delta V = -L\delta\zeta + W\delta\bar{z},$$

in which $\delta\zeta$ has the second of the above values. Thus

$$\frac{dV}{d\theta} = \frac{s}{2} \{(4a^2 - h^2 \sec^2 \theta) w - 4a^2 w'\} \sin \theta . \quad . \quad . \quad . \quad . \quad (a)$$

$$\therefore V = \frac{s}{2} \int_a^\phi \{4a^2 (w - w') - h^2 w \sec^2 \theta\} \sin \theta \, d\theta$$

$$= \frac{s}{2} \{-4a^2 (w - w') + h^2 w \sec a \sec \phi\} (\cos \phi - \cos a).$$

As regards stability or instability, a position of equilibrium is one of stability if the amount of work done by the forces in reaching it from a given position is a maximum; for, in that case, any other forces applied to disturb the system would have to do positive work. Hence any position of equilibrium will be stable or unstable according as it makes $\frac{d^2 V}{d\theta^2}$ negative or positive.

Assuming that the oblique position given by (3) exists, we have for it

$$\frac{d^2 V}{d\theta^2} = -swh^2 \sec^3\theta \sin^2\theta,$$

which is essentially negative; therefore the oblique position is stable. For the vertical position

$$\frac{d^2 V}{d\theta^2} = \frac{s}{2}\{4a^2(w-w') - h^2 w'\},$$

which is positive when the oblique position exists, and in this case the vertical position is therefore one of instability. When the oblique position does not exist, the above expression on the right is negative, and the vertical position is (as is evident *à priori*) stable.

The value of V could also be calculated from the moment, M, of the acting forces about A; for the work done by a couple of moment M for a small displacement of the body to which it is applied is $M \cdot \delta\theta$; hence

$$\delta V = M \cdot \delta\theta$$
$$= \frac{s}{2} \sin\theta \{4a^2(w-w') - h^2 w \sec^2\theta\} \cdot \delta\theta,$$

which is identical with (a).

As another example of the application of the principle of virtual work, consider the case of two thin uniform rods, AB, BC (Fig. 45), each of specific weight w', freely jointed together at the common extremity B, and resting partly immersed in a homogeneous liquid of specific weight w, the rod AB being freely moveable round a horizontal axis fixed at A at a given depth, $2h$, below the surface of the liquid. It is required to determine the position and nature of the equilibrium.

Pressure on Curved Surfaces. 141

Let $AB = 2a$, $BC = 2b$, $s =$ area of normal section of each rod, $\theta =$ inclination of AB and $\phi =$ inclination of BC to the vertical. Let G, G' be the centres of gravity of the rods, and H, H' their centres of buoyancy, H being the middle point of the immersed portion Am, and H' the middle point of Cn.

The positions of equilibrium can be easily found by elementary principles. Thus, considering the separate

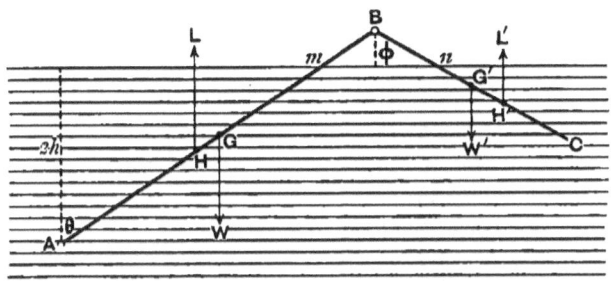

Fig. 45.

equilibrium of the rod BC, we see that the reaction of AB on BC is a vertical upward force at B equal to $W'-L'$; and since the moments of W' and L' about B are equal and opposite, we have at once

$$Cn = 2b\left\{1 - \left(1 - \frac{w'}{w}\right)^{\frac{1}{2}}\right\}, \quad \ldots \quad (1)$$

$= 2kb$, suppose.

Thus, L' is known; and then taking moments about A for the separate equilibrium of AB, we have

$$Am^2 = 4ka\{(a+2b)(2-k)-2b\}; \quad \ldots \quad (2)$$

and these two equations determine the position of the system.

To obtain the equation of virtual work, imagine θ to

142 *Hydrostatics and Elementary Hydrokinetics.*

increase by $\delta\theta$ and ϕ by $\delta\phi$, and calculate the vertical descents of the points H, G, G', H', the points H and H' being supposed not to shift their positions in the rods. If the vertical descents of these points are, respectively, $\delta\zeta$, δz, $\delta z'$, $\delta\zeta'$, the virtual work, δV, done by all the forces is given by the equation

$$\delta V = -L\delta\zeta + W\delta z + W'\delta z' - L'\delta\zeta', \quad \ldots \quad (3)$$

and this virtual work must be put equal to zero for the position of equilibrium.

Now $AH = h\sec\theta$, and if AB turns round A through $\delta\theta$, the vertical descent of H is $AH.\delta\theta.\sin\theta$, so that $\delta\zeta = h\tan\theta.\delta\theta$. To get the vertical descent of H', observe that if B did not move, the vertical descent of H' due to an increase of ϕ would be $-BH'.\delta\phi.\sin\phi$; but B descends through a distance $2a\sin\theta.\delta\theta$; hence

$$\delta\zeta' = 2a\sin\theta.\delta\theta - BH'\sin\phi\,\delta\phi.$$

But $BH' = b + (a\cos\theta - h)\sec\phi$; and we have

$\delta\zeta = h\tan\theta.\delta\theta$; $\delta z = a\sin\theta.\delta\theta$,

$\delta z' = 2a\sin\theta.\delta\theta - b\sin\phi.\delta\phi$,

$\delta\zeta' = 2a\sin\theta.\delta\theta - \{b + (a\cos\theta - h)\sec\phi\}\sin\phi.\delta\phi$.

Also $W = 2asw'$, $W' = 2bsw'$, $L = 2hsw\sec\theta$,

$L' = 2sw\{b - (a\cos\theta - h)\sec\phi\}$.

Substituting these values in (3), and collecting the coefficients of the independent variations $\delta\theta$ and $\delta\phi$, we have

$$\delta V = 2s\{(a^2 + 2ab)w' - 2abw - h^2w\sec^2\theta$$
$$+ 2a(a\cos\theta - h)w\sec\phi\}\sin\theta.\delta\theta$$
$$+ 2s\{b^2(w - w') - (a\cos\theta - h)^2 w\sec^2\phi\}\sin\phi.\delta\phi \quad . \quad (4)$$

Now since the position of equilibrium is obtained by putting $\delta V = 0$, we must have the coefficients of $\delta\theta$ and $\delta\phi$ each equal to zero; and thus we obtain both θ and ϕ.

Equating to zero the coefficient of $\delta\phi$, we obtain at once the result (1).

It is manifest that the expression (4) for δV is a perfect differential, since

$$\frac{dV}{d\theta} = 2s\{(a^2 + 2ab)w' - 2abw - h^2 w \sec^2\theta$$
$$+ 2a(a\cos\theta - h)w\sec\phi\}\sin\theta, \quad (5)$$

$$\frac{dV}{d\phi} = 2s\{b^2(w - w') - (a\cos\theta - h)^2 w \sec^2\phi\}\sin\phi, \quad (6)$$

and it is evident that we get identical results by differentiating (5) with respect to ϕ and (6) with respect to θ.

To find V, the potential work of the forces, in any position, integrate (6) with respect to ϕ and add an undetermined function of θ. Thus

$$V = -2s\{b^2(w-w')\cos\phi + (a\cos\theta - h)^2 w \sec\phi\}$$
$$+ 2s \cdot f(\theta). \quad (7)$$

Taking $\dfrac{d}{d\theta}$ of both sides of (7) and equating the result to $\dfrac{dV}{d\theta}$ in (5), we find

$$f'(\theta) = \{(a^2 + 2ab)w' - 2abw - h^2 w \sec^2\theta\}\sin\theta, \quad (8)$$

$$\therefore f(\theta) = -\{(a^2 + 2ab)w' - 2abw\}\cos\theta - h^2 w \sec\theta + C,$$

where C is a constant. Hence

$$\frac{V}{2s} = -\{(a^2 + 2ab)w' - 2abw\}\cos\theta - h^2 w \sec\theta$$
$$- b^2(w - w')\cos\phi - (a\cos\theta - h)^2 w \sec\phi + C. \quad (9)$$

This may be put into a more simple form by putting

$$Am = 2x \text{ and } Bn = 2y.$$

Thus $\cos\theta = \dfrac{h}{x}$; $\cos\phi = \dfrac{h(a-x)}{xy}$,

and if we put $-\dfrac{V}{2hs} = U$, and

144 *Hydrostatics and Elementary Hydrokinetics.*

$$(a^2 + 2ab) w' - 2abw = A, \quad b^2(w-w') = B,$$

$$U = \frac{A}{x} + wx + \frac{a-x}{x}\left(\frac{B}{y} + wy\right). \quad \ldots \quad (10)$$

Hence

$$\frac{dU}{dx} = w - \frac{1}{x^2}\left\{A + a\left(\frac{B}{y} + wy\right)\right\}; \quad \frac{dU}{dy} = \frac{a-x}{x}\left(w - \frac{B}{y^2}\right). (11)$$

For the position of equilibrium $\frac{dU}{dx} = 0$; and $\frac{dU}{dy} = 0$;

therefore $\quad y = \left(\frac{B}{w}\right)^{\frac{1}{2}}; \quad wx^2 = A + 2a(Bw)^{\frac{1}{2}}, \quad . \quad (12)$

which agree with the previous results, and the first shows that B must be positive, i.e. w must be $> w'$.

For the *stability* of equilibrium, V must be a maximum, and for a maximum or minimum (see Williamson's *Differential Calculus*, Chap. X.)

$$\frac{d^2V}{dx^2} \cdot \frac{d^2V}{dy^2} > \left(\frac{d^2V}{dxdy}\right)^2.$$

Now $\quad \dfrac{d^2U}{dx^2} = \dfrac{2}{x^3}\left\{A + a\left(\dfrac{B}{y} + wy\right)\right\}; \quad \dfrac{d^2U}{dy^2} = \dfrac{2B}{y^3}\dfrac{a-x}{x};$

$$\frac{d^2U}{dxdy} = -\frac{a}{x^2}\left(w - \frac{B}{y^2}\right),$$

and since in the oblique position (distinct from $\phi = 0$) this value of $\frac{d^2U}{dxdy}$ vanishes, and $\frac{d^2U}{dx^2}$ is essentially positive, the condition is simply that $\frac{d^2U}{dy^2}$ must be positive; i.e.

$$a - x \text{ must be } +, \quad \ldots \quad \ldots \quad (13)$$

and this is necessarily the case.

Since $\frac{d^2U}{dx^2}$ is $+$, $\frac{d^2V}{dx^2}$ is $-$, and the value of V which we

Pressure on Curved Surfaces. 145

are discussing is therefore (Williamson, above) a maximum and not a minimum. Hence from (12),

$$wa^2 > A + 2a(Bw)^{\frac{1}{2}}.$$

Restoring the values of A and B, this gives as the condition for stability

$$\frac{w'}{w} < \frac{a^2 + 4ab}{(a + 2b)^2}.$$

29. Positions of equilibrium of a freely floating body. A given body, provided that its weight is less than that of an equal volume of water, may be placed in several positions of equilibrium in the water.

We shall lay down a few definitions of terms in common use with regard to freely floating bodies.

The *section of floatation* of a floating body (Fig. 47) is the section of the body made by the surface, LM, of the liquid. In Fig. 35 the section of floatation is represented by the horizontal line AB.

The *area of floatation* is the area of the section of floatation.

The *displacement* is the volume of the displaced liquid, which is the volume included between the section of floatation and the surface of the immersed portion of the body.

In Fig. 35 the displacement is the volume represented in projection by the curve ADB.

In all possible positions of equilibrium of a given body floating freely in a given liquid the displacement is constant. For, if W is the weight of the body, V the displacement, and w the specific weight of the liquid (i.e. the weight per unit volume), the first condition of Cor. 2, Art 22, gives

$$Vw = W,$$
$$\therefore V = \frac{W}{w},$$

which shows that the displacement is constant.

L

Hence, without any reference to the second necessary condition of Cor 2, Art. 22, all possible positions of equilibrium are exhausted by describing planes so as to cut off a volume $\dfrac{W}{w}$ from the body. All planes which cut off this volume are (without reference to the second condition) possible planes of floatation; that is, if we mark the exterior surface of the body along the curve in which it is cut by any such plane, and we then place the body in the liquid so that this curve lies wholly in the free surface, LM, of the liquid, we shall obtain a position in which the body will float when left to itself, *provided the second condition of Cor. 2, Art. 22, is fulfilled in this position*. Of course, as a rule, this condition will not be fulfilled, so that of the (infinite) number of possible positions, as above defined, only a small number will satisfy both of the conditions that must hold.

All the planes which cut off the constant volume $\dfrac{W}{w}$ from the body envelop a surface called the *surface of floatation*, while the corresponding centres of buoyancy trace out a surface called the *surface of buoyancy*.

Now it is evident that, in order that a plane cutting off the volume $\dfrac{W}{w}$ should determine an actual area of floatation, the right line joining G, the centre of gravity of the body, to the corresponding centre, H, of buoyancy must be at right angles to this cutting plane, because in a position of equilibrium of the body the line GH is vertical, while the section of floatation $\left(\text{which cuts off the volume } \dfrac{W}{w}\right)$ is horizontal.

We shall now show that *in every position of equilibrium the line GH is normal to the surface of buoyancy at H.*

Pressure on Curved Surfaces. 147

Let AB (Fig. 46) be a plane cutting off a volume ADB from a homogeneous body; let H be the centre of gravity of this volume; let $A'B'$ be another plane differing slightly in position from AB and cutting off an equal volume, $A'DB'$; and let H' be the centre of gravity of this new volume. Then the line HH' is ultimately parallel to the plane AB. For, regard the volume ADB as consisting of the portion $A'DB$ and the thin wedge ACA'; and also regard the volume $A'DB'$ as consisting of the portion $A'DB$ and the thin wedge BCB'. Let n be the centre of gravity of the portion $A'DB$ which is common to both volumes, g the centre of gravity of the first wedge and g' that of the second. Then to find H we join g to n and divide gn at H, so that

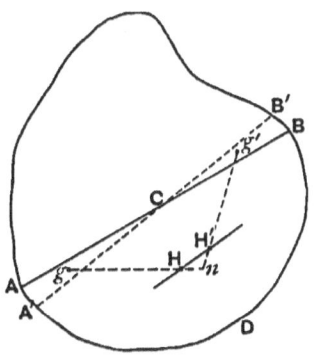

Fig. 46.

$$\frac{gH}{Hn} = \frac{\text{volume of } A'DB}{\text{volume of wedge}};$$

similarly $$\frac{g'H'}{H'n} = \frac{\text{volume of } A'DB}{\text{volume of wedge}}.$$

Hence $\frac{gH}{Hn} = \frac{g'H'}{H'n}$, and therefore the line HH' is parallel to gg', and therefore when the wedges are both made indefinitely thin, the line gg' then lying in the plane AB, the line HH', which then becomes a tangent at H to the locus of H, is parallel to the plane AB. Since this is true whatever be the *orientation* of the new cutting plane $A'B'$, the assemblage of lines HH' which touch the surface of buoyancy at H form the tangent plane to this surface at H, which plane is therefore parallel to the cutting plane AB.

148 *Hydrostatics and Elementary Hydrokinetics.*

It now follows that *all positions in which a given body can float freely in a homogeneous liquid are obtained by drawing normals, GH_1, GH_2, GH_3, ... from the centre of gravity, G, of the body to the surface of buoyancy, and placing the body so that any one of these normals is vertical.* For, the line GH must be vertical—that is, it must be perpendicular to the plane of floatation; and as the tangent plane at H to the surface of buoyancy has been proved to be parallel to the plane of floatation, the line GH must be the normal to the surface of buoyancy at H.

When the contour of the floating body is a surface of continuous curvature, the surface of buoyancy is, of course, a surface of continuous curvature; but when the contour of the body is not of continuous curvature (as in the case of a ship with a closed deck when all geometrically possible displacements—involving the submersion of the deck, the keel being above the surface of the water—are considered) the surface of buoyancy will be a broken or discontinuous surface.

As an example of a surface of buoyancy of discontinuous curvature, take the case in which the body is a triangular prism, the vertical section of which through its centre of gravity is a triangle ABC (Fig. 47), and consider all possible displacements of the body in the plane of this triangle.

If, for definiteness, we assume the specific weight, w, of the fluid to be to the specific weight, w', of the body as 16 to 9, the volume submerged will be $\frac{9}{16}$ of the volume of the body. We are therefore to draw all possible lines, such as LM, across the face of the triangle ABC cutting off areas, $LBCM$, equal to $\frac{9}{16}$ of the area ABC.

The immersed area will sometimes be triangular, and sometimes (as in the figure) quadrilateral. To trace out the locus of the centre of buoyancy when the vertex A alone is above the liquid, the line LM is to be drawn so as to cut off the triangular area $LAM = \frac{7}{16} ABC$; hence if $AM = y$, $AL = z$, $AC = b$, $AB = c$, we have

$$yz = \tfrac{7}{16} bc. \quad \ldots \ldots \ldots \quad (1)$$

Let the first position of the cutting line pass through B, and let the line revolve clockwise so that it assumes the position LM in the figure; then it will reach another position in which it passes through C. When the line passes through B, we have $z = c$, $\therefore y = \tfrac{7}{16} b = AQ$, suppose; the

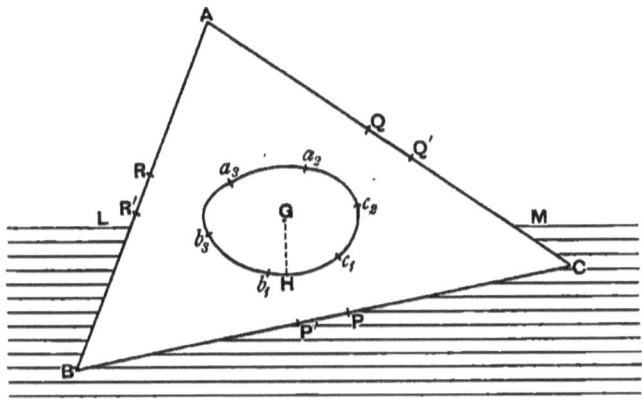

Fig. 47.

cutting line is then BQ, and the immersed area is the triangle QCB, whose centre of gravity is the point c_1.

In the second extreme position $y = b$, $\therefore z = \tfrac{7}{16} c = AR$, suppose; the cutting line is RC, and the immersed area is the triangle RCB, whose centre of gravity is b_1. The centres of gravity of all the intervening quadrilateral areas will lie on a curve, $c_1 H b_1$, between the points c_1 and b_1. This curve is easily proved to be a portion of a hyperbola having asymptotes parallel to AB and AC. For, AM and AL being y and z, the co-ordinates of the centre of gravity of the triangle LAM with reference to AC and AB

are ($\frac{1}{3}y$, $\frac{1}{3}z$), while those of ABC are ($\frac{1}{3}b$, $\frac{1}{3}c$); and if (η, ζ) are the co-ordinates of H, the centre of gravity of the quadrilateral $LBCM$, the *Theorem of Mass Moments* (Art. 10) gives

$$9\eta + \tfrac{7}{3}y = \tfrac{16}{3}b, \quad \ldots \ldots \quad (2)$$

$$9\zeta + \tfrac{7}{3}z = \tfrac{16}{3}c, \quad \ldots \ldots \quad (3)$$

which by (1) give

$$(\tfrac{16}{3}b - 9\eta)(\tfrac{16}{3}c - 9\zeta) = \frac{7^3}{144}bc, \quad \ldots \quad (4)$$

showing that the locus of H is a hyperbola whose centre is at the point ($\tfrac{16}{27}b$, $\tfrac{16}{27}c$).

Let the cutting line still revolve clockwise from the position RC, so that R moves towards A; then the immersed area will be a triangle whose vertex B is submerged, and the locus of the centre of gravity of this triangle will be a portion, $b_1 b_3$, of another hyperbola whose centre is B and asymptotes BA and BC, the point b_3 being the centre of gravity of the triangle ABP cut off when the revolving line is in the position AP, the point P being such that $BP = \tfrac{9}{16}BC$.

Thus there is an abrupt transition from one curve to another at the point b_1.

As the cutting line still revolves clockwise from the position AP, the locus of H will be a portion, $b_3 a_3$, of a hyperbola whose asymptotes are parallel to CA and CB, the immersed area being a quadrilateral until the line reaches the position $Q'B$, such that $CQ' = \tfrac{7}{16}b$. After this the immersed area will be a triangle with the vertex A immersed, the locus of H being a portion, $a_3 a_2$, of a hyperbola whose centre is A and asymptotes AB, AC, and the immersed area will continue triangular until the line reaches the position CR' such that $AR' = \tfrac{9}{16}c$; and so on.

Hence the vertical section of the surface of buoyancy consists of portions of six different hyperbolas whose points

Pressure on Curved Surfaces. 151

of intersection are $c_1, b_1, b_3, a_3, \ldots$. The number of normals that can be drawn from G, the centre of gravity of the prism, to this broken locus of H will determine the number of positions of equilibrium of the prism.

EXAMPLES.

1. A rectangular block of specific weight w' floats in a liquid of specific weight w with one face vertical; find the curve of buoyancy and the positions of equilibrium, the same face being always kept vertical.

Ans. Let the sides of the vertical face be $2b, 2c$, and suppose that in the initial position the side $2c$ is vertical; then, so long as the upper edge $2b$ is out of the liquid and the immersed portion a quadrilateral, the curve of buoyancy is a parabola, concave upwards, whose equation with reference to the horizontal and vertical lines through the initial centre of buoyancy as axes of x and y is

$$3 w' c x^2 = w b^2 y.$$

The initial position is one of equilibrium (which may or may not be stable), and other positions are obtained by drawing normals from the middle of the face to this parabola, provided that these normals fall within the relevant portion of the parabola. Now the relevant portion terminates at the point whose co-ordinates are $\left(\dfrac{b}{3}, \dfrac{w'}{w} \cdot \dfrac{c}{3}\right)$, this point being the centre of buoyancy when the immersed area begins to be triangular. In order that it should be possible to draw a normal within the limits the y of the point at which it is normal must be $< \dfrac{w'}{w} \cdot \dfrac{c}{3}$, and hence

$$\frac{b^2}{c^2} > 6 \frac{w'}{w}\left(1 - \frac{4}{3}\frac{w'}{w}\right).$$

To this portion of a parabola succeeds a portion of a hyperbola which is the curve of buoyancy so long as the immersed area is triangular; this, in turn, is succeeded by another portion of a parabola; and so on.

2. Find the surface of buoyancy in the case of a right circular cone immersed in a homogeneous liquid with its vertex downwards.

152 *Hydrostatics and Elementary Hydrokinetics.*

Ans. So long as no part of the base of the cone is submerged, the surface of buoyancy is a hyperboloid of revolution.

Let a be the semi-vertical angle of the cone, p the perpendicular from the vertex, V, on any plane cutting the cone, and ω the angle which p makes with the axis of the cone. Then the plane through p and the axis will cut the cone in two lines, VA, VB, which intersect the given cutting plane in the points A, B, which are the extremities of the major axis of the ellipse. If $VA = r_1$, $VB = r_2$, we have

$$VA = \frac{p}{\cos(\omega+a)}, \quad VB = \frac{p}{\cos(\omega-a)};$$

hence $AB = \dfrac{p \sin 2a}{\cos^2 a - \sin^2 \omega}$; and the semi-minor axis

$$= (VA . VB)^{\frac{1}{2}} \sin a = \frac{p \sin a}{(\cos^2 a - \sin^2 \omega)^{\frac{1}{2}}}.$$

If V is the volume of the displaced liquid, $V = \dfrac{p}{3} \times$ area of the ellipse cut off.

Hence $V = \dfrac{\pi}{6} \dfrac{p^3 \sin a \sin 2a}{(\cos^2 a - \sin^2 \omega)^{\frac{3}{2}}}$. From this it follows that the product $VA . VB$ is constant, and since the co-ordinates of the centre, C, of the ellipse are the halves of the sums of those of A and B, if we rotate the cutting plane so as to confine the motions of p to one plane, the locus of C is a hyperbola having the generators VA, VB for asymptotes. Hence for all possible positions of the plane, the locus of C is a hyperboloid generated by the revolution of this curve about the axis of the cone. But if H is the centre of buoyancy in any position, H lies on the line VC, and $VH = \frac{3}{4}VC$; hence the locus of H is a similar hyperboloid.

If l, m, n are the direction cosines of p with reference to any two rectangular axes of x and y through the vertex and the axis of the cone, we have $l^2 + m^2 = \sin^2 \omega$; and if x, y, z are the co-ordinates of C, $VA = r_1$, $VB = r_2$, we have

$$x = (r_1 - r_2) \frac{l \sin a}{2 \sin \omega},$$

$$y = (r_1 - r_2) \frac{m \sin a}{2 \sin \omega},$$

Pressure on Curved Surfaces.   153

$$z = (r_1 + r_2)\frac{\cos a}{2}.$$

Also, since $p = k(\cos^2 a - \sin^2 \omega)^{\frac{1}{2}}$, where k is a given constant, it follows that the locus of C is

$$\frac{z^2}{\cos^2 a} - \frac{x^2 + y^2}{\sin^2 a} = k^2.$$

30. Geometrical Theorem. In connexion with the question of the stability of floating bodies the following theorem is important.

A volume AKB, Fig. 48, being cut off from a solid body by a plane section $ALBL'$, any other plane, $A'LB'L'$, making a small angle with the first plane and cutting off an equal volume, $A'KB'$, must pass through the centroid (or 'centre of gravity'), C, of the area $ALBL'$.

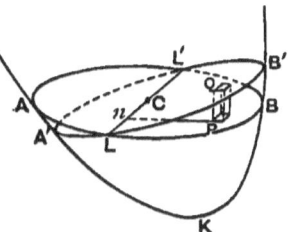

Fig. 48.

For, at any point, P, in the plane section $ALBL'$ describe a small element of area, dS; let the perpendicular, Pn, from P on the line, LL', of intersection of the two planes be denoted by x; let $\delta\theta$ be the angle between the two planes; and round the contour of dS draw perpendiculars to the plane of dS, these forming a prism which intersects the plane $A'LB'L'$ in a small area at Q. Then $\angle QnP = \delta\theta$, $QP = x\delta\theta$, and the volume of the small prism is very nearly $x\,dS.\delta\theta$. Hence the new volume $A'KB' = \text{vol. } AKB + \delta\theta \int x\,dS$, the prisms in the wedge $L'LBB'$ being taken positively, while those in the wedge $L'LA'A$ are taken negatively. If the two volumes cut off are the same, we must have

$$\int x\,dS = 0, \quad \ldots \ldots \quad (a)$$

the integration including all the elements of area of the

plane section $ALBL'$. Now, by the theorem of mass-moments the right-hand side of (a) is $A\bar{x}$, where A is the area of the plane section, and $\bar{x}$ the distance of its centroid from the line LL'; hence $\bar{x} = 0$, i.e. the centroid of the area must lie on LL'.

A visible representation of this fact is obtained by holding in the hand a tumbler partly filled with water and imparting to it small and rapid oscillations which cause the surface of the water to oscillate from right to left; the planes of the successive surfaces of the water can then be seen to pass always through the centre of the horizontal section.

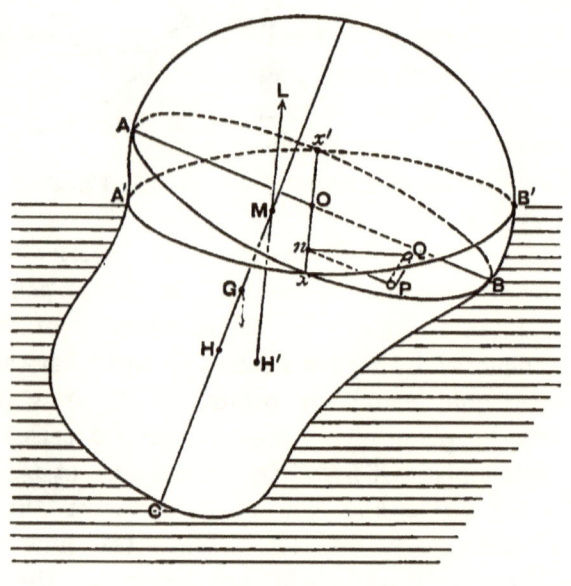

Fig. 49.

31. Small Displacements. Metacentre. Suppose a body, ACB, Fig. 49, floating in equilibrium in a homogeneous liquid to receive any small displacement; it is required to find whether the equilibrium is stable or unstable.

Pressure on Curved Surfaces.   155

Every displacement can be regarded as consisting of two kinds of displacement—viz. a vertical displacement of translation, upwards or downwards, which diminishes or increases the volume of the displaced liquid, and a rotatory or side displacement which leaves the volume of the displaced liquid unaltered.

If the displacement is small, these component displacements can be treated separately, and it is evident that equilibrium for the first kind of displacement is stable.

We shall confine our attention, then, to displacements of rotation which leave the volume of the displaced liquid, and therefore the magnitude of the force of buoyancy, unaltered.

Let G be the centre of gravity of the body, $AxBx'$ the section of floatation before displacement, H the centre of buoyancy (i. e. the centre of volume of the immersed portion ACB) before the displacement, and let this section of floatation be supposed to be marked on the surface of the body.

The body is represented as slighty displaced, the new section of floatation being $A'xB'x'$; the new centre of buoyancy (centre of volume of $A'CB'$) is H', which must be somewhere very close to H, the centre of volume of ACB.

If the body, having been displaced to the new position, is then left to itself, it will be acted upon by two forces, viz. its weight, W, acting through G, and the force of buoyancy, L, (which is equal to W, since the volume of the displaced liquid is constant,) acting vertically upwards through H'. These forces form a couple.

Now it is clear that if the line $H'L$ cuts the line GH above G, in a point M, the body will be acted upon by a couple which tends to destroy the displacement; while, if M is below G, the moment of the couple, being in the sense of the displacement, will cause the body to fall farther from the position of equilibrium, which is therefore unstable.

It may happen, on account of the shape of the body and the position of the axis, $x'x$, of displacement, that the vertical line through H' does not intersect the old line GH of centres of gravity. At present we shall confine our attention to cases in which it does intersect GH, and subsequently we shall find the condition that such intersection shall take place.

Manifestly if G is below H, the equilibrium will be stable, and the consideration of this case may be dismissed. The case in which G is above H is very important inasmuch as it is the case of ships generally, and especially that of large ironclads, in which so much of the mass is in the upper portion.

If p is the length of the perpendicular from G on the line $H'L$, the moment

$$L \cdot p \quad \ldots \quad \ldots \quad \ldots \quad (1)$$

of the new force of buoyancy about G is called the *moment of stability*.

To calculate p, or the position of the point M—which point is called the *metacentre*—replace the actual force of buoyancy due to liquid $A'CB'$ by a force of buoyancy consisting of three components, viz.,

(a) an upward force due to liquid ACB,
(b) an upward force due to liquid $Bxx'B'$,
(c) a downward force due to liquid $Ax'xA'$.

The two sections of floatation intersect in the line xx', and since the volumes ACB and $A'CB'$ are equal, this line xx' must pass through the centroid, O, of the section of floatation (Art. 30).

Since the volume of the wedge $Bxx'B'$ = the volume of the wedge $Ax'xA'$, the forces (b) and (c) being the weights

Pressure on Curved Surfaces. 157

of these wedges of liquid, form a couple, each acting through the centre of gravity of the corresponding wedge; while force (a) is L acting up through H.

Also $L \cdot p =$ the sum of the moments of these forces about the axis through G perpendicular to the plane of displacement. Now since the forces (b) and (c) form a couple, the sum of their moments about all parallel axes is the same, and hence the sum of their moments about the horizontal axis through $G =$ the sum of their moments about xx', which latter we shall take. The wedges may be broken up into an indefinitely great number of slender prisms perpendicular either to the plane $A'xB'x'$ or to the plane $AxBx'$. Taking the latter mode, at any point P in the area $AxBx'$ describe an indefinitely small area dS, and round its contour erect perpendiculars which will cut off a small area at Q on the plane $A'xBx'$. Let AOB be the diameter at O perpendicular to $x'x$; take Ox and OB as axes of x and y; let the perpendicular Pn from P on $x'x$ be y, and let θ be the small angle, QnP, through which the body is displaced round $x'x$. Then the volume of the prism PQ is $\theta y dS$; its weight (reversed for buoyancy) acts at the middle point of PQ, and may be resolved into the components $\theta wy dS \cos\theta$ parallel to PQ and $\theta wy dS \sin\theta$ parallel to Pn; i.e. into components $\theta wy dS$ and $\theta^2 wy dS$ in these directions.

The moment of the latter, being of the second order in θ, may be neglected, while the moment of the former is

$$\theta w y^2 dS. \qquad \qquad (2)$$

By integrating (2) throughout both wedges, we obtain the sum of the moments of the forces (b) and (c), since they both give moments of the same sign about xx'.

Hence the moment of buoyancy due to the wedges is

$$\theta w \int y^2 dS, \qquad \qquad (3)$$

the integration extending all over the section $AxBx'$ of floatation.

If A is the area of this plane section, and k its radius of gyration about the axis xx' of displacement, (3) is

$$\theta w . Ak^2. \qquad (4)$$

The moment of the force (a) about G, in the sense of the previous moment, is

$$-L . GH . \theta. \qquad (5)$$

Hence, if V is the volume of the displaced liquid, since $L = Vw$, the moment of the whole of the forces of buoyancy about the horizontal axis through G perpendicular to the plane of displacement is

$$\theta w (Ak^2 - V . GH); \qquad (6)$$

and, equating this to (1), we have, since $p = \theta . GM$,

$$V . GM = Ak^2 - V . GH, \qquad (7)$$

$$\therefore HM = \frac{Ak^2}{V}, \qquad (8)$$

which determines the position of the metacentre, M.

For stability, therefore,

$$\frac{Ak^2}{V} > HG. \qquad (9)$$

Displacements of constant volume may take place round any diameter of the section, $AxBx'$, of floatation provided that the diameter passes through the centroid of this section (Art. 30); and since for all such displacements both A and V are constant, equation (8) shows that the metacentre will be highest when the displacement takes place round that diameter about which the moment of inertia of the section of floatation is greatest, and lowest if it takes place round the diameter about which the moment of inertia is least. These two diameters are the *principal axes* of the section of floatation at its centroid. If k_2 and k_1 are the greatest and least radii of gyration of the section of floatation about its principal axes, and M_2, M_1, the corresponding metacentres for displacements round them,

$$HM_2 = \frac{Ak_2^2}{V}, \quad HM_1 = \frac{Ak_1^2}{V}. \quad . \quad . \quad . \quad (10)$$

The equilibrium will, then, be least stable when the displacement takes place round the diameter of least moment of inertia, which in the case of a ship is the line from stem to stern.

Since $Vw = W$, (8) can be written

$$HM = \frac{Awk^2}{W}. \quad . \quad . \quad . \quad . \quad (11)$$

32. Experimental determination of Metacentre. The height of the metacentre above the centre of gravity of a ship can be found experimentally by means of a plumb-line and a moveable mass on the deck. Suppose one end of a long string fastened to the top of one of the masts and let a heavy particle hang from the other end of the string. Now if a considerable mass, P, be shifted from one side of the deck to the other, the ship will be tilted through a small angle which can be measured by means of the pendulum if the bob of the pendulum moves in front of a vertical sheet of paper on which the amount of displacement of the bob can be marked. If l is the length of the string and s the distance traversed on the paper by the bob while the mass P is shifted across the deck, $\dfrac{l}{s}$ is the circular measure of the whole angle of deflection of the ship.

Let G be the centre of gravity of the ship, W the weight of the ship and moveable mass together, $2b$ the breadth of the deck, a the perpendicular from G on the plane of the

160 *Hydrostatics and Elementary Hydrokinetics.*

deck, and 2θ the whole angle, $\dfrac{l}{s}$, of deflection. Then, on account of the symmetry of the ship, we can in Fig. 49 take the line HG as passing through O.

Let the mass P be at B, and take moments of the forces acting about G; then

$$W.GM.\theta = P(b+a\theta),$$

$$\therefore\ GM = \frac{P}{W}\left(\frac{b}{\theta} + a\right).$$

The value of a is usually much smaller than $\dfrac{b}{\theta}$, so that, with sufficient accuracy, we have

$$GM = \frac{P}{W}\cdot\frac{b}{\theta},$$

where $\theta = \dfrac{l}{2s}$.

Thus, in a ship of 10,000 tons the breadth of whose deck is 40 feet, if a mass of 50 tons moved from one side to the other causes the bob of a plumb-line 20 feet long to move over 10 inches, the metacentric height is about $4\frac{4}{5}$ feet.

The metacentric heights of large war vessels vary from about $2\frac{1}{2}$ feet to 6 feet.

Examples of the Metacentre.

1. A uniform rectangular block, of specific weight w', floats, with one of its edges vertical, in a liquid, of specific weight w; find the relation between its linear dimensions so that the equilibrium shall be stable.

Let $2a$, $2b$ be the lengths of the horizontal edges, and $2c$ the length of the vertical edge, and let $b < a$. Then the equilibrium is most unsafe when a displacement is made round the longest diameter of the section of floatation. If x is the length of the vertical edge immersed,

$$x = 2c\frac{w'}{w},$$

and therefore
$$HG = c\left(1 - \frac{w'}{w}\right).$$

Also $k^2 = \tfrac{1}{3}b^2$ round the axis of most dangerous displacement, and $V = \dfrac{W}{w}$, where $W = 8abcw' =$ weight of body. Hence
$$HM = \frac{b^2 w}{6cw'},$$
so that for stability $\dfrac{b^2 w}{6cw'} > c\left(1 - \dfrac{w'}{w}\right)$, that is
$$\frac{b}{c} > \sqrt{6\frac{w'}{w}\left(1 - \frac{w'}{w}\right)}.$$

2. If the floating body is a solid cylinder, floating with its axis vertical, find the condition for stability.

Ans. If r is the radius of the base and h the height,
$$\frac{r}{h} > 2\sqrt{\frac{w'}{w}\left(1 - \frac{w'}{w}\right)}.$$

3. If the floating body is a solid cone, floating with its axis vertical and vertex downwards, find the condition for stability.

Ans. If r is the radius of the base and h the height,
$$\frac{r}{h} > \sqrt{\left(\frac{w}{w'}\right)^{\frac{1}{3}} - 1}.$$

4. If the floating body is a solid isosceles prism whose base is uppermost, find the condition for stability.

Ans. If $2b$ is the length of the shorter side of the base and h the height of the prism,
$$\frac{b}{h} > \sqrt{\left(\frac{w}{w'}\right)^{\frac{1}{2}} - 1}.$$

5. If the cone in example 3 floats with its vertex uppermost, find the condition for stability.

Ans.
$$\frac{r}{h} > \sqrt{\left(\frac{w}{w-w'}\right)^{\frac{1}{3}} - 1}.$$

We have assumed that H', Fig. 49, lies in the plane of displacement, and we can easily see that this will not be the case unless the axis, $x'x$, of displacement is a principal axis of the section, $AxBx'$, of floatation. For, if we seek the co-ordinates of H' (which is the centre of volume of the new volume $A'CB'$) we may regard, as before, the volume $A'CB'$ as 'resolved' into the original volume ACB, the positive wedge $B'xBx'$, and the negative wedge $A'xx'A$. Hence if x is the distance of the point P from the line OB and ξ the distance of H' from the vertical plane containing OB, we have
$$V.\xi = \theta \int xy\,dS,$$
since the volume of the prism PQ is $\theta y\,dS$, and its volume-moment about OB is $\theta xy\,dS$, the integration extending all over the area $AxBx'$.

This shows that $\xi = 0$ only when Ox and OB are principal axes at O. In the case of a square or circular section of floatation, every axis through O is a principal axis, and hence H' always lies in the plane of displacement.

In general, therefore, a small angular displacement round a diameter of the section of floatation produces a moment of the forces not only round this axis but also round the perpendicular axis in the plane of floatation, the effect of which would be to produce small oscillations of the body about this axis.

The question of stability, however, is not affected by this consideration, since any small angular displacement, θ, round an axis $x'x$ could be resolved into two separate small angular displacements

$$\theta \cos \alpha \quad \text{and} \quad \theta \sin \alpha$$

round the two *principal* axes at O, where α is the angle made by $x'x$ with one of these principal axes; and if the equilibrium were stable for small displacements round the most dangerous of the principal axes, it would be so for the given displacement round $x'x$. On the other hand, if the equilibrium is unstable round one of the principal axes, it will be unstable round all axes in the section at O, unless these axes are inclined at indefinitely small angles to the other principal axis—supposing the equilibrium to be stable for displacements round this axis.

33. Surfaces of Revolution. When the figure of the floating body is that of a surface of revolution, take the origin, O, of co-ordinates at its lowest point, the axis of x being vertically upwards and that of y horizontal. Then if (x, y) are the co-ordinates which determine the surface of floatation in the erect position, and (x', y') those belonging to any other parallel section, we have

$$A = \pi y^2, \quad k^2 = \tfrac{1}{4} y^2, \quad V = \pi \int_0^x y'^2 \, dx';$$

hence
$$HM = \frac{y^4}{4 \int_0^x y'^2 \, dx'} \quad \ldots \quad (1)$$

Also, by mass-moments,

$$OH \times \pi \int_0^x y'^2 \, dx' = \pi \int_0^x x' y'^2 \, dx';$$

therefore
$$OM = \frac{y^4 + 4 \int_0^x x' y'^2 \, dx'}{4 \int_0^x y'^2 \, dx} \quad \ldots \quad (2)$$

Thus, to determine the figure of the floating body when HM is of constant length whatever be the depth of immersion, let $HM = m$ in (1),

$$\therefore \quad \tfrac{1}{4} y^4 = m \int_0^x y'^2 \, dx'.$$

Differentiate both sides with respect to x; then (see Williamson's *Integral Calculus*, Chap. VI)

$$y^3 \frac{dy}{dx} = my^2,$$

$$\therefore\ y^2 = 2mx,$$

which shows that the generating curve is a parabola; hence when a paraboloid of revolution floats in a liquid the height of the metacentre above the centre of buoyancy is constant for all depths of immersion.

EXAMPLE.

Find the nature of the generating curve so that for the surface of revolution and for all depths of immersion the height of the metacentre above the lowest point shall be any assigned function of the co-ordinates of the section of floatation.

Let $OM = \phi(x, y)$ in (2); then, writing ϕ instead of $\phi(x, y)$ for shortness,

$$\tfrac{1}{4}y^4 + \int_0^x x' y'^2 \, dx' = \phi \int_0^x y'^2 \, dx'. \quad \ldots \quad (1)$$

Differentiating with respect to x, and putting p for $\frac{dy}{dx}$,

$$py^3 + xy^2 = y^2 \phi + \left(\frac{d\phi}{dx} + p\frac{d\phi}{dy}\right) \int_0^x y'^2 \, dx. \quad \ldots \quad (2)$$

Dividing out and again differentiating,

$$\frac{d}{dx}\left\{\frac{py^3 + (x-\phi)y^2}{\frac{d\phi}{dx} + p\frac{d\phi}{dy}}\right\} = y^2. \quad \ldots \quad (3)$$

This is the differential equation of the required generating curve.

If, for instance, the metacentre is at a constant height, a, above the lowest point, we know that the curve is a circle, and this appears at once from (2), since $\phi = a$,

$$\therefore\ py + x = a, \quad \therefore\ y \, dy + (x-a) \, dx = 0, \quad \therefore\ (x-a)^2 + y^2 = a^2.$$

Pressure on Curved Surfaces.

34. Metacentric Evolute. Suppose a body of given mass to float in a liquid; then if we consider all possible displacements—and not merely small displacements—in which the volume of the displaced liquid is constant, the lines of action of the forces of buoyancy will envelope a certain surface fixed in the body. This surface is called the *metacentric evolute* for the given displaced volume.

As a particular case, consider the displacements of a square board, $ABCD$, Fig. 50, floating in a liquid of double its own specific weight. The displacement is always half the volume of the board; and when the board floats with AB horizontal, the centre of buoyancy is H, the metacentre being M, such that $HM = \frac{1}{3}a$, where $2a = AB$. In this position the equilibrium is unstable.

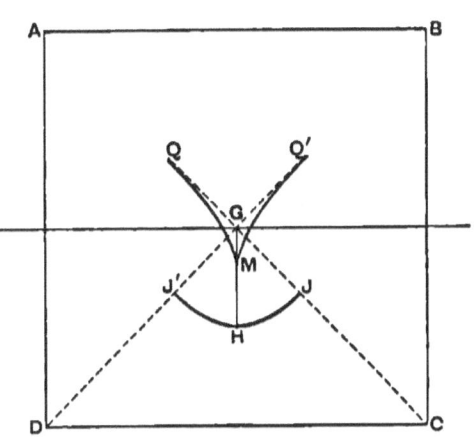

Fig. 50.

The curve of buoyancy for positions intermediate to those in which the surfaces of floatation are DB and CA is the portion $J'HJ$ of a parabola whose parameter is $\frac{2}{3}a$. The lines of action of the forces of buoyancy are always normals to this parabola, and their envelope is the evolute, QMQ', of the parabola. The positions in which DB and CA are in the surface of the fluid are positions of stable equilibrium, the metacentric heights GQ and GQ' being each $\dfrac{\sqrt{2}}{3}a$.

166 *Hydrostatics and Elementary Hydrokinetics.*

In general, for the displacements of any body in one plane—the volume of the displaced liquid being constant—the metacentric evolute is the evolute of the curve of buoyancy in the plane.

35. Stability in two Fluids. Let $DAOB$, Fig. 51, represent a body floating partly in a homogeneous fluid of

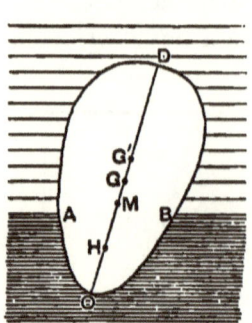

Fig. 51.

specific weight w' and partly in one of specific weight w, the latter being the lower, and suppose the position of equilibrium to be found. We may evidently imagine the volume, DAB, of the upper fluid completed by adding the portion AOB, and all the forces in play will be those due to an immersion of the whole volume in a fluid of specific weight w' and an immersion of the portion AOB in one of specific weight $w-w'$.

Let G be the centre of gravity of the body; G' its centre of volume, i.e. the centre of gravity of the whole volume supposed to be homogeneously filled; H the centre of volume of the portion in the lower fluid before displacement; M the metacentre corresponding to this lower fluid (of specific weight $w-w'$); V the volume of the lower and V' that of the upper fluid displaced. The position of M is given by the equation $HM = \dfrac{Ak^2}{V}$.

For simplicity we have assumed G, G' and H in the original position to lie on the same vertical line; but the method of investigating any case in which they are not thus simply situated will be readily understood from the simple case supposed.

The points G', G, M, H may in any individual case have relative positions different from those represented in the figure.

We may evidently suppose the displacement to be made round some diameter of the section AB through its centroid, in which case the wedge forces of buoyancy at the section are equivalent to a couple, whose moment in the present instance is $Ak^2(w-w')$.

The equilibrium will be stable if the sum of the moments of the forces acting on the body in its position of displacement round an axis perpendicular to the plane of displacement, drawn through G or through any other convenient point, is in a sense opposed to that of the displacement.

Now, if W = weight of body, the forces in action are W acting down through G, together with $V(w-w')$ acting up through M, and $(V+V')w'$ up through G'. The sum of their moments about H in the sense opposed to the angular displacement is

$$\theta\{V(w-w').HM+(V+V')w'.HG'-W.HG\},$$

and if the expression in brackets is positive, the equilibrium is stable.

It is sometimes more convenient to take the restoring moment about the lowest point, O, of the axis of the body.

In the above expression we may put $W = Vw + V'w'$; and it is evident that if the centre of gravity, G, of the body coincides with its centre of volume, G', the condition becomes simply $HM > HG$—as is evident à priori.

Example.

A circular cone the length of whose axis is 20 inches is formed of two substances whose specific gravities are 3 and 8, the denser forming a cone whose axis is 12 inches long and the other forming the frustum which completes the whole cone; it is immersed, vertex downwards, in a liquid whose specific gravity is 14, on top of which rests a liquid whose specific

168 *Hydrostatics and Elementary Hydrokinetics.*

gravity is 1, the whole cone being immersed; show that for stable equilibrium the radius of the base must be greater than 11·639 inches.

36. Floating Vessel containing Liquid. Suppose a vessel, represented in Fig. 52, to contain a given volume of

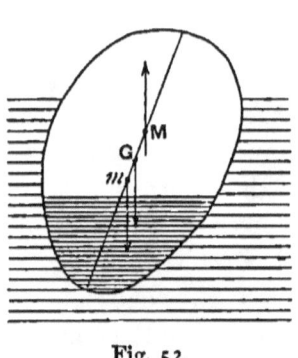

Fig. 52.

liquid of specific weight w' and to float in a liquid of specific weight w. If the vessel receives a small angular displacement, there will be a force of buoyancy due to the external fluid acting upwards through its metacentre M; the line of action of the weight of the contained fluid acts through its new centre of gravity and it intersects the line GM in m, the metacentre of this contained fluid.

This force acts downwards, and W, the weight of the vessel acting through its centre of gravity, G, also acts downwards. The weight of the internal fluid may assist either in promoting stability or in promoting instability according to the position of m. If, as in the figure, m is below G, this force promotes stability. If $V =$ volume of displaced external fluid, $V' =$ volume of internal fluid, the restoring moment is

$$\theta(wV \cdot GM + w'V' \cdot Gm).$$

EXAMPLES.

1. Find the least height to which a uniform heavy cylindrical vessel of negligible thickness can be filled with water so that when it is placed with its axis vertical in water the equilibrium may be stable.

Ans. Let h be the distance of the centre of gravity of the vessel from the base, W the weight of the vessel, and A the area of the base; then the least height to which it can be filled is

$$h - \frac{W}{2Aw}.$$

2. If the cylinder contains a liquid of specific weight w' and floats in a liquid of specific weight w, with its axis vertical, find the condition of stability.

Ans. Let $w' = n.w$, $W = Acw$, and $x =$ the height to which the cylinder is filled; then, for stability, the expression

$$2n(n-1)x^2 + 4ncx + 2c^2 - (n-1)r^2 - 4ch$$

must be positive.

3. If a uniform hollow cone of negligible thickness contains a liquid of specific weight w' and floats in a liquid of specific weight w with its axis vertical and vertex downwards, find the condition of stability.

Ans. If x is the length of the axis occupied by the internal fluid, y the length occupied by the external fluid, h the whole length of the axis, l the distance of the centre of gravity of the cone from the vertex, $r =$ radius of base, $w' = nw$, $W =$ weight and $V =$ volume of cone, and if $W = m.Vw$, we have

$$y^3 - nx^3 = mh^3,$$

and for stability the expression

$$3\left(1 + \frac{r^2}{h^2}\right)(y^4 - nx^4) - 4mh^3 l$$

must be positive.

4. A thin vessel in the form of a surface of revolution contains a given quantity of homogeneous liquid and rests with its vertex at the highest point of a rough curved surface, find the condition of stability for small lateral displacements.

Ans. Let W be the weight of the vessel (without the liquid), h the distance of its centre of gravity from the vertex, V the volume of the liquid, w its specific weight, z the distance of its centre of gravity from the vertex, A the area of the free surface of the liquid, k the radius of gyration of this area about its diameter of displacement, ρ and ρ' the radii of curvature of the vessel and the fixed surface in the plane of displacement; then the restoring moment is proportional to

$$(Vw + W)\rho - \left(1 + \frac{\rho}{\rho'}\right)(Vwz + Ak^2w + Wh),$$

and if this expression is positive, the equilibrium is stable. The restoring moment is equal to this expression multiplied by $\frac{\rho'\theta}{\rho + \rho'}$, where θ is the small angular displacement of the vessel. (See *Statics*, vol. ii., Art. 279, 4th ed.)

5. In the last example find the position of the metacentre.

Ans. If H is the centre of gravity of the contained fluid,

$$HM = \frac{Ak^2}{V} + \frac{W}{Vw}\left(h - \frac{\rho\rho'}{\rho + \rho'}\right).$$

6. If the vessel is a paraboloid of revolution resting on a horizontal plane, the weight of the liquid being P and the latus rectum of the parabola $4a$, the condition for stability is

$$W(2a - h) > \tfrac{2}{3}P\left(\frac{P}{2\pi aw}\right)^{\frac{1}{2}}.$$

37. Stability in Heterogeneous Fluid. We shall now suppose that a body floats in a fluid of variable density which is subject to the action of gravity. The level surfaces of the external force being horizontal planes, these planes will also be surfaces of constant density. Hence if w is the specific weight of the fluid at any point whose depth below the free surface, LN, Fig. 53, of the fluid is ζ, we have

$$w = f(\zeta). \quad \ldots \ldots \ldots \quad (1)$$

Suppose the dotted curve to represent the original position of the floating body, and that the full curve ACB represents its position when it has received a slight angular displacement, θ, round any assigned horizontal line Ox—which we suppose to be perpendicular to the plane of the paper.

Pressure on Curved Surfaces.

Take the vertical plane through the original line joining G and H, the centres of gravity of the body and of buoyancy, which is perpendicular to Ox as plane of yz, the point, O, in which this plane cuts Ox being taken as origin, the vertical Oz as axis of z, and the horizontal line, Oy, perpendicular to Ox as axis of y. Thus the displacements of all points of the body take place in planes parallel to the plane of yz.

The section of floatation of the body in the displaced position is represented by $A'B'$.

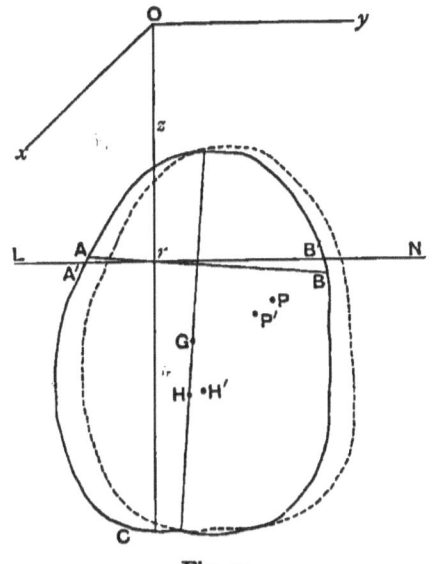

Fig. 53.

Suppose AB to be the section of the body made by the plane LN in the original position; and in this position let b be the distance between the line GH and the axis Oz. Let h be the height of O above LN.

The equation of the plane $A'B'$ is $z - h = 0$, and this was the equation of AB in the original position; but by rotation in the sense indicated in the figure the equation of AB in its displaced position becomes $z - \theta y - h = 0$, and therefore the old and new positions of the plane AB—and of every plane horizontal section of the body—intersect on the axis Oz.

Suppose P' to be any point in the body whose original position was P (the latter point being supposed to be marked in fixed space and not in the body); and let x, y, z be the co-ordinates of P with reference to the fixed axes at O.

172 *Hydrostatics and Elementary Hydrokinetics.*

Then the co-ordinates of P' are $(x, y-\theta z, z+\theta y)$, so that the density of the fluid which would exist at P' if the body were removed would be, by (1),

$$f(z+\theta y - h), \text{ or } f(\zeta + \theta y), \text{ or } w + \theta y \frac{dw}{dz}, \quad . \quad . \quad (2)$$

since the depth of P' below the surface LN is $z+\theta y-h$. Hence when the element of volume $dx\,dy\,dz$ at P is carried to P', it will experience a force of buoyancy

$$\left(w + \theta y \frac{dw}{dz}\right) dx\,dy\,dz; \quad . \quad . \quad . \quad . \quad (3)$$

and since the points P are all those included within the original volume, BCA, immersed, the corresponding forces of buoyancy will omit the wedge $B'rB$ and include the wedge ArA'—the latter not, in reality, contributing any force of buoyancy at all in the displaced position, while the former does. We must therefore specially include the wedge $B'rB$ and exclude ArA'.

Let w_0 be the specific weight of the fluid at the surface LN; let dS be the area of any element of the surface AB (such as that represented at P in Fig. 49); then if y_0 is the distance of this element from the line through r parallel to Ox, the volume of the small cylinder standing on dS, as in Fig. 49, is $\theta y_0\,dS$. Also let x_0 be the x co-ordinate of the element dS, and let c be the original depth of G below the horizontal plane xOy. Then we have, in their new positions,

the co-ordinates of P ...... x, $y-\theta z$, $z+\theta y$,
,, ,, ,, G ...... 0, $b-\theta c$, $c+\theta b$,
,, ,, ,, dS ...... x_0, $y_0-\theta h$, $h+\theta y_0$.

Now we shall calculate the sum, L, of the moments of the forces of buoyancy round the horizontal axis through G parallel to Ox in the sense opposite to that of the displacement, i.e. counterclockwise as we view the figure. If a force having components X, Y, Z acts at the point (x, y, z), its moments round axes through the point (a, β, γ) parallel to the axes are $Z(y-\beta) - Y(z-\gamma)$, and two similar expressions (*Statics*, vol. ii., Art. 202).

In the present case only the z-component of force exists, and this at P' is the expression (3) with a negative sign, while at the new position of the surface element dS it is

$$-\theta w_0 y_0 dS. \quad \ldots \quad \ldots \quad (4)$$

Hence we have

$$L = \iiint \left(w + \theta y \frac{dw}{dz}\right) \{y - b - \theta(z-c)\} \, dx \, dy \, dz$$

$$+ \theta w_0 \int y_0 (y_0 - b) \, dS. \quad . \quad (5)$$

Now observe that we neglect θ^2; also if W is the weight of the volume ACB of fluid originally displaced,

$$\iiint wy \, dx \, dy \, dz = W \cdot b,$$

since the y of H was originally b. Hence the term independent of θ in (5) disappears, as it must, of course; and we have

$$L = \theta \iiint (y^2 - by) \frac{dw}{dz} \, dx \, dy \, dz - \theta \iiint w(z-c) \, dx \, dy \, dz$$

$$+ \theta w_0 \int (y_0^2 - by_0) \, dS. \quad . \quad (6)$$

Observe also that w is a function of z alone, so that the first triple integral can be written in the form

$$\int \left[\iint (y^2 - by) \, dx \, dy\right] \frac{dw}{dz} dz, \quad \ldots \quad (7)$$

and if A denotes the area of any section for which z is constant (i.e. any section of the body parallel to AB), k the radius of gyration of this section round the line in its

plane parallel to Ox, at the point where Oz cuts the section, and $\bar{y}$ the distance of the 'centre of gravity' of the area from this same line, the double integral in the brackets in (7) is
$$A(k^2 - l\bar{y}), \quad \ldots \ldots \quad (8)$$
so that the first integral in (6) is
$$\theta \int A(k^2 - l\bar{y}) \frac{dw}{dz} dz. \quad \ldots \ldots \quad (9)$$

The second integral in (6) is $-\theta W \cdot HG$, and the last is $\theta w_0 A_0 (k_0^2 - l\bar{y}_0)$, where k_0 is the radius of gyration of the section, AB, of floatation (whose area is A_0) round the line through r parallel to Ox, and $\bar{y}_0$ is the distance of the 'centre of gravity' of the section from this line. Hence (6) becomes

$$\frac{L}{\theta} = \int A(k^2 - l\bar{y}) \frac{dw}{dz} dz + w_0 A_0 (k_0^2 - l\bar{y}_0) - W \cdot HG. \quad (10)$$

For stability this must be a positive moment; and in the particular case in which w is constant and the displacement is made round a diameter of the section AB, it is obvious that we get the same condition as in Art. 31.

But the forces of buoyancy will also, in general, produce a moment round the horizontal axis through G parallel to Oy, i.e. a moment tending to turn the body across the plane of displacement. If M is this moment, we have

$$M = \iiint \left(w + \theta y \frac{dw}{dz} \right) x\, dx\, dy\, dz + \theta w_0 \int x_0 y_0 \, dS. \quad (11)$$

Let P denote the product of inertia, $\iint xy\, dx\, dy$, of any section round axes in its plane parallel to Ox and Oy at the point where the section is cut by Oz; then

$$\frac{M}{\theta} = \int P \frac{dw}{dz} dz + w_0 P_0. \quad \ldots \ldots \quad (12)$$

Pressure on Curved Surfaces. 175

This moment will not exist if P is zero for all sections, or if the fluid is homogeneous and P is zero for the surface of floatation.

Let us now calculate the *work done* in the displacement of the body round Ox.

The work which would be done on a material system by force the components of whose intensity at (x,y,z) are X, Y, Z for any small displacement whose typical components are $\delta x, \delta y, \delta z$ is

$$\int (X\delta x + Y\delta y + Z\delta z)\,dm; \quad \ldots \quad (13)$$

and if the displacement is produced by small rotations, $\delta\theta_1, \delta\theta_2, \delta\theta_3$, round the axes of co-ordinates, we have $\delta x = y\delta\theta_1 - x\delta\theta_2$, with similar values of δy and δz. Hence, if L, M, N are the typical moments of the force intensity about the axes, the work is

$$L\delta\theta_1 + M\delta\theta_2 + N\delta\theta_3 \ldots \quad (14)$$

In the present case the only rotation is that about Ox. $\therefore \delta\theta_2 = \delta\theta_3 = 0$. Consider the moment L as that of the forces of buoyancy in the displaced position ACB, and calculate the element of work done by these forces in any *further* small displacement by which the angle θ is increased by $d\theta$. Then the infinitesimal element of work done in this further displacement is

$$L\,d\theta. \quad \ldots \quad (15)$$

But (taking the forces of buoyancy alone),

$$L = -\iiint \left(w + \theta y \frac{dw}{dz}\right)(y - \theta z)\,dx\,dy\,dz - \theta w_0 \int y_0^2\,dS. \quad (16)$$

$$= -Wb - \theta\left\{\int A k^2 \frac{dw}{dz}\,dz - W(c + HG) + w_0 A_0 k_0^2\right\}. \quad (17)$$

$$= -Wb - K\theta, \text{ suppose}; \quad \ldots \quad (18)$$

and the integral of this expression from $\theta = 0$ to $\theta = \theta$

expresses the work done by the forces of buoyancy in the displacement from the initial position of the body (represented by the dotted contour) to that represented by ACB. Hence the work is
$$-Wb\theta - \tfrac{1}{2}K\theta^2. \quad \ldots \quad (19)$$

The work done by the weight of the body is simply $W\delta\bar{z}$, in which $\delta\bar{z}$ must be accurate as far as θ^2; i.e. $\delta\bar{z} = b\theta - \tfrac{1}{2}c\theta^2$. Hence the work done by all the forces is

$$-\frac{\theta^2}{2}\left\{\int Ak^2\frac{dw}{dz}dz - W\cdot HG + w_0 A_0 k_0^2\right\}, \quad \ldots \quad (20)$$

and this, with reversed sign, is the work which must be done *against* the forces to produce the displacement.

EXAMPLES.

1. If a solid homogeneous cone float, vertex down, in a fluid in which the density is directly proportional to the depth, find the condition of stability.

Ans. If, as at p. 131, h' is the length of the axis immersed and h is the height of the cone, the equilibrium will be stable if $\cos^2 a < \dfrac{4h'}{5h}$, where a is the semivertical angle of the cone.

2. Determine the condition of stability of a solid homogeneous cylinder under the same circumstances.

Ans. If r is the radius, h the height of the cylinder, and h' the length of the axis immersed (see p. 131), the condition of stability is
$$r^2 > h'(h - \tfrac{2}{3}h').$$

3. If a spherical balloon of weight B is held at a given height by a rope made fast to the ground, find the work done in displacing it about the ground end of the rope through a small angle.

Ans. If h is the height of the centre of the balloon and W the weight of the displaced air, the work is
$$\tfrac{1}{2}\theta^2(W-B)h,$$
where W has the value given in example 3, p. 131.

38. Green's Equation. Let ABC, Fig. 54, be any closed surface; let U and V be any functions of x, y, z, the co-ordinates of any point P at which an element of volume $d\Omega$ is taken; and let ∇^2 stand for the operation

$$\frac{d^2}{dx^2} + \frac{d^2}{dy^2} + \frac{d^2}{dz^2};$$

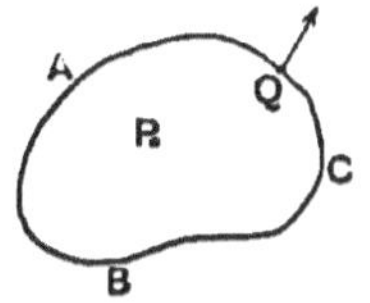

then if we take the integral

$$\int U \nabla^2 V d\Omega$$

Fig. 54.

throughout the volume enclosed by ABC, the result can be expressed in terms of another volume-integral taken through the same space and of a surface-integral taken over the bounding surface ABC. Thus, let Q be any point on the surface, at which an element of area dS is taken and let dn be an element of the normal at Q drawn outwards into the surrounding space (in the sense of the arrow). Then we have (see *Statics*, vol. ii, chap. xvii, Section iv.)

$$\int U \nabla^2 V . d\Omega = \int U \frac{dV}{dn} . dS$$
$$- \int \left(\frac{dU}{dx} \frac{dV}{dx} + \frac{dU}{dy} \frac{dV}{dy} + \frac{dU}{dz} \frac{dV}{dz} \right) d\Omega . \quad (1)$$

In exactly the same way, if ϕ is any other function of x, y, z, we have

$$\int U \left(\frac{d}{dx} . \phi \frac{dV}{dx} + \frac{d}{dy} . \phi \frac{dV}{dy} + \frac{d}{dz} . \phi \frac{dV}{dz} \right) . d\Omega$$
$$= \int U \phi \frac{V}{dn} . dS - \int \phi \left(\frac{dU}{dx} \frac{dV}{dx} + \frac{dU}{dy} \frac{dV}{dy} + \frac{dU}{dz} \frac{dV}{dz} \right) . d\Omega. \quad (2)$$

The first of these is known as Green's equation; the second is a modification made by Sir W. Thomson.

By assigning to U various values (such as a constant value, the value V, &c.) we obtain (as shown in *Statics* above) various remarkable theorems with physical applications.

A most remarkable consequence of (2) is this. If ϕ and V are any two functions satisfying the equation

$$\frac{d}{dx}\cdot\phi\frac{dV}{dx} + \frac{d}{dy}\cdot\phi\frac{dV}{dy} + \frac{d}{dz}\cdot\phi\frac{dV}{dz} = 0 \quad . \quad . \quad (3)$$

at all points within a closed surface, ABC, and if the value of V is assigned at every point, Q, on the surface itself, its value at each internal point, P, is determinate.

For, if possible, let there be two different functions, viz.,

$$V \equiv f(x, y, z),$$
$$V' \equiv f'(x, y, z);$$

each satisfying (3) and such that $V = V'$ at each point, Q, on the surface, while V is, of course, not equal to V' at each internal point P. Denote $V - V'$ by ξ; then ξ satisfies (3). Now employ (2) for the volume and surface of ABC, and, moreover, choose for U the value ξ. Then

$$\int \xi \left(\frac{d}{dx}\cdot\phi\frac{d\xi}{dx} + \frac{d}{dy}\cdot\phi\frac{d\xi}{dy} + \frac{d}{dz}\cdot\phi\frac{d\xi}{dz} \right) \cdot d\Omega$$
$$= \int \xi\phi\frac{d\xi}{dn}\cdot dS - \int \phi \left[\left(\frac{d\xi}{dx}\right)^2 + \left(\frac{d\xi}{dy}\right)^2 + \left(\frac{d\xi}{dz}\right)^2 \right] \cdot d\Omega. \quad (4)$$

But each term under the integral on the left-hand side vanishes, and the surface-value of ξ which enters into each term of the first integral on the right also vanishes; therefore the second integral on the right vanishes; but since each term of this integral is a square, we must have each term equal to zero, i. e.,

$$\frac{d\xi}{dx} = 0, \quad \frac{d\xi}{dy} = 0, \quad \frac{d\xi}{dz} = 0,$$

must hold for all points inside ABC; and this requires that

ξ is constant for all internal points, and ∴ zero, since it is zero at the surface points.

Hence there cannot be two functions, V, V', satisfying (3), agreeing at each surface point, while differing at internal points. If, therefore, any *one* function, $f(x, y, z)$, of the co-ordinates is known to satisfy (3) and to have at each point on the surface an assigned particular value, it is the only one applicable to the points enclosed by the surface.

The application of this result to the case of fluid pressure is obvious. If at each point of any fluid-mass the external forces satisfy the equation

$$\frac{dX}{dx} + \frac{dY}{dy} + \frac{dZ}{dz} = 0, \quad \ldots \quad (5)$$

~~i.e., if the external forces have a potential~~, equations (1), (2), (3), p. 82, give

$$\frac{d}{dx} \cdot \frac{1}{\rho}\frac{dp}{dx} + \frac{d}{dy} \cdot \frac{1}{\rho}\frac{dp}{dy} + \frac{d}{dz} \cdot \frac{1}{\rho}\frac{dp}{dz} = 0. \quad \ldots \quad (6)$$

Hence if ABC is the surface of a foreign body immersed in the fluid, the distribution of the fluid which could, under the influence of the given external forces, statically replace the body is determinate since the value of the pressure intensity is assigned at each surface point, Q. At each internal point, P, the pressure intensity is determinate, and if ρ is, for the given fluid, a given function of p—say $f(p)$—the value of p at P is given by the equation

$$\frac{dp}{f(p)} = dV, \quad \ldots \quad (7)$$

where V is the potential function of the external forces, involving the co-ordinates of P. This is the result referred to in p. 113.

EXAMPLE.

If in the midst of a mass of fluid which is not self-attracting there is a solid body which attracts its own particles and those of the fluid according to the law of inverse square of distance, and if the surface, A, of this body is one of constant potential, prove that the intensity of pressure, p, of the fluid at any point, P, is less than the intensity of pressure, p_0, at any point on A by an amount given by the equation

$$p = p_0 - \frac{1}{4\pi\gamma M} \cdot \int \rho R^2 \, d\Omega,$$

where γ is the constant of gravitation (*Statics*, vol. ii., Art. 315), M is the mass of the solid body, ρ is the density of the fluid at any point at which the attraction per unit mass due to the body is R, $d\Omega$ is an element of volume, and the integration extends over the space included between the surface A and the equipotential surface, S, described through P.

In Green's equation (1) for U choose $p - p_0$ and let V be the potential at any point due to the solid body. Then we have

$$\int (p-p_0) \nabla^2 V \, d\Omega = \int (p-p_0) \frac{dV}{dn} dS - \int \left(\frac{dp}{dx} \frac{dV}{dx} + \ldots \right) d\Omega, \quad (1)$$

in which the surface-integral on the right is taken over the surface A and over the surface S, and the element of normal dn is drawn into the space *outside* the volume enclosed by A and S; this space is therefore the *interior* of the solid body and the exterior of S, so that dn in the integration over A is measured towards the interior of M.

Now we know that (*Statics*, vol. ii., Art. 329) $\nabla^2 V = 4\pi\gamma\rho'$, where ρ' is the density of the *attracting matter* (to which V is due) at the point to which V applies; and as there is none of this attracting matter at any of the points within the volume (that included between A and S) included in the integration, $\nabla^2 V = 0$. Again, at every point on the surface of A we have $p - p_0 = 0$, therefore the part of the surface-integral on the right which relates to the surface A is zero. Further at every point on S p is constant; hence the surface-integral is simply

$$(p - p_0) \int \frac{dV}{dn} dS,$$

and is confined to the surface S. Moreover, at every point in the fluid $\frac{dp}{dx} = \rho X$, &c., and $X = \frac{dV}{dx}$; hence (1) becomes

$$0 = (p-p_0)\int \frac{dV}{dn} dS - \int \rho R^2 d\Omega. \quad \ldots \quad (2)$$

Now (*Statics, ibid.*)

$$\int \frac{dV}{dn} dS = -4\pi\gamma M, \quad \ldots \quad \ldots \quad (3)$$

so that the required result follows at once from (2) and (3).

39. Line-Integrals and Surface-Integrals. If any directed magnitude, or vector, has for components u, v, w along three fixed rectangular axes, the magnitude which has for components λ, μ, ν along these axes, where

$$\frac{dw}{dy} - \frac{dv}{dz} = \lambda, \quad \ldots \quad \ldots \quad (1)$$

$$\frac{du}{dz} - \frac{dw}{dx} = \mu, \quad \ldots \quad \ldots \quad (2)$$

$$\frac{dv}{dx} - \frac{du}{dy} = \nu, \quad \ldots \quad \ldots \quad (3)$$

has been called the 'curl' of the given vector by Clerk Maxwell. (In the theory of Stress and Strain, and in the motion of a fluid, it is convenient to define the curl as having the *halves* of the above components.)

Any vector and its curl possess the following fundamental relation: *the line-integral of the tangential component of any vector along any closed curve is equal to the surface-integral of the normal component of its curl taken over any curved surface having the given curve for a bounding edge.* (See *Statics*, vol. ii, Art. 316, *a*.)

If l, m, n are the direction-cosines of the normal at any point of such a surface, and dS the area of a superficial element at that point, while ds denotes an element of length of the bounding edge of the surface, the theorem is expressed by the equation

$$\int (l\lambda + m\mu + n\nu)\, dS = \int \left(u\frac{dx}{ds} + v\frac{dy}{ds} + w\frac{dz}{ds} \right) ds. \quad (4)$$

Now we are often given the components of curl, λ, μ, ν, and from these we require to determine the vector from which they arise. In view of such a problem, the following fact is useful. If we can find, by any means, some particular values, u_0, v_0, w_0, of the components of the required vector which will satisfy the equations (1), (2), (3), the *general* values of u, v, w are simply

$$u = u_0 + \frac{d\phi}{dx}, \quad \ldots \ldots (5)$$

$$v = v_0 + \frac{d\phi}{dy}, \quad \ldots \ldots (6)$$

$$w = w_0 + \frac{d\phi}{dz}, \quad \ldots \ldots (7)$$

where ϕ is any function whatever of x, y, z. This is evident, because if we substitute u_0, v_0, w_0 for u, v, w in (1), (2), (3), we have, by subtraction,

$$\frac{d(w - w_0)}{dy} = \frac{d(v - v_0)}{dz}$$

and two analogous equations; and these signify that the expression

$$(u - u_0)\, dx + (v - v_0)\, dy + (w - w_0)\, dz$$

is a perfect differential of some function of x, y, z. If this function is denoted by ϕ, we have the results (5), (6), (7).

Of course it is a necessity from (1), (2), (3) that any possible components of curl of a vector should satisfy the identity

$$\frac{d\lambda}{dx} + \frac{d\mu}{dy} + \frac{d\nu}{dz} \equiv 0. \quad \ldots \ldots (8)$$

Thus, it is not possible to determine a vector the components of whose curl are x, y, z; but it is possible to determine one whose components of curl are x, y, $-2z$. The values $u_0 = \frac{3}{2}yz$, $v_0 = -\frac{1}{2}zx$, $w_0 = \frac{1}{2}xy$ will give these latter components of curl; so will the values $u_1 = yz$, $v_1 = -zx$, $w_1 = 0$. But it is obvious that if ϕ denotes the function $\frac{1}{2}xyz$, these two sets of components are related thus

$$u_0 = u_1 + \frac{d\phi}{dx}, \quad v_0 = v_1 + \frac{d\phi}{dy} \quad \&c.$$

In the line-integral along the closed curve the vector whose components are $\dfrac{d\phi}{dx}$, $\dfrac{d\phi}{dy}$, $\dfrac{d\phi}{dz}$ may be rejected, if ϕ is a single-valued function.

Stokes's method of determining values of u, v, w from the given values of λ, μ, ν will be found in Lamb's *Treatise on the Motion of Fluids*, p. 150.

Examples.

1. Given any unclosed curved surface in a heavy homogeneous liquid, is it possible to express the total component of pressure, on one side of the surface, parallel to—
    (a) a horizontal line,
    (b) a vertical line,
by an integral taken along the bounding edge of the surface?

Ans. The first is possible, but not the second. If a horizontal line is drawn at the surface of the liquid, which is taken as the plane of x, y, the component in the first case is $\int l z\, dS$, and this $= -\frac{1}{2}\int z^2 \dfrac{dy}{ds} ds$. This result is evident from elementary principles; because, if through the edge of the surface we describe a horizontal cylinder whose generators are parallel to the axis of x, and take a section of this cylinder perpendicular to its axis, the horizontal component of the pressure on the curved surface is equal to the pressure on this plane section, and is therefore independent of the shape and size of the given surface.

184 *Hydrostatics and Elementary Hydrokinetics.*

2. Given any unclosed curved surface in a heavy homogeneous liquid, is it possible to express the sum of the moments of the pressures, on one side of the surface, about—

    (*a*) a horizontal line,
    (*b*) a vertical line,

by an integral taken along the bounding edge of the surface?

Ans. The second is possible, but not the first; for,

$$\int (mzx - lyz)\, dS = \tfrac{1}{2} \int z^2 \left(x \frac{dx}{ds} + y \frac{dy}{ds} \right) ds;$$

and the result also follows from elementary principles, by closing the surface with a fixed cap described on the bounding edge, and then imagining the given surface to vary in size and shape, while retaining its bounding edge.

CHAPTER VI.

GASES

40. Definition of a Perfect Gas. When defining the modulus of cubical compressibility of a substance (Chap. I) the law which regulates the compressibility of gases was given. We may accept this law as an adequate definition and say that—

A perfect gas is a fluid whose resilience, or compressibility, of volume, when its temperature is constant, is numerically equal to its intensity of pressure.

It was shown (Art. 8) that the expression for the resilience of volume of any substance is

$$-v\frac{dp}{dv},$$

and that if this is equal to p, we have by integration

$$pv = \text{const.} \quad \ldots \ldots \ldots (a)$$

If the volume v was v_0 when the intensity of pressure was p_0, we must therefore have

$$pv = p_0 v_0. \quad \ldots \ldots \ldots (\beta)$$

If we take a given mass of gas and assume that at no point in it is there any intensity of pressure due to the weight of the gas, the pressure-intensity must be regarded as the same at all points and equal to whatever value it has at the surface. Let us suppose the gas contained in a cylindrical tube fitted with a gas-tight piston which can be weighted with various loads; then the volume of the gas will vary with each pressure intensity; and if its temperature remains unaltered, we can graphically represent its various states as expressed in the fundamental equation (β), thus: draw any two rectangular axes, Ov, Op, and let the volumes assumed by the gas be measured, on any scale, along Ov, while the intensities of pressure are measured on any scale along OP.

If, on these scales, OM and ON represent respectively any volume and the corresponding intensity of pressure, the point, P, whose co-ordinates are OM and ON will graphically represent the state of the gas; and all points, such as P, whose co-ordinates satisfy (β) will be found on a rectangular hyperbola passing through P and having the axes Ov and Op for asymptotes.

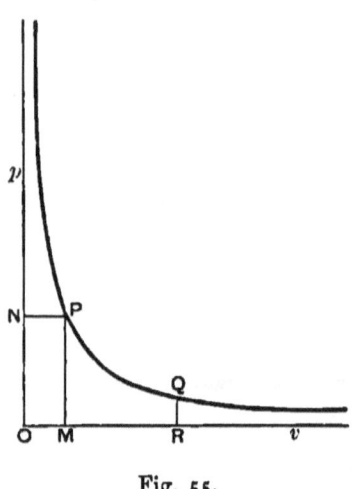

Fig. 55.

Thus, then, *the curve of transformation of a given mass of gas at constant temperature is a rectangular hyperbola.* Such transformation is called an *isothermal transformation.*

The figure exhibits the fact that when the intensity of pressure is infinitely increased the volume of the gas becomes infinitely small, and that when the intensity of pressure is infinitely reduced, the volume becomes infinitely great.

The first result would be strictly true for a substance whose transformations strictly follow the law (a) for all values of p; but it will be readily understood that there exists no gas for which (a) holds indefinitely, and that when the intensity of pressure is very greatly increased the gas may approximate to, and actually become, a liquid.

The law (a) was arrived at experimentally by Boyle and by Mariotte, and was not deduced as a mathematical consequence of the physical definition of a gas which we have given at the beginning of this Article. It is generally known in this country as *the law of Boyle and Mariotte*, and it may be formally enunciated as follows—*the temperature remaining constant, the volume of a given mass of gas varies inversely as its intensity of pressure*.

The experimental verification is as follows. Let $HABK$ (Fig. 56) be a bent glass tube of uniform section—at least in the leg AH which is closed at the top. Let the gas to be experimented upon be enclosed in the branch AH by means of a column, ALB, of mercury, the branch LBK of the tube being open to the atmosphere. Suppose matters arranged so that when the gas in AH is in equilibrium of temperature with the surrounding air, after the pouring in of the mercury has ceased, the surfaces A and B of the mercury are at the same level in both branches. Then the intensity of pressure at any point in the surface A is equal to that at any point in B; so that if p_0 is the atmospheric intensity of pressure, p_0 is also the intensity of pressure of the gas in AH.

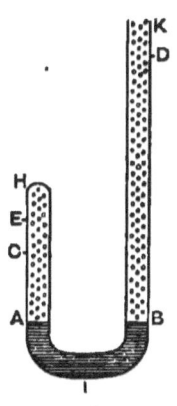

Fig. 56.

For simplicity denote p_0 by the height of the barometer at the time of the experiment.

Let this height be h (inches or millimètres), and let v_0 (cubic inches or cubic millimètres) be the volume of the gas AH. If w is the weight of a unit volume (cubic inch or cubic millimètre) of the mercury, we have $p_0 = w \cdot h$.

Let us now, by pouring mercury slowly into the open

branch at K, reduce the volume of the gas in AH to half its value. If $CH = \frac{1}{2} AH$, the mercury is to be poured in until its level in the closed branch stands at C after all disturbance and heating effect due to the pouring in of the mercury have subsided. If we now read the difference of level between C and the surface of the mercury in the branch LK, we shall find it exactly equal to h, the height of the barometer. Equating the intensity of pressure at C due to the imprisoned gas to the intensity of pressure due to the mercury and the atmosphere, we see that the former must be equal to $p_0 + wh$, i.e., the new intensity of pressure $= 2p_0$, while the new volume is $\frac{1}{2} v_0$.

Again, let $EH = \frac{1}{4} AH$, and let us pour mercury in at K until the volume of the imprisoned gas is EH, i.e., $\frac{1}{4} v_0$. We shall then find that the difference of level between E and the surface, F, of the mercury (not represented in the figure) in the open branch is 3 times the height of the barometer, i.e., $3h$, so that the intensity of pressure of the gas in EH is $p_0 + 3wh$, or $4p_0$.

Hence we have the following succession of volumes and pressure intensities for the gas, its temperature being the same all through,

$$(v_0, p_0), \quad (\tfrac{1}{2} v_0, 2p_0), \quad (\tfrac{1}{4} v_0, 4p_0).$$

If, in the same way, the volume is reduced to $\frac{1}{n} v_0$, the difference of level of the mercury in the two branches is found to be $(n-1)h$, so that the new intensity of pressure is np_0; and from these results we see that in each case the volume of the gas is inversely proportional to its intensity of pressure, as expressed by equation (β).

The law of Boyle and Mariotte may also be verified in the following simple manner by means of a single straight tube, about 2 mm. in diameter.

Let AD be a tube of uniform section closed at the end A

and open at D; let a portion, AB of the tube be filled with air or other gas, and let a thread of mercury, BC, of length l, separate this gas from the external air.

When the tube is held horizontal and all disturbance has subsided, let the volume, v_0, of the gas AB be read; its intensity of pressure is the same as that at C, i.e., p_0, the atmospheric in‑ tensity.

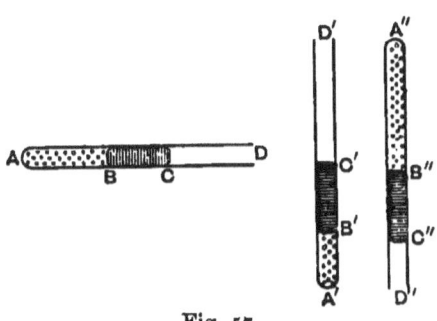

Fig. 57.

Now let the tube be held in a vertical position with the closed end A' downwards and let the gas occupy the volume $A'B'$, or v'. Its intensity of pressure is now equal to that at B' due to everything above B', i.e., $p_0 + wl$, where w = weight of unit volume of mercury. If h is the height of the barometer during the experiment, $p_0 = wh$, and if p' is the intensity of pressure in $A'B'$,
$$p' = w(h+l).$$

Finally, let the tube be held vertically with the closed end A'' uppermost, and let the volume of the gas be $A''B''$, or v''. If its intensity of pressure is p'', the intensity at C'' is $p'' + wl$ due to everything above C''; but p_0 is also the intensity of pressure at C'' since that is a point in the external air. Hence
$$p'' = w(h-l).$$
Hence, as regards volume and intensity of pressure, we have the succession of states
$$\{v_0, wh\}, \ \{v', w(h+l)\}, \ \{v'', w(h-l)\},$$
and we find on trial that
$$v_0 h = v'(h+l) = v''(h-l),$$
according to the requirements of Boyle's law.

Reference has already been made to the change of a gas into a liquid by compression. To this we shall subsequently return. At present we shall merely remark that Boyle's law is not accurately obeyed by any known gas, but the approximation is very close in the case of all gases when they are not near the state in which, either by increase of pressure or by diminution of temperature, liquefaction begins. When any gas is near the state of liquefaction, its volume decreases more rapidly with increased pressure than it would if it followed Boyle's law. 'When it is actually at the point of condensation, the slightest increase of pressure condenses the whole of it into a liquid.' (Clerk Maxwell's *Theory of Heat*, Chap. I.)

41. Law of Dalton and Gay-Lussac. The volume of a given mass of gas may be altered by heat as well as by pressure. The law relating to this change was discovered independently by Dalton in 1801 and by Gay-Lussac in 1802; and, apparently, it was discovered fifteen years previously by M. Charles, although not published by him. It is this—

The intensity of pressure being constant, the volume of a given mass of gas, when its temperature is raised from the freezing to the boiling point of water, increases by a fraction of the volume at the first temperature, which fraction is the same for all gases.

In short, the law is that all gases have the same coefficient of expansion, and that this is independent of the magnitude of the (constant) intensity of pressure under which they expand.

The fraction in question is, with certain reservations to be mentioned presently,

$$\cdot 3665,$$

so that if we measure degrees of heat by the Centigrade thermometer, this is the fractional increase of volume for

100°. The rate of expansion per degree is also found to be uniform, and is ·003665, which we shall use in the form $\frac{1}{273}$*.

Hence if v_0 denotes the volume of a given mass of any gas at 0° C, and v its volume at $t°$ C, we have

$$v = v_0 \left(1 + \frac{t}{273}\right), \quad \ldots \quad (1)$$

whatever be the intensity of pressure; and if v' is its volume at t', we have

$$\frac{v}{273+t} = \frac{v'}{273+t'}. \quad \ldots \quad (2)$$

If the point from which the temperature is reckoned on the Centigrade thermometer is removed 273° below the ordinary zero, i.e., the point at which water freezes when its surface intensity of pressure is that due to a standard atmosphere (indicated by a mercurial column 760 mm. in height), the expression $273 + t$ indicates the newly measured temperature, and is always denoted by T, and called the *absolute temperature* of the substance, the new point of reckoning being called the *absolute zero* of temperature.

If it were possible to have $T = 0$, that is $t = -273°$, for the gas—supposing the substance to remain a gas at all temperatures with constant coefficient of expansion, $\frac{1}{273}$—equation (1) would give

$$v = 0,$$

i.e., the gas would be reduced to zero volume. As the substance does not satisfy the above supposition, but alters its state in the process of lowering the temperature, the

* The fraction is more accurately $\frac{1}{272 \cdot 85}$, but the above is usually taken for simplicity. Clerk Maxwell (*Theory of Heat*) gives various values; thus, $\frac{1}{273\frac{1}{3}}$ and $\frac{1}{273 \cdot 7}$, the latter deduced from experiments of Thomson and Joule.

consequence is not realised, and it would thus appear that the notion of an *absolute zero* of temperature at $-273°$ C is a gratuitous error. Indeed, if the conception of *absolute temperature* rested on no other foundation, we might similarly argue from the coefficient of expansion of platinum, for instance, that since for this body $v = v_0 \left(1 + \dfrac{t}{37699}\right)$, nearly, where v_0 is its volume at zero and v its volume at $t°$, if we make $t = -37699$ we shall arrive at the absolute zero of temperature.

The truth is that the measure of absolute temperature rests on quite another basis, that it is intimately connected with the coefficient of expansion of a perfect gas, and that $273 + t$ is properly to be regarded as measuring the absolute temperature of a body whose temperature indicated by a Centigrade thermometer is t. This will be shown later on.

Adopting absolute temperature, then, equation (2) gives

$$\frac{v}{T} = \frac{v'}{T'}. \quad \ldots \ldots \ldots \quad (3)$$

Of course in the expression of the law of Dalton and Gay-Lussac it is not necessary to signalise the particular temperature corresponding to the *freezing of water* as possessing any special reference to the expansion of gases.

The law may be stated thus: *all gases expand, per degree, by the same fraction of their volumes at any common temperature.* This is obvious because their volumes at *any* temperature, τ, will all be the same multiple of their volumes at $0°$, and a constant fraction of the latter will give a constant fraction of the former.

In symbols, for any gas let u be the volume at τ, v that at t, v_0 that at zero, and a the coefficient of expansion with reference to the volume at zero; then

$$v = v_0(1 + at); \; u = v_0(1 + a\tau);$$

and therefore
$$v = u\frac{1+at}{1+a\tau} = u\frac{1+a\tau+a(t-\tau)}{1+a\tau}$$
$$= u\left\{1 + \frac{a}{1+a\tau}(t-\tau)\right\}$$
$$= u\{1 + \beta(t-\tau)\},$$

where $\beta = \dfrac{a}{1+a\tau}$, so that β is obviously the rate of expansion of the gas reckoned as a fraction of the volume u; and if a is the same for all gases, so is β.

It is remarkable that a is the same for all gases when far removed from their condensing points, i. e., from the liquid states, and that it is independent of the intensity of pressure under which the expansion takes place.

Clerk Maxwell (*Theory of Heat*) points out that *if the law of Dalton and Gay-Lussac is true for any one intensity of pressure, and if the law of Boyle holds, it follows that the former law holds for all intensities of pressure.*

This is very easily proved thus. Let v_0 be the volume of a given mass of gas at $(0°, p)$, i. e., p is its intensity of pressure, and let the law of Dalton hold for this pressure intensity; then if v is its volume at (t, p)
$$v = v_0(1 + at).$$

Now, keeping t constant, alter p to p'; then by Boyle's law the new volume, u, is given by the equation
$$u = v_0(1 + at)\cdot\frac{p}{p'}.$$

But if v_0 at $(0°, p)$ were altered by keeping its temperature zero and changing its intensity of pressure to p', its value, u_0, would be $v_0 \cdot \dfrac{p}{p'}$ by Boyle's law; so that the last equation gives
$$u = u_0(1 + at),$$
and therefore Dalton's law holds for p' if it holds for p.

With regard to the *accuracy* of the law of Dalton and Gay-Lussac, M. Regnault has found that, a being the coefficient of expansion per degree Centigrade,

for Carbonic acid, $\quad a = \cdot 003710$,

,, Protoxide of Nitrogen ,, $= \cdot 003719$,

,, Sulphurous acid, $\quad$,, $= \cdot 003903$,

,, Cyanogen $\quad$,, $= \cdot 003877$;

the last two of which are notably greater than the coefficient of expansion of air; but these are precisely the gases that can be most easily liquefied, while it is found that for all gases which can be liquefied only with great difficulty, a has very nearly the same small value, $\cdot 003665$, that it has for air. Hence M. Regnault modifies the law of Dalton and Gay-Lussac by saying that the coefficients of expansion of all gases approach more nearly to equality as their intensities of pressure become more feeble; so that it is only when gases are in a state of great tenuity that they have the same coefficient of expansion.

42. General Equation for the Transformation of a Gas. Given the volume, v, of a mass of gas at the temperature t, and pressure intensity p, find its volume at t' and p'.

First let the temperature be altered from t to t', the pressure intensity remaining p; then the volume v becomes u, where

$$u = v \frac{273 + t'}{273 + t},$$

by equation (2) of last Art.

Now keep the temperature constantly equal to t' and alter p to p'; then u becomes v', where

$$v' = u \cdot \frac{p}{p'},$$

by Boyle's law. Hence we have

$$v' = v\frac{273+t'}{273+t} \cdot \frac{p}{p'}, \qquad \ldots \quad (1)$$

or $\quad \dfrac{v'.p'}{273+t'} = \dfrac{v.p}{273+t}, \qquad \ldots \ldots \quad (2)$

or $\quad \dfrac{v'.p'}{T'} = \dfrac{v.p}{T}, \qquad \ldots \qquad \ldots \quad (a)$

where T and T' are the *absolute* temperatures of the gas.

Hence, whatever changes of pressure and temperature may be made in a *given mass* of gas, we have the result

$$\frac{v.p}{T} = \text{constant} \quad \ldots \ldots \quad (\beta)$$

between its volume, pressure intensity, and absolute temperature.

This most important result is the general equation for the transformation of a given mass of gas.

43. Formula in English Measures. Since the freezing point of water is marked 32° on Fahrenheit's thermometer, and the boiling point 212°, the fractional expansion of gas per degree Fahrenheit is $\dfrac{·3665}{180}$, or about $\dfrac{1}{491·13}$, *of the volume at* 32°. This fraction is usually taken as $\frac{1}{492}$; and this, as will presently be seen, would place the absolute zero of temperature 460 Fahrenheit degrees below the zero of the Fahrenheit scale. The experiments of Joule and Thomson indicate $-460·66$ as the position of the absolute zero; but for all practical purposes we can take -460.

If a given mass of gas has a volume u at 32° F, and its temperature is raised to t, we have, if the intensity of pressure is unaltered,

$$v = u\left(1 + \frac{t-32}{492}\right).$$

196 *Hydrostatics and Elementary Hydrokinetics.*

Hence $v = u\dfrac{460+t}{492}$; and if v' is its volume at t', we have

$$\frac{v}{460+t} = \frac{v'}{460+t'}.$$

If the intensity of pressure, estimated in any way, alters from p to p',

$$\frac{vp}{460+t} = \frac{v'p'}{460+t'}. \quad \ldots \ldots (a)$$

It thus appears that if the temperature of the gas were reduced to $-460°$ F, its volume would vanish, supposing that it obeys the laws of a gas during the whole process.

If we denote by T the absolute temperature, $460+t$, of the gas, we have the general equation of transformation of a given mass

$$\frac{vp}{T} = \frac{v'p'}{T'} = \text{constant}. \quad \ldots \ldots (\beta)$$

44. Law of Avogadro. One of the fundamental laws of gases is known as the Law of Avogadro. It is the following: *equal volumes of all substances when in the state of perfect gas, and at the same temperature and intensity of pressure, contain the same number of molecules.*

This law enables us to find the relative molecular weights of all substances by converting these substances into vapours, and then measuring the weights of known volumes of the vapours at known temperatures and intensities of pressure. Thus, it is found that a cubic foot of oxygen weighs 16 times as much as a cubic foot of hydrogen under like conditions of temperature and pressure; hence we conclude that the mass of each molecule of oxygen is 16 times that of a molecule of hydrogen.

45. Air Thermometer. A long capillary glass tube, AC, terminating in a bulb, B, is filled with air, and a short thread, m, of mercury is inserted into it, the end of the tube (beyond *C*) being open. In order to fill the bulb and portion of the tube to the left of *m* with air deprived of moisture, the tube and bulb are first filled with mercury which is boiled in the bulb. The open end is then inserted into a cork fitting into the neck of a tube, *D*, filled with chloride

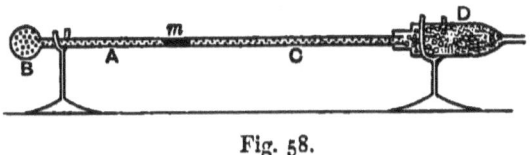

Fig. 58.

of calcium, which has the property of absorbing aqueous vapour from air, a fine platinum wire having been inserted into the stem *CA* through the tube *D*. If the instrument is supported in a position slightly inclined to the horizon on two stands and the platinum wire is agitated, air enters through the chloride of calcium, and gradually displaces the mercury from the bulb and stem, the process being stopped when only a very short thread of mercury is left.

The air in the instrument may now be considered to be dry.

Detach the stem from the drying tube *D*, and place it in a vertical position with the bulb *B* in a vessel filled with melting ice. Suppose the barometer to stand at 760 mm., thus indicating the standard atmospheric pressure. Then when the air has assumed the temperature of the melting ice, mark o on the stem *AC* at the under limit of the mercury index *m*.

If then the instrument is placed vertical with the bulb *B* surrounded by the steam of boiling water—close to the surface of the water, but not *in* it—the index *m* will move up towards *C*, and at its lower limit let 100 be marked on the stem. Thus the two standard Centigrade temperatures are marked on the stem, and the intervening space is to be divided into 100 equal parts if the stem has previously been ascertained to be of uniform bore. The graduations may be carried then below zero and beyond 100°.

If the tube of the air thermometer is made cylindrical all through—so that the bulb B is simply a uniform continuation of the stem—and we continue the graduations to 273 parts below the zero, we shall here reach the bottom, B, of the tube.

Hence the definition of the *absolute temperature* of a body, which we are so far justified in giving, is simply, in the words of Clerk Maxwell, its *temperature reckoned from the bottom of the tube of the air thermometer*.

The upper end of the stem of an air thermometer necessarily remains open to the atmosphere, otherwise the index, m, would not move or would scarcely move at all: if the end were closed and the air uniformly heated, m would not move.

Hence the air thermometer cannot be used to indicate temperature except in conjunction with the barometer. If the latter stands at p instead of p_0, the standard height (which we have above supposed to be 760 mm.) and the temperature indicated by the index m is t, the real reading is not t but that at which the index would stand if the intensity of pressure were altered to p_0. To find the point at which the index would stand in this case, let s be the area of the cross-section of the tube, c the length of the tube between two successive degrees, and B the volume of the bulb and tube up to the zero mark. Then when the index m stands at the mark t', the volume of the gas is $B + cst'$. But since at the absolute zero the volume of the gas would vanish, $B = 273\,cs$; hence

$$v = (273 + t')\,cs,$$

and this is at the intensity of pressure p, its true temperature being t. If p were altered to p_0 without any change in the true temperature, the index would stand at t and the volume would be $(273 + t)\,cs$. Now since these volumes are inversely as the intensities of pressure, we have

$$273 + t = (273 + t') \frac{p}{p_0},$$

which gives the true reading.

46. Work done in Expansion. If a gas is contained in an expansible envelope, the pressure of the gas on each element, dS, of area of the envelope is continually driving the element dS outwards along the normal to it, and hence the pressures on the various elements perform a certain amount of work in the increase of the total volume of the gas from one value to another.

Thus if p pounds' weight per square foot is the intensity of pressure of the gas at any instant and we take dS square feet at any point, P, Fig. 59, of the envelope, ABC, the force acting on dS is pdS. If the point P is moved by the pressure to a close position, P', along the normal, the work done by this force is

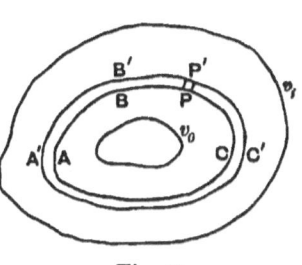

Fig. 59.

$$pdS \times PP'.$$

Now if the new position of the envelope is $A'B'C'$, $dS \times PP'$ is the volume of the small cylinder standing on dS and terminated by the new surface $A'B'C'$; and the sum of the works done by the pressures on all the elements of ABC in moving this surface to $A'B'C$ is

$$\Sigma (pdS \times PP'), \text{ or } p \cdot \Sigma (dS \times PP'),$$

since p has the same value at all points of ABC. But $\Sigma (dS \times PP')$ is the sum of all such small cylinders as that above described, i.e., it is the volume of the whole space between ABC and $A'B'C'$. If, then, v cubic feet is the volume of ABC, and $v + dv$ the volume of $A'B'C'$, the volume included is dv, and the work, dW, of expansion from ABC to $A'B'C'$ is given by

$$dW = pdv. \qquad (1)$$

Hence the work done by the pressure of the gas on its envelope in expanding from an initial volume v_0 to any final volume, v_1, is given by the equation

$$W = \int_{v_0}^{v_1} pdv. \qquad (2)$$

If p is measured in dynes per square centimètre, and v in cubic centimètres, the work will be in ergs.

The amount of work may be represented in a diagram by describing the curve (such as PQ in Fig. 55) whose abscissæ represent the volumes of the gas and whose ordinates represent the corresponding intensities of pressure.

In the particular case of *isothermal expansion*, $pv = p_0 v_0$, so that

$$W = p_0 v_0 \int_{v_0}^{v_1} \frac{dv}{v}$$

$$= p_0 v_0 \log_e \frac{v_1}{v_0}, \qquad (3)$$

from which it appears that the work done by the pressure of the gas in expanding isothermally from one given volume to another is independent of the temperature.

The work done by the pressure of an expanding gas on its envelope may or may not be equal to the work done against any external pressure which acts on the surface of the envelope. Thus, if the gas is contained in a horizontal cylinder and kept in by means of a piston on which the atmosphere presses, when the piston is released the work done by the pressure of the gas on the piston is equal to the work done against the atmospheric pressure plus the gain of kinetic energy of the piston.

Gases.

EXAMPLES.

1. A circular cone, hollow but of great weight, is lowered into the sea by a rope attached to its vertex; find the volume of the compressed air in the cone when the vertex is at a given depth below the surface.

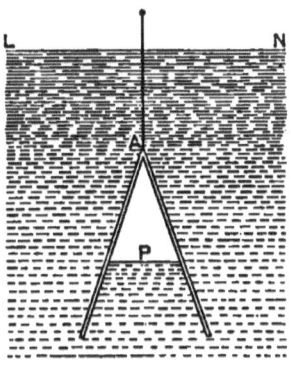

Fig. 60.

Let Fig. 60 represent a section of the cone; let c be the depth of the vertex below the surface, LN, of the water, $h =$ height of cone, $V =$ its volume, $t =$ the temperature of the air at the surface, $t' =$ temperature of the water, and therefore of the air in the cone; let P be the surface of the water within the cone, and let k be the height of a column of sea-water in a water barometer.

If these quantities are in English measure, we may regard the lengths as measured in feet, and the temperature as Fahrenheit; then k will be about 33 feet.

Now if x is the depth of P below A, the volume of the air in the cone is $V\dfrac{x^3}{h^3}$. The intensity of pressure of this air measured by a column of water is $k+c+x$. Hence the following diagrams represent the history of this mass of air as regards volume, temperature, and intensity of pressure:

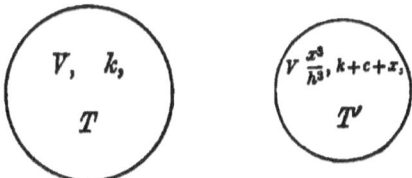

in which T and T' are *absolute* temperatures.

From Art. 42 or Art. 43 we have, then,

$$\frac{Vk}{T} = \frac{Vx^3(k+c+x)}{h^3\,T'},$$

$$\therefore\ x^4 + (k+c)x^3 - kh^3\frac{T'}{T} = 0,$$

from which x can be found.

A vessel used in this manner is called a *diving bell*. The above is a conical diving bell.

2. If in the above position of the cone it is desired to free the interior of water completely by pumping the air above the surface into the cone, find the volume of this surface air that will be required.

Let U be the volume required, and h the height of the cone; then suppose the cone to be wholly filled with air of the temperature t' of the surrounding water, and write down the history of this air, thus:

$V+U, k,$
T

$V, k+c+h,$
T'

$$\therefore\ \frac{(V+U)k}{T} = \frac{V(k+c+h)}{T'},$$

$$\therefore\ U = V\left\{\left(1 + \frac{c+h}{k}\right)\frac{T}{T'} - 1\right\}.$$

(It is not improbable that the student will fall into the error of supposing that U can be calculated as the volume of the surface air which is required to occupy the lower portion of the cone in Fig. 60, i.e., the portion occupied by water.)

Of course the result is the same whether the vessel is conical or of any other figure.

3. If a conical diving bell of height h feet contains a mercurial barometer the column of which stands at p_0 inches when the bell is above the surface of the water, and at a height p when below, infer the depth of the top of the bell below the surface.

Ans. $\dfrac{13\cdot596}{12}(p-p_0) - h\left(\dfrac{T'\,p_0}{T\,p}\right)^{\frac{1}{3}}$ feet.

4. Deduce the depth for a cylindrical or prismatic bell.

Ans. $\dfrac{13\cdot596}{12}(p-p_0) - h\dfrac{T'\,p_0}{T\,p}$.

Gases.

5. Find the tension of the suspending chain in a diving bell which occupies any position in water.

Ans. The weight of the bell and its appurtenances diminished by the weight of the water which is displaced from all causes.

(The water is displaced by the chain, the thickness of the bell, and the air within the bell; the weight of this water is the force of buoyancy. In strictness, the weight of the contained air should be added to that of the bell.)

6. If at the bottom of a river 40 feet deep, when the temperature is $40°$ F, a bubble of air has the volume $\frac{1}{10^5}$ of a cubic inch, what will be its volume on reaching the surface where the temperature is $50°$ F, and the height of a water barometer is 34 feet?

Ans. $\frac{2\cdot 22}{10^5}$ cubic inches.

7. If an open vessel (such as a tumbler) made of a substance whose specific gravity is greater than that of water is forced, mouth downwards, into water, show that its equilibrium becomes unstable after a certain depth has been reached.

(If the volume of the solid substance of the vessel is v, and in any position of the vessel if X is the volume of its compressed air, the downward force, P, required to hold it in equilibrium is given by the equation

$$P = Xw - v(w'-w),$$

where $w =$ specific weight of water, $w' =$ specific weight of substance of vessel.

Hence when X is so far diminished by forcing the vessel down that $Xw = v(w'-w)$, the pressure P vanishes, and after this an upward pull would be required.)

8. If v is small (i.e., if the thickness of the vessel is small), and if V is the volume of the interior of the vessel, prove that when the position of instability is reached, the depth of the top of the vessel below the surface of the water is approximately

$$k\left\{\frac{Vw}{v(w'-w)} - 1\right\},$$

where k is the height of a water barometer at the surface.

9. If v cubic inches of the external air at the absolute temperature T are inserted into the Torricellian vacuum of a uniform cylindrical barometer tube, calculate the depression produced in the column of mercury if the absolute temperature changes to T'.

Ans. Let h inches be the height of the barometer at first, $a =$ length of Torricellian vacuum, s square inches $=$ area of cross-section of tube, $x =$ length of tube finally occupied by the air; then
$$x(x-a) = \frac{T'}{T} \cdot \frac{vh}{s}.$$

10. A diving-bell of any shape occupies a given position below the surface of water; the bell has a platform inside; if a large block of wood falls from the platform into the water, prove that the water will rise inside the bell, but that the bell now contains less water than before.

Let the depth of the top be c, let h be the height of a water barometer at the surface, put $k = c + h$, let $B =$ volume of the block of wood, w' its specific weight, $w =$ specific weight of water, $V =$ whole volume of the interior of the bell, let x be the depth of the water in the bell below the top of the bell, and let X be the volume of the interior of the bell above this surface. Then
$$(X-B)(x+k) = Vh. \qquad \ldots \ldots (1)$$

When the wood falls the volume of it which remains above the surface is $B\left(1 - \dfrac{w'}{w}\right)$. Let x' be the new depth of the water surface in the bell below the top of the bell, and X' the volume of the interior of the bell above this new surface. Then
$$\left\{X' - B\left(1 - \frac{w'}{w}\right)\right\}(x'+k) = Vh. \quad \ldots \quad (2)$$

Now since X obviously increases with x, we must have $x' < x$, since in the opposite case each of the factors at the left-hand side of (2) would be greater than the corresponding factor in (1), and the equations would be inconsistent.

Hence the surface of the water rises.

Again, if $\Omega =$ the first volume of water in the bell, and Ω' the second
$$\Omega = V - X,$$
$$\Omega' = V - X' - B\frac{w'}{w};$$

$$\therefore \quad \Omega - \Omega' = B\frac{w'}{w} - (X - X')\ldots \quad \ldots \quad (3)$$

Now from (1) and (2)
$$X - X' = B\frac{w'}{w} - \frac{Vh(x-x')}{(x+k)(x'+k)},$$
$$\therefore \quad \Omega - \Omega' = V\frac{h(x-x')}{(x+k)(x'+k)},$$

which shows that Ω' is less than Ω.

47. Weight of Gas. It is obvious that the weight of a cubic-foot of air, or any other gas, is not the same when its temperature is 20° or 100°, as when it is 0°, supposing the intensity of pressure the same. In other words, the weight of a cubic foot of air depends on the temperature and pressure intensity at which it is taken.

Taking the units of the Metric System, let us enquire what is the weight of v litres (i. e., cubic decimètres) of dry air when its temperature is $t°C$ and its intensity of pressure denoted by a column of mercury p millimètres high.

Supposing that we knew the weight of 1 litre of air when its temperature is 0° and its intensity of pressure that of a standard atmosphere, denoted by a column of mercury 760 mm. high, we could answer the question by finding the number of litres which would be occupied by the given v litres if its state were changed from (p, t) to $(760, 0°)$. But by (1) or (2) of Art. 42, if we put $t'=0$, $p'=760$, we have
$$v_0 = \frac{273}{760} \cdot \frac{vp}{T}. \quad \ldots \quad \ldots \quad (1)$$

Now M. Regnault found that the mass of
1 litre of dry air at $(760, 0°) = 1\cdot293187$ grammes . (2)

Hence the mass of v litres at (p, t) is v_0 multiplied by this number. Denoting the mass in grammes by W, we have then
$$W = \cdot 4645 \frac{v \cdot p}{T}, \quad \ldots \quad \ldots \quad (a)$$

in which, be it remembered, T is the absolute Centigrade temperature of the air, p its pressure intensity estimated in millimètres of mercury, v its volume in litres, and W its mass in grammes.

For any other gas, if its specific gravity at $(760, 0°)$ is denoted by s, the mass of a litre of it in this state is $1\cdot293187 \times s$ grammes, and evidently if W is the mass of v at (p, t), we have simply

$$W = \cdot 4645 \frac{v \cdot p \cdot s}{T}. \quad \ldots \ldots (\beta)$$

The specific gravity of a gas is above assumed to be the ratio of the weight of any volume of the gas to the weight of an equal volume of dry air at $(760, 0°)$; but it is easy to see that we get exactly the same result by taking the ratio of the weight of a volume of the gas at (p, t) to the weight of an equal volume of air also at (p, t), whatever the pressure intensity, p, and the temperature, t, may be, if it be true that all gases have the same coefficient of expansion; for, equal volumes, v, of the two gases at (p, t) will become equal volumes, v_0, at $(760, 0°)$ since

$$v_0 = \frac{v \cdot p}{1 + at} \cdot \frac{1}{760},$$

and a is the same for both gases.

48. The equation $p = k\rho$. From (β) we see that if for $\dfrac{W}{v}$ we write ρ, where ρ is the mass, in grammes, of the gas per litre, we have

$$p = \frac{1}{\cdot 4645} \cdot \frac{T}{s} \cdot \rho, \quad \ldots \ldots (1)$$

p being measured in millimètres of mercury.

Let p be measured in grammes' weight per square centimètre, and let ρ be the mass, in grammes, per cubic centimètre; then, if we have 1 cubic cm. of the gas at

(p, t), and this becomes x cubic cms. at zero and an intensity of pressure of $76 \times 13\cdot596$ grammes' weight per sq. cm., we have
$$\frac{x \times 76 \times 13\cdot596}{273} = \frac{1 \times p}{T}. \quad \ldots \quad (2)$$

Now the mass of 1 cubic cm. at the latter temperature and pressure being $\frac{1\cdot293187}{1000} s$ grammes, the mass of x cubic cms. is obtained by multiplying this by the value of x in (2), and this mass is ρ, the density of the gas at (t, p), i.e., the mass of 1 cubic cm. Hence we have
$$p = 2926\cdot9 \frac{T}{s} \rho, \quad \ldots \quad (3)$$
where p is in grammes' weight per sq. cm. and ρ in grammes per cub. cm. Hence if we write the relation between p and ρ in the form $p = k\rho$, we see that
$$k = 2926\cdot9 \frac{T}{s}.$$

If p is measured in dynes per sq. cm., we must multiply this value of k by the value of g in dynes, i.e., by 981 (about). In this case, then,
$$p = 2926\cdot9 \frac{gT}{s} \rho. \quad \ldots \quad (4)$$

(Observe that here s is the specific gravity of the gas referred to air.)

If v is the volume of the gas at (p, T), and w its mass, we have, by multiplying both sides of (3) by v,
$$pv = 2926\cdot9 \frac{w}{s} . T. \quad \ldots \quad (5)$$

It is sometimes useful to express p in kilogrammes' weight per square decimètre, v in cubic decimètres, and w in kilogrammes; in which case (5) becomes
$$pv = 292\cdot69 \frac{w}{s} . T. \quad \ldots \quad (6)$$

208 *Hydrostatics and Elementary Hydrokinetics.*

It is usual to write (5) or (6) in the form
$$pv = RwT, \quad \ldots \ldots \ldots (7)$$
where R stands for the constant $\dfrac{2926 \cdot 9}{s}$ in the first case.

49. Formulae in English Measures. The equation connecting the volumes, &c., of a given mass of gas in English measures is
$$\frac{vp}{460+t} = \frac{v'p'}{460+t'}. \quad \ldots \ldots (1)$$

To obtain the formula for the mass of air, analogous to (a), Art. 47, we may either convert the metric formula into English measures, or deduce a formula from special observations on the mass of a given volume of air under standard conditions. Dr. Prout found that the mass of 100 cubic inches of dry air at the temperature 60° F at an intensity of pressure indicated by 30 inches of mercury in a barometer tube is 31·0117 grains; in other words,

the mass of 1 cubic foot at (60°, 30″) is ·0765546 pounds. (a)

Now if we have v cubic feet of dry air at $(t°, p)$, where p is in inches of mercury, this would, by (1), become
$$\frac{52}{3} \cdot \frac{vp}{460+t}$$
cubic feet at (60°, 30), and multiplying this by the number (a), we have
$$W = 1 \cdot 326946 \, \frac{vp}{460+t}, \quad \ldots \ldots (2)$$
for the mass, in pounds, of the given v cubic feet at $(t°, p)$, the intensity of pressure, p, being supposed taken in inches of mercury.

For a gas of specific gravity s (referred to air),
$$W = 1 \cdot 326946 \, \frac{vps}{460+t}. \quad \ldots \ldots (3)$$

Gases.

If p is estimated in pounds' weight per square foot, and $460 + t$ is denoted by T, we have

$$W = \frac{1}{53 \cdot 30222} \cdot \frac{vps}{T}, \quad \ldots \quad (4)$$

$$\therefore \; p = 53 \cdot 30222 \frac{T}{s} \cdot \rho, \quad \ldots \quad (5)$$

where ρ is the density of the gas in pounds per cubic foot, p is its intensity of pressure in pounds' weight per square foot, &c.

To obtain the analogue of (6), Art. 48, multiply both sides of this equation by v; then, with sufficient accuracy,

$$pv = 53 \cdot 3 \frac{w}{s} T, \quad \ldots \quad (6)$$

in which w is the mass of the gas in pounds; and if we write this in the form

$$pv = RwT, \quad \ldots \quad (7)$$

R stands for $\frac{53 \cdot 3}{s}$.

50. Barometric Formula. We are now in a position to deduce a formula for the height of a mountain, by neglecting the variation of gravity between the base and the summit, and by assuming the temperature of the air to be constant within these limits. The latter assumption would often be far from the truth; but we shall presently see how it can be corrected.

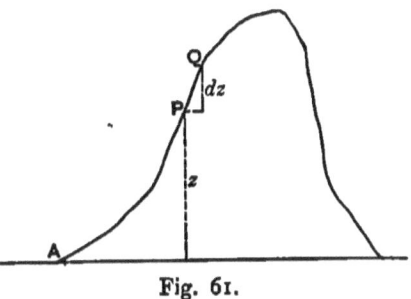

Fig. 61.

Let A, Fig. 61, be a point at the base of the mountain where the height of the barometer is p_0 inches; let P be a

210 *Hydrostatics and Elementary Hydrokinetics.*

point at a height z feet above A, and at P let the height of the barometer be p inches; let Q be a point very near P, the difference of level between P and Q being dz feet; and let $t°\,F$ be the temperature of the air at P.

Imagine a horizontal area of 1 square foot at P; then the atmospheric pressure on this area is the weight of the column of air standing on it and terminated by the limit of the atmosphere. Hence the difference of the pressures on this area at P and Q is the weight of the vertical column of air between P and Q standing on 1 square foot, i. e., the weight of dz cubic feet.

But if $p-dp$ is the height of the barometric column at Q, the difference of the pressures on 1 square foot at P and at Q is the weight of a column of mercury standing on 1 square foot having the height of $-dp$ inches. Now the mass of 1 cubic foot of mercury $= 848\frac{3}{4}$ pounds at the temperature of melting ice; but if the temperature of the mercury is $t°\,F$, this requires correction. The coefficient of absolute expansion of mercury per degree Fahrenheit is very nearly $\frac{1}{9915}$; hence, if w is the weight of a cubic foot at $32°$, the weight of a cubic foot at $t°$ is

$$\frac{w}{1+\dfrac{t-32}{9915}}, \text{ or } w\left(1-\frac{t-32}{9915}\right),$$

nearly; so that the weight of the column of mercury corresponding to the barometric fall between P and Q is

$$-\frac{848.75}{12}\left(1-\frac{t-32}{9915}\right)dp.$$

If, then, p is the height of the barometric column at the temperature $t°\,F$, the *corrected height* is

$$p\left(1-\frac{t-32}{9915}\right). \quad\quad\quad\quad (a)$$

To save a multiplicity of symbols, we shall suppose that

the heights of the mercury at the stations $A, P, Q, \ldots$ are thus corrected; in other words, we shall suppose in the subsequent work that p, p_0, dp, &c. are *corrected heights*. With this understanding, if we equate the weight of the column of air obtained by writing dz for v in (2) of last Article to the weight of the column of fall of the barometer between P and Q, we have

$$1\cdot 326946 \frac{p\,dz}{460+t} = -\frac{848\cdot 75}{12}\cdot dp \quad \ldots \quad (1)$$

(If p is not a corrected height, the temperature coefficient which multiplies p in (a) must be considered as part of the variable dp in the right-hand side of this equation, so that it does not disappear by division, unless the temperature is constant all the way up the mountain.) Hence

$$dz = -53\cdot 3022\,(460+t)\cdot\frac{dp}{p}, \quad \ldots \quad (2)$$

which is the differential relation between z and p, from which the relation between them can be obtained only by taking the temperature t constant in the term $460+t$. If we do this, and integrate from A to P, we have

$$\int_0^z dz = -53\cdot 3022\,(460+t)\int_{p_0}^p \frac{dp}{p},$$

$$\therefore\; z = 53\cdot 3022\,(460+t)\,\log_e\frac{p_0}{p}. \quad \ldots \quad (3)$$

Now $\log_e n = \dfrac{1}{\cdot 4343}\log_{10} n$; hence (3) becomes

$$z\,(\text{feet}) = 122\cdot 73\,(460+t)\,\log_{10}\frac{p_0}{p}, \quad \ldots \quad (4)$$

which is sometimes put into the form

$$z = \{60383 + 122\cdot 73\,(t-32)\}\,\log_{10}\frac{p_0}{p}. \quad \ldots \quad (5)$$

The constant value of t in this expression is usually taken to be the arithmetic mean, $\dfrac{t_0+t}{2}$, between the air temperatures t_0 at A and t at P. In the case of a very high mountain, several observations might be made at different levels, taking the temperature constant in each successive stage and equal to half the sum of the temperatures of the air at the beginning and end of the stage, and then adding the calculated heights of the successive stages together.

The result (5) can also be deduced from the general equation of equilibrium, (2) or (3) of Art. 17. If p is the intensity of pressure of the air at P in pounds' weight per square foot, and ρ is the mass of the air in pounds per cubic foot, observing that z is measured *upwards* and the force ρ pounds' weight *downwards*,

$$\frac{dp}{dz} = -\rho, \qquad (6)$$

and by (5) of last Art.,

$$p = 53 \cdot 30222\, T \cdot \rho, \qquad (7)$$

$$\therefore dz = -53 \cdot 30222\, T \frac{dp}{p}, \qquad (8)$$

which is (2) above.

If the mountain is so high that there is a sensible variation of gravity between the base and the summit, and if p is still measured with reference to the weight of a pound on the earth's surface at the sea level, the force acting on ρ pounds mass at the height z is $\rho \left(\dfrac{r}{r+z}\right)^2$ pounds' weight at the sea level, where r is the radius of the earth; so that (6) is to be replaced by

$$\frac{dp}{dz} = -\rho \frac{r^2}{(r+z)^2}, \qquad (9)$$

while (7) still holds. Hence

$$\frac{r^2}{(r+z)^2} dz = -53 \cdot 30222\, T \frac{dp}{p}, \qquad (10)$$

$$\therefore r^2 \int_{z_1}^{z} \frac{dz}{(r+z)^2} = -53\cdot30222\, T \int_{p_1}^{p} \frac{dp}{p} \quad \ldots \quad (11)$$

Assuming the station A to be at the height z_1 above the sea level, and that the intensity of pressure at A is p_1 pounds' weight per square foot. The integral of (11) is

$$r^2 \left(\frac{1}{r+z_1} - \frac{1}{r+z} \right) = 122\cdot73\, T \log_{10} \frac{p_1}{p} \quad \ldots \quad (12)$$

In this equation p and p_1 are measured in pounds' weight per square foot; but they must be inferred from the readings of the barometer at the two stations; and if the heights of the barometer at the stations are h and h_1 and the weights of a unit volume of mercury at these stations are w and w_1, respectively, we have $p = wh$ and $p_1 = w_1 h_1$, therefore $\dfrac{p_1}{p} = \dfrac{h_1}{h} \cdot \dfrac{w_1}{w} = \dfrac{h_1}{h} \left(\dfrac{r+z}{r+z_1} \right)^2$; hence (12) becomes

$$r^2 \left(\frac{1}{r+z_1} - \frac{1}{r+z} \right) = 122\cdot73\, T \left\{ \log_{10} \frac{h_1}{h} + 2 \log_{10} \frac{r+z}{r+z_1} \right\}, \quad (13)$$

in which the barometric heights may, of course, be measured in inches, millimètres, or any other units of length.

Observe that $\dfrac{z}{r}$ and $\dfrac{z_1}{r}$ are very small fractions whose squares may be neglected, so that if $z - z_1$ is denoted by Δ, since $\log_{10}\left(1 + \dfrac{z}{r}\right)$ is approximately equal to $\cdot 4343 \dfrac{z}{r}$, we have

$$\Delta = 122\cdot73\, T \left(1 + \frac{z + z_1}{r} \right) \left\{ \log_{10} \frac{h_1}{h} + \cdot 8686 \frac{\Delta}{r} \right\}. \quad (14)$$

Let $A = 122\cdot73\, T \log_{10} \dfrac{p_1}{p}$ and $B = 122\cdot73 \times \cdot 8686\, T$; then, putting $z = \Delta + z_1$, we have

$$\Delta = \left(A + B \frac{\Delta}{r} \right) \left(1 + \frac{\Delta + 2z_1}{r} \right) \quad \ldots \quad (15)$$

214 *Hydrostatics and Elementary Hydrokinetics.*

If the variation of gravity were neglected, we should have $\Delta = A$, as in (4), and this value may be put for Δ in the terms of (13) which involve r, so that finally

$$\Delta = A\left(1 + \frac{A+B+2z_1}{r}\right), \quad \ldots \quad (16)$$

an equation which gives the difference of level between the top and base of the mountain when the height of the base above the sea level is known.

It is understood, as before explained, that in this expression p_1 and p are the *corrected heights* of the barometer at the two stations.

51. Metric Formulae. By the same method—viz., the equating of the weight of the vertical column of air between P and Q, Fig. 61, standing on a horizontal square decimètre to the weight of the column of fall of mercury standing on the same area—we obtain the height of a mountain in metric measures.

Thus, neglecting the variation of gravity, since a litre is a cubic decimètre, if z is the height of P above A in decimètres and the *corrected* barometric height at P is p millimètres, the weight of the fall of mercury is $-135 \cdot 96\, dp$ grammes' weight; hence from (a) of Art. 47,

$$-135 \cdot 96\, dp = \cdot 4645 \frac{p\, dz}{273 + t}, \quad \ldots \quad (1)$$

$$\therefore z = 673 \cdot 962\,(273 + t) \log_{10} \frac{p_0}{p},$$

z being in decimètres. If z is taken in metres, this becomes

$$z = 18399 \cdot 2\left(1 + \frac{t}{273}\right) \log_{10} \frac{p_0}{p} \quad \ldots \quad (2)$$

If the variation of gravity is taken into account, let p be measured in grammes' weight per square centimètre, &c.,

as in (3) of Art. 48; then the equations are

$$\frac{dp}{dz} = -\rho \frac{r^2}{(r+z)^2},$$

$$p = 2926 \cdot 9\, T\rho,$$

$$\therefore r^2 \left(\frac{1}{r+z_1} - \frac{1}{r+z}\right) = 673 \cdot 962\, T \log_{10} \frac{p_1}{p},$$

in which again we can put $\dfrac{p_1}{p} = \dfrac{h_1}{h}\left(\dfrac{r+z}{r+z_1}\right)^2$, and deduce a result similar to (14).

EXAMPLES.

1. If a cubic inch of water is converted into steam at $212°$ F, find the volume of the steam.

Ans. 1696 cubic inches. Hence it is approximately true that 1 cubic inch of water yields 1 cubic foot of steam.

2. Calculate the mass of air in a room whose dimensions are 18, 18, and 10 feet, the temperature being $60°$ F and the barometer standing at 30 inches.

Ans. 248 pounds.

3. If 1 pound of water is converted into steam at $212°$ F under the intensity of pressure of 15 pounds' weight per square inch, prove that it will yield 26.66 cubic feet.

4. If any volume of water is converted into steam at the temperature $t°$ F under the intensity of pressure p pounds' weight per square inch, prove that the ratio of the volume of the steam to that of the water from which it has been formed is about

$$37 \cdot 1 \times \frac{460 + t}{p}.$$

[This is called the *relative volume* of steam at the given temperature and pressure.]

5. At the foot of a mountain the temperature of the air is $66°$F, and the height of the barometer $29 \cdot 35$ inches; at the top the temperature is $50°$ F, and the barometric height $24 \cdot 81$; find the height of the mountain, assuming the coefficient of expansion of mercury to be $\frac{1}{10000}$ per degree Fah.

Ans. About 4590 feet.

216 *Hydrostatics and Elementary Hydrokinetics.*

52. Nature of Gas Pressure. According to the Kinetic Theory of Gases, when a gas is contained in a vessel, the pressure exerted by the gas on each element of area of the vessel is due to the incessant impacts of the molecules of the gas on the element of area.

These molecules are, of course, extremely small: at each instant a certain number of them will be in actual contact; but there will be a certain *average* distance between them, and it is assumed that this distance is vastly greater than the diameter, or greatest linear dimension, of a molecule. Thus we are to imagine the space inside the vessel as being comparatively void of gaseous matter. Nevertheless, this space is what we mean by the *volume of the gas*, which, therefore, is something very different from the sum of the volumes of its material particles; i. e., from the aggregate number of the cubic centimètres occupied by its material: the volume of the gas is the volume of the space within which the excursions of all its molecules are confined.

Again, when the gas has settled down to constant conditions of temperature and pressure, it is clear that, in every respect, the state of affairs is the same at any one instant as at any other. Not only so, but if we imagine any area—say one square centimètre—placed at any point, P, inside the vessel and occupying any position (orientation) at this point, the number of molecules passing through this area in any time—say one second—whether from right to left or left to right, is always the same.

It is easy to calculate the intensity of pressure produced at each point of a vessel containing a system of molecules moving in this manner; but we shall confine our attention here to the consideration of a very simple case from which the state of affairs in the general case may be inferred. The discussion of the general case will be found in Watson's *Treatise on the Kinetic Theory of Gases*. The development

Gases. 217

of the subject is, in the first instance, mainly the work of Clerk Maxwell.

Imagine a cylinder whose base is AB, Fig. 62, to contain a large number of molecules, the mass of each being μ grammes, all moving in lines parallel to the axis of the cylinder, each with a velocity of v centimètres per second. Then if we measure a length AA' equal to $v \cdot \Delta t$ and draw a plane $A'B'$ parallel to AB, all the molecules within the space $ABB'A'$ which are moving in the sense $A'A$ will in the time Δt strike the base AB, and each will be reflected with a velocity which is assumed to be equal to v, i.e., the coefficient of restitution for each particle and the base AB is assumed to be unity. Let the area of the base AB be S square centimètres.

Fig. 62.

Now if there are n molecules in each cubic centimètre, there are $Snv\Delta t$ within the space $ABB'A'$, and since there are as many moving in the sense AA' as in the sense $A'A$, the number striking the base AB in the time Δt is

$$\tfrac{1}{2} Snv\Delta t.$$

Hence the momentum incident on AB is $\tfrac{1}{2}Sn\mu v^2\Delta t$; and since the same quantity of momentum is generated in the opposite sense by impact on the base, the total change of momentum generated by impact on the base in the time Δt is

$$Sn\mu v^2 \Delta t.$$

But if P dynes is the mean value of the pressure exerted by the base on the column of molecules, $P\Delta t$ is the impulse of this force in the time Δt, and this is equal to the change of momentum in the same time. Hence

$$P = Sn\mu v^2 \quad \ldots \quad \ldots \quad (1)$$

Of course n is enormously great and μ indefinitely small;

each is unknown, but $n\mu$ is the mass, in grammes, of one cubic centimètre of the gas. Denote this by ρ, and let p denote $\dfrac{P}{S}$, the intensity of the pressure; then

$$p = \rho v^2. \qquad \qquad (2)$$

This expresses p in dynes per square centimètre. If we divide it by 981, or rather by g, the acceleration of a freely falling body in centimètres per second per second, we have

$$p = \frac{\rho v^2}{g}, \qquad \qquad (3)$$

which gives the intensity of pressure in grammes' weight per square cm.

If v is in feet per second, ρ in pounds per cubic foot, and g in feet per second per second, (3) gives the intensity of pressure in *pounds' weight per square foot*.

This formula would apply to the case of a waterfall which strikes the ground, the water not being reflected by impact, as the student will easily see on re-examining the details. If h is the height of the waterfall, the velocity of the molecules striking the plane is given by the equation $v^2 = 2gh$, so that

$$p = 2\rho h, \qquad \qquad (4)$$

which equation gives p in gravitation measure—either grammes' weight per sq. cm., or pounds' weight per square foot; for the latter units ρ is about $62\tfrac{1}{2}$ pounds per cubic foot. Hence (4) shows that the intensity of pressure on the plane is twice as great as that produced by a statical column of water of the same height.

But the case of a waterfall is in other respects different from that of a gas; for, in the latter case the molecules in any column are not all moving in the same direction, so that the value of p in (2) or (3) must be greatly superior to that of the intensity of pressure actually produced by

a gas contained in the cylinder. We must remember that v in these equations is the mean value of the velocity in a fixed direction; and that, if v_1, v_2, v_3 are the components of velocity of a molecule in three fixed rectangular directions, and v is the resultant velocity of the molecules,
$$v^2 = v_1^2 + v_2^2 + v_3^2.$$
This shows that when we consider an indefinitely great number of molecules, if $\overline{v^2}$ is the mean value of the squares of their resultant velocities,
$$\overline{v^2} = 3v_x^2, \quad \ldots \ldots \quad (5)$$
where v_x^2 is the mean value of the square of velocity *in a fixed direction*.

Now v_x^2 in (5) must take the place of v^2 in (2) or (3).

Hence, then, for a system of molecules, moving in *all* directions inside a vessel,
$$p = \tfrac{1}{3} \rho \overline{v^2} . \quad \ldots \ldots \quad (6)$$
or
$$p = \tfrac{1}{3} \frac{\rho \overline{v^2}}{g}, \quad \ldots \ldots \quad (7)$$
according as p is measured in absolute units (dynes or poundals) or in gravitation units (grammes' weight or pounds' weight).

This elementary method of treating the question is not satisfactory. The following is more thorough and scientific.

Imagine the molecules of a gas contained within any vessel divided at any instant into groups, the velocities of all those in the same group being nearly the same in magnitude and nearly the same in direction. No one of these groups is localised in a definite portion of the volume of the vessel—the grouping is not with reference to *place* but with reference to the *characteristics of velocity*, so that each group occupies the whole space within the vessel.

Now we can graphically represent the velocities in any

group thus. Take any fixed origin, O, and draw a line OP to represent in magnitude and direction any velocity q; let Ox, Oy, Oz be any three rectangular axes at O, and let the direction of OP be expressed by means of the usual angles of colatitude and longitude (see *Statics*, vol. i., Art. 176), i. e., let θ be the angle zOP and let ϕ be the angle between the plane zOP and the plane of xz; describe a sphere with centre O and radius OP; let the axis Oz meet this sphere in z; on the surface of this sphere take a point Q near P, its colatitude ZOQ being $\theta + d\theta$, and its longitude $\phi + d\phi$; through P draw a parallel of latitude PR meeting the meridian zQ in R, and through Q draw a parallel of latitude QS meeting the meridian zP in S; we thus obtain a small spherical quadrilateral, $PRQS$ whose area is $q^2 \sin\theta \, d\theta \, d\phi$; produce OP to P' so that PP' represents dq, and with OP' as radius describe another sphere; produce OR, OQ, OS to meet the surface of this second sphere in R', Q', S', respectively; then we obtain a small element of volume between the two quadrilaterals $PRQS$ and $P'R'Q'S'$, and the volume of this element is

$$q^2 \sin\theta \, dq \, d\theta \, d\phi, \text{ or } d\Omega. \quad \ldots \quad (8)$$

The number of molecules in the group whose velocities are represented by radii vectores drawn from O to points contained within this elementary volume will be proportional to the product of this element and some function of q, θ, ϕ. But since any given speed is found in all directions indifferently within the vessel, it is clear that if from O to points contained within any element, $d\Omega$, of volume radii be drawn and taken to represent a velocity group of nearly the same speed, q, and nearly the same direction, the quantity by which $d\Omega$ must be multiplied to give the number of molecules in the group must be a function of q alone and not of its direction (θ, ϕ.)

Gases.

Hence the number of molecules within the vessel and contained in the selected group is

$$f(q) \cdot q^2 \sin \theta \, dq \, d\theta \, d\phi, \quad \ldots \quad (9)$$

where $f(q)$ is some unknown function of q.

It is not impossible that the following objection will be raised by the reader to the assumption that the number of molecules in the group is $f(q) \cdot d\Omega$: this expression vanishes if $d\Omega$ is zero, i. e., if we consider the number of molecules moving with *exactly the same* velocity, represented as above by the radius vector OP, whereas every molecule in the vessel might be moving with one and the same velocity—as in the case of the elementary example first treated, viz., that in which a stream of particles moved along the axis of a cylinder.

To this the reply is as follows: the fundamental assumption of the Kinetic Theory is that *all possible speeds*, from 0 to ∞, *in all possible directions* characterise the state of affairs within the vessel, and therefore that the conception of the existence of only one speed among the molecules is wholly inadmissible. Hence, there being an enormously great number of molecules, the number of those moving with the *same* speed—whether in the same direction or not is immaterial—is relatively zero.

The number of the group (9) contained within a unit volume of the space within the vessel is obtained by dividing the expression (9) by the number of units of volume in the vessel. We may, then, assume that (9) expresses the number of the group per unit volume.

To calculate the intensity of pressure at any point, M, in the gas, take a small plane area, dS, at the point perpendicular to the direction of the axis Oz; on dS describe a cylinder the length of whose axis is $q \cos \theta \cdot dt$, where q expresses the velocity characterising any group, and dt

222 *Hydrostatics and Elementary Hydrokinetics.*

is an infinitesimal element of time. Then all the molecules within this small cylinder, whose volume is $q \cos \theta \, dS \, dt$, will impinge on the plane dS within the time dt. Now it matters not whether dS is a small *material* plane from each face of which the impinging molecules are reflected, their normal velocities being all restored to them in the reverse sense, or an *imagined* plane area through which two streams of molecules pass in opposite senses; for in either case the quantity of momentum which in the time dt travels from the plane in the same sense is the same.

The number of molecules of the (q, θ, ϕ) group contained within this cylinder is

$$f(q) \cdot q^2 \sin \theta \, dq \, d\theta \, d\phi \times q \cos \theta \, dS \, dt, \quad . \quad . \quad (10)$$

half of them moving towards the plane and half from it. Let μ be the mass of each molecule; then, since the velocity of each of these perpendicular to the plane is $q \cos \theta$, the quantity of momentum passing from the plane in one sense within the time dt is

$$\mu f(q) \cdot q^4 \sin \theta \cos^2 \theta \, dq \, d\theta \, d\phi \cdot dS \, dt. \quad . \quad . \quad (11)$$

If p is the mean value of the force exerted, per unit area, on the plane, the impulse of the force exerted by the plane is $p \, dS \, dt$, and this must be equated to the integral of (11) for all possible velocity groups. All such groups are included by taking the variables between the limits indicated thus:
$$\left. q \right|_0^\infty, \quad \left. \theta \right|_0^\pi, \quad \left. \phi \right|_0^{2\pi}.$$

Hence

$$p = \mu \int_0^\infty f(q) \cdot q^4 \, dq \int_0^\pi \sin \theta \cos^2 \theta \, d\theta \int_0^{2\pi} d\phi$$

$$= \frac{4\pi\mu}{3} \int_0^\infty f(q) \cdot q^4 \, dq. \quad . \quad . \quad . \quad . \quad . \quad . \quad (12)$$

To see how this is connected with the mean square of velocity of the molecules, let n be the number of molecules per unit volume; then from (9)

$$n = \int_0^\infty f(q) \cdot q^2 \, dq \int_0^\pi \sin\theta \, d\theta \int_0^{2\pi} d\phi$$

$$= 4\pi \int_0^\infty f(q) \cdot q^2 \, dq. \quad \ldots \quad (13)$$

To get the mean square of velocity, $\overline{v^2}$, multiply (9) by q^2, integrate and equate the result to $n\overline{v^2}$; then

$$n\overline{v^2} = 4\pi \int_0^\infty f(q) \cdot q^4 \, dq. \quad \ldots \quad (14)$$

Hence

$$p = \tfrac{1}{3} n\mu \cdot \overline{v^2}$$

$$= \tfrac{1}{3} \rho \overline{v^2}, \quad \ldots \quad (15)$$

as before obtained. (Of course p is here expressed in absolute units; in gravitation units $p = \tfrac{1}{3}\dfrac{\rho \overline{v^2}}{g}$, as before explained.)

To get the mean velocity, $\bar{v}$ (which, of course, is not equal to the square root of $\overline{v^2}$) multiply (9) by q, integrate, and equate the result to $n\bar{v}$; then

$$n\bar{v} = 4\pi \int_0^\infty f(q) \cdot q^3 \, dq. \quad \ldots \quad (16)$$

Thus we see that the result (15) is true whatever may be the form of the function $f(q)$. In the Kinetic Theory of Gases this function must have a particular form in order that the incessant collisions between the molecules may render the state of affairs *statistically* the same at all times. It does not belong to our present purpose to find the form of this function; its determination involves only the most elementary principles of the collision of elastic spheres, and it will be found in detail in Watson's *Kinetic Theory of Gases*. The statistical permanence of the state of the molecules requires that

$$f(q) \equiv A e^{-\frac{q^2}{c^2}}, \quad \ldots \ldots \quad (17)$$

where A is a constant, and c is also a constant of the nature of velocity.

Thus
$$n = 4\pi A \int_0^\infty e^{-\frac{q^2}{c^2}} \cdot q^2 \, dq, \quad \ldots \quad (18)$$

$$n\overline{v^2} = 4\pi A \int_0^\infty e^{-\frac{q^2}{c^2}} \cdot q^4 \, dq, \quad \ldots \quad (19)$$

$$n\overline{v} = 4\pi A \int_0^\infty e^{-\frac{q^2}{c^2}} \cdot q^3 \, dq. \quad \ldots \quad (20)$$

Now it is known (see Williamson's *Integral Calculus*, Art. 116, or Price's *Infin. Cal.*, vol. ii., chap. iv) that

$$\int_0^\infty e^{-kx^2} dx = \tfrac{1}{2} \left(\frac{\pi}{k}\right)^{\frac{1}{2}}. \quad \ldots \quad (21)$$

Hence differentiating both sides with respect to k,

$$\int_0^\infty e^{-kx^2} \cdot x^2 \, dx = \tfrac{1}{4} \left(\frac{\pi}{k^3}\right)^{\frac{1}{2}}. \quad \ldots \quad (22)$$

and again differentiating this with respect to k,

$$\int_0^\infty e^{-kx^2} \cdot x^4 \, dx = \tfrac{3}{8} \left(\frac{\pi}{k^5}\right)^{\frac{1}{2}}. \quad \ldots \quad (23)$$

Also it is obvious that $\int_0^\infty e^{-kx^2} \cdot x \, dx = \dfrac{1}{2k}$, and by differentiating this with respect to k,

$$\int_0^\infty e^{-kx^2} \cdot x^3 \, dx = \dfrac{1}{2k^2}. \quad \ldots \quad (24)$$

Gases.

From these results we have

$$n = A\pi^{\frac{3}{2}} c^3, \quad \ldots \ldots \ldots (25)$$

$$\overline{v^2} = \tfrac{3}{2} c^2, \quad \ldots \ldots \ldots (26)$$

$$\bar{v} = \frac{2}{\sqrt{\pi}} c. \quad \ldots \ldots \ldots (27)$$

From the last two we have

$$\overline{v^2} = \frac{3}{8}\frac{\pi}{}(\bar{v})^2, \quad \ldots \ldots (28)$$

which shows the relation between the mean square of the velocities and the square of the mean velocity, the former being the greater. In terms of the latter, the intensity of pressure is given by the equation

$$p = \frac{\pi}{8}\rho(\bar{v})^2, \text{ or } p = \frac{\pi}{8}\frac{\rho}{g}(\bar{v})^2, \quad \ldots \ldots (29)$$

according as p is taken in absolute or in gravitation measure.

EXAMPLE.

It is found that the mass of 1 cubic foot of hydrogen at 0°C when its intensity of pressure is 2116·4 pounds' weight per square foot is ·005592 pounds, the value of g being 32·2 feet per second per second; find the mean velocity of the hydrogen molecules.

From the second value of p in (29) we have

$$2116\cdot 4 = \frac{\pi}{8} \cdot \frac{\cdot 005592}{32\cdot 2} \cdot (\bar{v})^2,$$

$$\therefore \quad \bar{v} = 5570\cdot 8 \text{ feet per second, nearly.}$$

The value of $\sqrt{\overline{v^2}}$, which is that velocity whose square is equal to the mean square of the velocities, and which is called the *velocity of mean square*, is found from (15) to be 6046·5 feet per second, nearly.

The kinetic energy of the molecules in a unit volume is obtained by multiplying (9) by $\frac{\mu q^2}{2g}$. If E denotes this kinetic energy (in gravitation measure) we thus find

$$E = \frac{\mu}{2g} \cdot 4\pi \int_0^\infty f(q) \cdot q^4 \, dq = \frac{\mu n}{2g} \overline{v^2}$$

$$= \frac{\rho \overline{v^2}}{2g} \qquad \qquad \qquad (30)$$

Thus we have

$$p = \tfrac{2}{3} E, \qquad \qquad \qquad (31)$$

expressing the intensity of pressure in terms of the kinetic energy per unit volume.

53. Mixture of Gases. When two or more different gases are present in the same space, *each of them produces exactly the same intensity of pressure as if none of the others were present*. This fact is a result of the kinetic theory—'when several different sets of spheres are present together in the region under consideration, the distribution of the centres and of the velocities of the spheres of each set is independent of the coexistence of the remaining sets.' (Watson's *Kinetic Theory of Gases*, Prop. VI.) The *à priori* possibility of such a result is manifest if we remember that the void spaces in a vessel which contains even several gases are very much greater than the occupied spaces.

The result may be proved thus: suppose two masses of gas whose volumes, intensities of pressure, and absolute temperatures are represented in the following figures,

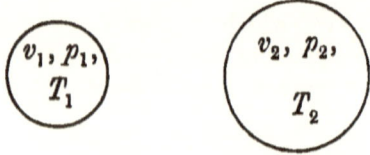

to be mixed and contained within a given volume V; and when the mixture becomes homogeneous, let its intensity of pressure become P and its absolute temperature T. Then if we take the first gas and alter its intensity of pressure to p_2 and its absolute temperature to T_2, its volume will become

$$v_2 \frac{p_1}{p_2} \cdot \frac{T_2}{T_1},$$

and in this state if it is added to the second gas, the volume of the mixture will be

$$v_1 \frac{p_1}{p_2} \cdot \frac{T_2}{T_1} + v_2$$

at (p_2, T_2). Denote this volume by U. Then if (U, p_2, T_2') is altered to (V, P, T),

$$\frac{VP}{T} = \frac{Up_2}{T_2},$$

$$\therefore \frac{VP}{T} = \frac{v_1 p_1}{T_1} + \frac{v_2 p_2}{T_2}. \quad \ldots \ldots \quad (1)$$

Now if the first gas filled the volume V alone, at the absolute temperature T, its intensity of pressure would be

$$\frac{v_1}{V} \cdot \frac{T}{T_1} \cdot p_1,$$

and if the second occupied V alone, its intensity of pressure would be

$$\frac{v_2}{V} \cdot \frac{T}{T_2'} \cdot p_2,$$

and the sum of these is

$$\frac{T}{V} \left(\frac{v_1 p_1}{T_1} + \frac{v_2 p_2}{T_2} \right),$$

which is precisely the value of P given by (1).

If several gases of given volumes, intensities of pressure, absolute temperatures, and specific gravities, are mixed together

228 *Hydrostatics and Elementary Hydrokinetics.*

in a vessel of given volume at a given absolute temperature, find the intensity of pressure and specific gravity of the mixture.

Let the specific gravities of the gases be $s_1, s_2, s_3, \ldots$, the specific gravity of the mixture S, its absolute temperature T, and its volume V; then by the laws of Boyle and Dalton, we have

$$\frac{VP}{T} = \frac{v_1 p_1}{T_1} + \frac{v_2 p_2}{T_2} + \frac{v_3 p_3}{T_3} + \ldots; \quad . \quad . \quad (2)$$

and since the weight of the mixture is equal to the sum of the weights of the constituents,

$$\frac{VPS}{T} = \frac{v_1 p_1 s_1}{T_1} + \frac{v_2 p_2 s_2}{T_2} + \frac{v_3 p_3 s_3}{T_3} + \ldots . \quad . \quad (3)$$

54. Vapours. Many liquids, such as water, mercury, ether, the various alcohols, and acetones, gradually pass into a gaseous condition—i. e., continuously give off vapours—at ordinary temperatures. Some of these liquids are much more volatile than others. For example, a small quantity of alcohol if left in an unclosed vessel will disappear in a short time, whereas the same mass of water would, under like circumstances, take a very long time to pass away as a gas.

These vapours are essentially the same as the bodies which we have defined as gases. In fact, all gases can be regarded as the vapours of liquids, although the liquids from which some of them come can be obtained or produced only with extreme difficulty. Thus, it is now known that even hydrogen and oxygen are the vapours of two liquids; and it was known a long time ago that nearly all the gases with which we are familiar could be converted into liquids, and even solids.

Every vapour—whether it be the gas which we call

hydrogen or the vapour of alcohol or of any ordinary liquid —obeys the typical law of a gas (that of Boyle and Mariotte), provided that the vapour, as regards temperature or intensity of pressure, is not near its liquid state; and, of course, none of these bodies obey this law when they are near this state.

The first fundamental characteristic of a vapour which we shall signalise is this: *at a given temperature, there is a limit to the intensity of pressure which a given vapour can exert, or which can by any means be exerted upon that vapour.*

For example, if we take the vapour of water at the temperature of 100° C, or 212° F, we cannot produce on it a greater intensity of pressure than about that of 15 pounds' weight per square inch. If we attempt to exceed this limit, some of the vapour will at once become water. Again, if we take the vapour of water at the temperature of 107° F, it cannot sustain a greater intensity of pressure than that of about 1 pound weight per square inch.

The result is different for different liquids. Thus, if we take the vapour of mercury at the temperature of 100° C, the greatest intensity of pressure that can be exerted upon it is that of about $\frac{1}{88}$ of a pound weight per square inch.

On the other hand, the vapour of bisulphide of carbon at the temperature of 100° C can sustain an intensity of pressure of about 65.7 pounds' weight per square inch; and the vapours of alcohol and ether, at this temperature, can exert intensities of about $33\frac{1}{2}$ and $97\frac{1}{2}$ pounds' weight per square inch, respectively.

To verify experimentally the above characteristic of a vapour, take a long glass tube (barometric tube), ABC, Fig. 63, dipping into a vessel, DE, of mercury; let A be the point to which the mercury in the tube reaches, the Torricellian vacuum being the portion of the tube above A.

Then *AC* is the height of the mercurial barometer and marks the intensity of atmospheric pressure.

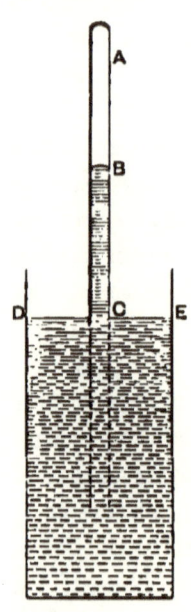

Fig. 63.

Suppose now that a drop of ether is inserted into the tube through the open end below the mercury in the vessel. The ether will rise up through the mercury in the tube and reach the Torricellian vacuum, where it will be at once converted into vapour whose pressure will depress the surface of the column of mercury in the tube. Let another drop of ether be similarly inserted; then a further depression of the mercury column takes place. Add, again, another drop of ether, and so on, until the mercury column refuses to be further depressed. *It is an experimental fact that such a limit to depression is reached*; and if we continue to insert drops of ether into the tube, they are not converted into vapour on reaching the top, but remain in the liquid state on the top, *B*, of the column of mercury. It is supposed that the point *B* marks the depression of the mercury when the mercury refuses to be further depressed, and when the ether ceases to be further converted into vapour.

Now it is found experimentally that the position of this limiting point *B* depends simply on the temperature of the surrounding air, i. e., on the temperature of the ether vapour at the top of the tube. For example, if the temperature of the ether is

$-20°$ C, the depression $AB =$ 68 millimètres,

$\quad\quad 0°$,, ,, ,, $= 182$,,

$\quad\quad 60°$,, ,, ,, $= 1728$,,

Supposing now that for any given temperature, t, the point B is the lowest to which the mercury is depressed, the tube ABC being held fixed, let us consider what happens—

1°. if the tube is forced down farther into the vessel DE,

2°. if the tube is raised.

In the first case we find that the effect is to liquefy some of the ether in AB and to keep the height, BC, of B above the level of the mercury DE constant.

In the second case, if there was any of the liquid ether resting on the mercury at B, some of this would be vapourised, and we should still observe that the height BC is constant.

Now if the heights AC and BC are estimated in millimètres, the intensity of pressure of the vapour is measured by

$AC - BC$ millimètres of mercury.

Thus, if the temperature of the vapour is 60° C, the distance AB being 1728 millimètres, the point B will be 968 millimètres below C, assuming the height, AC, of a barometric column to be 760 mm.

A vapour which, at a given temperature, t, cannot submit to *increase* of pressure (becoming in part liquid if an attempt to increase the pressure is made) is called a *saturated vapour* at the temperature t. Its intensity of pressure can, of course, be diminished, and then it ceases to be a *saturated* vapour. Thus we may define the saturation intensity of pressure of any vapour at the temperature t as *the greatest intensity of pressure that the vapour at this temperature can bear without liquefying.*

As has been said, the vapours of different liquids, all at the same temperature, have very different intensities of pressure corresponding to their saturation states. There is no known mathematical formula which gives the value of the intensity of pressure, p, corresponding to the state of saturation of a given vapour at a given temperature, t.

In the case of *aqueous vapour*, i. e., the vapour of water, it is most important to have a knowledge of the maximum intensities of pressure corresponding to a long range of temperatures. M. Regnault by a simple method of experiment was enabled to observe these saturation pressures for temperatures ranging from below zero to above 100° C; and from the table which he compiled an empirical formula connecting p and t may be constructed. One such formula will be given presently.

It will be useful to regard the matter from a different point of view, and, as it were, to deduce the liquid from the vapour instead of the vapour from the liquid.

Thus, suppose that we propose this problem: given ether vapour at $-20°$ C, at $0°$ C, at $60°$ C, and at $100°$ C, successively, how shall we convert it into liquid ether?

The answer would be: produce on it an intensity of pressure of about $1\frac{1}{3}$, $3\frac{1}{2}$, 34, and $97\frac{1}{2}$ pounds' weight per square inch, respectively.

It looks as if in the case of any gas whatever *at any assigned temperature* the answer would be the same kind— i. e., simply produce a certain intensity of pressure; but we shall subsequently see that in the case, for example, of hydrogen at any ordinary temperature this process would not suffice.

55. Mixture of Gas and Vapour. *If a given space is saturated with the vapour of any liquid, at a given temperature, the intensity of pressure of the vapour is the same whether this space is a vacuum or contains any gas.*

This follows from the result which has been already given for the mixture of any gases in a given space, viz., that each gas behaves as a vacuum to all the others;

and it may be experimentally verified by the following method.

A glass tube HD, Fig. 64, is attached to another, LB, both being vertical, the system being provided with a stopcock s. The tube LB is fitted with another stopcock b, and terminates at A where the funnel, F, or a glass globe, G, fitted with a stopcock, c, can be screwed on. At first the two tubes are filled with mercury, the tube LB being filled up to b and the stopcock s closed. The globe G contains dry air, or any other gas.

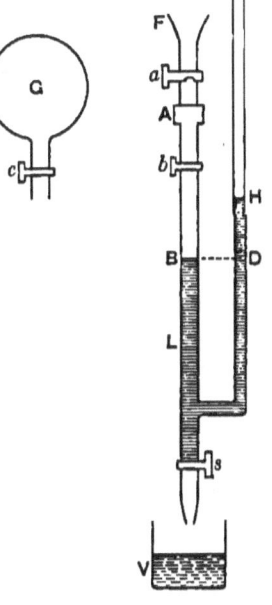

Fig. 64.

Let the globe be screwed on at A and the stopcocks c, b, s opened, and let a little mercury flow into a vessel V so that dry air fills the top of the tube LB. Close s, and pour mercury into the tube HD until the level of the mercury is the same in both tubes. The tube HD being open to the atmosphere, when the level is the same, the air at the top of LB is at the atmospheric pressure. Let the common level of the mercury at this stage be BD. Now remove the globe from A, and screw on the funnel, F. This funnel is fitted with a stopcock, a, which is not perforated but has a small cavity at the side, as seen in the figure. Let some of the liquid whose vapour we are considering be poured into F, and let a be turned so that liquid falls down to b; and let b be turned so that the liquid falls into the air space bB and becomes vapourised, when its pressure, added to that of the air, depresses the level of the mercury below B. Repeat this operation until the level of the mercury ceases to sink; and let the final level be L; then the space bL is saturated with the vapour.

Now pour mercury into the tube DH until its level in the other tube is restored from L to B. This will liquefy some of the vapour, but will leave its intensity of pressure unaltered. Suppose that the level of the mercury in DH is now at H. Then if h is the height of the barometer during the experiment, the intensity of pressure of the mixed gases in bB is represented by $h + HD$; but as the air in bB has been restored to its original volume, its intensity of pressure is the same as it was at first, i.e., it is represented by a column of mercury of height h; hence the intensity of pressure of the vapour is represented by HD.

But if we now take a barometer tube, such as that represented in Fig. 63, and insert a few drops of the liquid in question into the Torricellian vacuum until the space becomes saturated, we shall find that the depression of the mercury column is equal to HD, which shows that the intensity of pressure of the saturated vapour of the given liquid at the given temperature is uninfluenced by the presence of air with the vapour. Hence in a given volume which is saturated with any vapour there is the same mass of vapour whether the given space is a vacuum or contains any gas or gases.

56. Moist Air. It follows from Art. 53 that when the atmosphere contains aqueous vapour the intensity of pressure which is observed by a barometer is the sum of the intensities due to the air itself and to the vapour which it contains.

If, then, h (inches or millimètres) is the height of the barometer, and if the intensity of pressure of the vapour present in the air is measured by f (inches or millimètres) of the liquid which the barometer contains, the intensity of pressure of the air itself is measured by

$$h-f.$$

The weight of a given volume of moist air is affected by this consideration.

Thus, assuming the specific gravity of aqueous vapour to be ·622, or $\frac{5}{8}$ nearly, the weight of the air in a volume of v litres, with the notation of Art. 47, is $·4646\dfrac{v(p-f)}{T}$, and that of the vapour in this volume is $·4645\dfrac{vf}{T} \times \frac{5}{8}$; so that the weight of the whole is

$$·4645\frac{v(p-\frac{3}{8}f)}{T}. \quad \ldots \quad \ldots \quad (1)$$

Similarly, formula (2) of Art. 49 will be replaced by

$$W = 1·326946\frac{v(p-\frac{3}{8}f)}{460+t}. \quad \ldots \quad (2)$$

The accurate measurement of the intensity of pressure of the aqueous vapour present in the air at any temperature is a matter of difficulty; but we shall presently see how it can be approximately effected by means of a hygrometer.

57. Vapour of Water. A knowledge of the maximum intensity of pressure of the vapour of water for any given temperature is important; and, as said before, this knowledge must be derived from experiment. Water may be made to boil at any temperature whatever by producing a suitable intensity of pressure on its surface.

If any given intensity of pressure is by any means produced on the surface of a liquid, and heat is continuously applied to the liquid, *the liquid will boil when the intensity* of pressure of its vapour becomes equal to the intensity of pressure on the surface of the liquid.

The given intensity of pressure on the surface is then the maximum intensity of pressure of the vapour at the temperature of the liquid.

This principle is the basis of the experiments of Regnault for the determination of the maximum intensities of pressure of water vapour at various temperatures.

His method was to make water boil under a continuous series of surface pressures and to read the corresponding temperatures of boiling.

The figure of the apparatus used will be found in treatises on Experimental Physics (see, for example, Ganot's *Physics*, Art. 351).

Experiments with the same object had a short time previously been carried out by Dulong and Arago, at the request of the French Academy, and the results were expressed by an empirical formula known as the *French Commissioners' Formula*. If t is the centigrade temperature at which water is made to boil by adjusting the intensity of pressure on its surface, this intensity is expressed by the equation

$$n = \left(1 + \cdot 7153 \frac{t-100}{100}\right)^5, \quad \ldots \quad (1)$$

in which n denotes the number of atmospheres, of 760 mm., under which the water boils; that is, the intensity of pressure on the surface (and therefore of the vapour) is represented by a column of mercury $760\,n$ millimètres high.

When compared with the tabulated results of M. Regnault, this formula is fairly accurate for pressures ranging from 1 atmosphere to 24 atmospheres; but for very low pressures the error in the formula is great.

The following table gives a few temperatures below

Gases. 237

100° C at which water is caused to boil, together with the corresponding intensities of pressure on its surface (or of its vapour), measured in mm. of mercury, as calculated from the formula and as observed by Regnault:

t	mm. calculated	mm. observed
0°	1·42	4·60
20°	10·88	17·39
60°	140·90	148·79
70°	227·14	233·09
80°	351·21	354·64
90°	524·39	525·45

Thus the disagreement at low temperatures is very marked.

From (1) we can deduce a formula in English measures. Let intensity of pressure be measured in pounds' weight per square inch, and temperature on the Fahrenheit scale. Putting $\frac{5}{9}(t-32)$ for t in (1), the number of standard atmospheres becomes

$$\left(1 + \frac{t-212}{251\cdot643}\right)^5, \text{ or } \left(\frac{39\cdot643 + t}{251\cdot643}\right)^5;$$

and taking a column of mercury 760 mm. high as equivalent to 14·697 pounds' weight per square inch, the intensity of pressure of the vapour in pounds' weight per square inch is the product of this expression and 14·697; that is $\left(\frac{39\cdot643 + t}{147\cdot03}\right)^5$; or, as it is usually taken

$$p = \left(\frac{40 + t}{147}\right)^5 \quad \ldots \ldots \quad (2)$$

238 *Hydrostatics and Elementary Hydrokinetics.*

If intensity of pressure is measured in pounds' weight per square inch, while temperature is measured on the Centigrade scale, the formula gives

$$p = \left(\frac{39.8 + t}{81.67}\right)^5. \quad \ldots \quad (3)$$

The height of a mountain may be deduced from the temperatures at which water boils at the base and at the summit, on the usual assumption of a constant mean temperature of the air, without the aid of a barometer; for, from the observed temperatures of boiling, we can deduce the corresponding atmospheric pressures by formula (1) or (2), and make use of these pressures in the barometric formulas of Arts. 50 and 51.

EXAMPLES.

1. If at the base and the summit of a mountain water boils at 212° and 190°, Fahrenheit, respectively, and the mean temperature of the air is 40°, find the height of the mountain.

Ans. About 12172 feet.

2. At the base of a mountain 28000 feet high the atmospheric intensity of pressure is 14 pounds' weight per square inch; assuming the temperature of the air to be uniformly 32° F, find the temperature at which water boils at the summit.

Ans. About 161° F.

58. Principles of Thermodynamics. It is now an accepted principle that a quantity of heat is the same thing as a quantity of kinetic energy—that, in fact, heat is the kinetic energy of molecular motions in a body. Now as kinetic energy is the equivalent of work, and can be measured in ergs, foot-pounds' weight, mètre-kilogrammes' weight, and many other analogous forms, it follows that a quantity of heat can be expressed as so many ergs, or so many foot-pounds' weight, &c.

For a long time this was not the mode adopted for

expressing a quantity of heat, because until the experiments of Joule were made, it was not recognised that *heat* and *work* are equivalent things. To take an example, when by burning coal under a vessel containing 1 pound of water the temperature of this water was raised 1° F (supposing none of the imparted heat to be radiated from the water) it was said that a quantity of heat called *one thermal unit* was imparted to the water. This mode of speaking is still adopted: it merely amounts to a definition, and indicates one particular way of measuring quantities of heat. But at the present time we should also describe the result thus—*a quantity of kinetic energy equivalent to about 772 foot-pounds' weight has been imparted to the pound of water.* The measurement of quantities of heat in thermal units and in foot-pounds' weight may be compared with the measurement of areas in acres and in square yards.

The British thermal unit is defined as *the quantity of heat (or molecular kinetic energy) which must be imparted to one pound of water at its temperature of maximum density (about 39° F) to raise this temperature 1° F.*

The Metric thermal unit is defined as *the quantity of heat (or molecular kinetic energy) which must be imparted to one kilogramme of water at its temperature of maximum density (about 4° C) to raise its temperature 1° C.*

It is not quite true that the quantity of heat necessary to raise by one degree the temperature of a given quantity of water is the same at all temperatures. Thus, it is found that the heat necessary to raise the temperature of one pound of water from 32° F to 212°, instead of being 180 times that required to raise it from 32° to 33°, is 180·9 times as great. The quantity of heat necessary to raise the temperature by one degree increases very slightly as the temperature of the water increases; and this fact may be otherwise expressed thus: the specific heat of water increases slightly as the temperature of the water increases.

The Metric thermal unit is very frequently called a *calorie*, and the quantity of kinetic energy in it is about 425 mètre-kilogrammes' weight (usually called *kilogramme-mètres*).

In the C. G. S. system, the thermal unit is *the quantity of heat* (or molecular kinetic energy) *which must be imparted to one gramme of water at its temperature of maximum density to raise its temperature* $1°$ C, and it amounts to about 42×10^6 ergs.

In any of these systems the number of work-units which is equivalent to the thermal units is called *the dynamical equivalent of heat*, and it is usually denoted by J (with reference to the name of Joule). Thus, in the British system $J = 772$, in the Metric gravitation system $J = 425$, and in the C. G. S. system $J = 42 \times 10^6$, all these numbers being only approximate.

Let us now consider what happens when a quantity of heat is communicated to any body. The result can be comprised under two heads—

$1°$. Kinetic energy is imparted to the molecules of the body (i. e., the body is heated) and at the same time work is done against the attractive forces of these molecules (i. e., a certain amount of internal static energy is generated in the body);

$2°$. The body at its bounding surface does work against external pressures.

Thus the first heading comprises both *heat in the body* and *internal work*. The sum of these two is called the *energy of the body*, and is denoted by the symbol U.

Suppose now that a small quantity of heat, dQ, which we shall estimate in dynamical units, is communicated to a body, and that the effect of this is to generate a small

quantity, dU, of energy in the body, and to make the body perform a small quantity, dW, of external work. Then we have the equation
$$dQ = dU + dW. \quad \ldots \ldots \quad (1)$$

As it is not our object to consider the principles of thermodynamics in their application to every kind of body, we shall suppose the body in question to be a gas or a vapour.

Then there are three things which determine the state of such a body—viz., its *temperature*, T, its *volume*, v, and its *intensity of pressure*, p. These three are not independent; for, in the case of a given mass of gas, we have the equation $\dfrac{vp}{T}$ = a constant, or as in Art. 48,
$$vp = wR \cdot T; \quad \ldots \ldots \quad (2)$$
so that the state of the body at any instant may be considered as depending on any *two* of the quantities, v, p, T. Now the internal energy, U, of the body is always the same in the same state of the body. Hence we can consider either
$$U = f_1(v, p) \text{ or } U = f_2(v, T), \text{ or } U = f_3(p, T), \quad (3)$$
where f_1, f_2, f_3 are symbols of functionality.

Now, confining our attention to gases, the following is a fundamental result:

the energy, U, of a given mass of gas is a function of its temperature, T, alone, i.e., it does not depend on either v or p.

This result was verified by the following experiment of Joule.

A and B (Fig. 65) are two copper vessels which communicate with each other by a tube in which is a stopcock, c. These vessels are placed inside a vessel, CD, which is in the form of two nearly closed cylinders communicating by means of a narrow passage between two walls. The reservoir A is filled with air, or any gas, at a high intensity of pressure (22 atmospheres in Joule's experiment), while B has been exhausted of air. Water is poured into the vessel CD, and since A and B have nearly the capacity of the cylinders in which they are placed, the quantity of this water is small, and thus any alteration of its temperature would be the more easily observed.

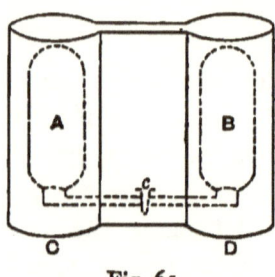

Fig. 65.

Now let the stopcock c be suddenly opened. The gas rushes out of A, increasing its volume and diminishing its intensity of pressure; and when the whole mass of this gas is considered, we see that it has altered its state without doing any external work. At the same time no heat has been communicated to it.

If the water is kept stirred the result of the experiment will be that the water neither rises nor falls in temperature. Actually, it was found that if the water was at rest, the temperature of that surrounding A was lowered and the temperature of that surrounding B was raised; but, on the whole, these local alterations of temperature balanced each other, and the body (i.e., the mass of gas) underwent change without development of heat.

Now, applying to this case equation (1), we see that $dQ = 0$ and $dW = 0$, therefore $dU = 0$; and since both v and p have altered, but not T, we see that U cannot be a function of v or p. Hence for a gas

$$U = f(T), \quad \ldots \ldots \ldots (4)$$

i.e. the energy of the gas is a function of its temperature alone.

This amounts to the statement that *the molecular forces in a gas are zero*; for, if they existed, and were *attractive*, they would perform *negative* work when the volume expanded; and if they were *repulsive*, they would perform *positive* work in the same case.

Subsequent experiments by Joule and Thomson revealed, however, small changes of temperature, showing that the internal forces are not wholly evanescent; but these changes became smaller, for any gas, as its temperature was higher; and the changes of temperature which, under the above circumstances, take place are smaller the more nearly the gas approaches to the condition of a *perfect* gas. Thus, for air and for hydrogen they are much smaller than for carbonic acid gas. Such a result would naturally be expected; and hence we are to regard the result (4) as holding accurately in the case of a *perfect* gas only; for other gases the result is approximately true.

Supposing, then, that we are dealing with a perfect gas only, equation (1) can be written

$$dQ = \frac{dU}{dT} dT + dW. \quad \ldots \ldots \quad (5)$$

When the gas does external work by overcoming resistance applied over its bounding surface, the intensity of this resistance being equal to the intensity of pressure, p, the element, dW, of work done during a small increment, dv, of volume of the gas, is given by the equation (Art. 46)

$$dW = pdv. \quad \ldots \ldots \quad (6)$$

Hence (5) becomes

$$dQ = \frac{dU}{dT} dT + pdv. \quad \ldots \ldots \quad (7)$$

If w is the mass of the quantity of gas with which we are dealing, we have equation (2), which enables us to express dv in (7) in terms of dT and dp, thus

$$dv = \frac{wR}{p} dT - \frac{wRT}{p^2} dp$$

$$= \frac{wR}{p} dT - \frac{v}{p} dp; \quad \ldots \ldots \quad (8)$$

hence (7) becomes

$$dQ = \left(\frac{dU}{dT} + wR\right) dT - v\,dp. \quad \ldots \quad (9)$$

It will be observed that we have been all through measuring heat in work-units.

We now lay down two definitions:

(*a*) The specific heat of any gas at constant volume is the limiting value of the ratio of an infinitesimal quantity of heat to the infinitesimal rise of temperature produced, when this heat is imparted to a unit of mass which is not allowed to expand.

(*b*) The specific heat of any gas at constant pressure is the limiting value of the ratio of an infinitesimal quantity of heat to the infinitesimal rise of temperature produced, when this heat is imparted to a unit of mass which is allowed to expand under constant pressure.

Thus (*a*) is the value of $\frac{1}{w}\frac{dQ}{dT}$ when v is constant, and if this is denoted by c, (7) shows that

$$c = \frac{1}{w}\frac{dU}{dT}. \quad \ldots \ldots \ldots \quad (10)$$

Also (*b*) is the value of $\frac{1}{w}\frac{dQ}{dT}$ when p is constant, and if this is denoted by C, (9) shows that

$$C = \frac{1}{w}\frac{dU}{dT} + R \quad \ldots \ldots \quad (11)$$

Gases.

It is evident *à priori* that $C > c$, because when the gas is allowed to expand, it uses some of the imparted heat to do work against external resistance, so that all of the heat does not go towards raising the temperature, whereas, when no expansion is allowed, all this heat is used in raising the temperature.

Now one of the fundamental laws of a perfect gas is that *for any given gas each of these specific heats is a constant at all temperatures.* Of course, since

$$C = c + R, \quad \ldots \ldots \quad (12)$$

if either of them is constant, the other must be constant. This law was proved experimentally by M. Regnault.

Since the specific heats are constant, we may say (as is usually done) that the specific heat of a gas at constant volume is the quantity of heat necessary to raise its temperature one degree when the heat is imparted to the unit of mass at constant volume; or the quantity of heat necessary to raise its temperature by any number of degrees, if we divide by the number of degrees.

We have supposed that Q is measured in work-units: if Q is measured in thermal units (whether English or Metric), and J is the corresponding work-unit value of the thermal unit, the fundamental equation (1) becomes

$$JdQ = dU + dW; \quad \ldots \ldots \quad (13)$$

also, c being measured with reference to thermal units, $c = \dfrac{1}{w} \dfrac{dQ}{dT}$, when v is constant; so that

$$Jc = \frac{1}{w}\frac{dU}{dT}; \quad JC = \frac{1}{w}\frac{dU}{dT} + R \quad \ldots \quad (14)$$

$$\therefore J(C - c) = R. \quad \ldots \ldots \quad (15)$$

Thus, Regnault found that for dry air $C = \cdot 2375$, and if we use English measures, $R = 53\cdot 3$ (Art. 49), $J = 772$; hence $c = \cdot 1685$, which gives $\dfrac{C}{c} = 1\cdot 409$.

From (10) we have the value of the internal energy of a given mass, w, of a perfect gas at the absolute temperature T, viz.,
$$U = cwT \quad \ldots \ldots \ldots (16)$$
in units of work if c is expressed with reference to these units; or
$$U = JcwT \ldots \ldots \ldots (17)$$
if c is expressed with reference to thermal units.

It appears, therefore, that when a gas is expanded isothermally by the continuous application of heat, all the heat supplied is utilised in doing work against external resistance.

59. Adiabatic Expansion. Carnot's Cycle. If a mass of gas is contained within a cylinder closed by a moveable piston, the cylinder and the piston being both impermeable to heat, so that no heat can be communicated to the gas from without and no heat can escape from the gas itself, while the gas may drive out the piston before it or may be compressed by means of the piston, the transformation is called *adiabatic*. If the gas is compressed by means of the piston, heat is generated in the gas; but no heat is gained or lost *by conduction or radiation* in an adiabatic transformation.

We propose to find the relation existing between the volume and the temperature, and the volume and intensity of pressure, of the gas during this transformation.

These are found by using (7) and (9) of last Article and putting $dQ = 0$.

Hence
$$cw\,dT + p\,dv = 0, \ \ldots \ldots \ldots (1)$$
$$Cw\,dT - v\,dp = 0, \ \ldots \ldots \ldots (2)$$
in which the specific heats c, C are in work-units.

From these we derive the equation
$$Cpdv + cvdp = 0, \quad \ldots \ldots (3)$$
which gives by integration
$$pv^{\frac{C}{c}} = \text{constant} \quad \ldots \ldots (4)$$

It has been found that $\frac{C}{c} = 1.408$, nearly, for air. Denoting by n the value of this ratio for any gas, if (p_0, v_0) are the intensity of pressure and volume when the adiabatic transformation begins, the equation of the curve which is analogous to the hyperbola in Fig. 55, and which exhibits the relation between p and v in the adiabatic transformation, is
$$pv^n = p_0 v_0^n, \quad \ldots \ldots (5)$$
which, combined with the fundamental equation
$$\frac{pv}{T} = \frac{p_0 v_0}{T_0}, \quad \ldots \ldots (6)$$
gives the relation between p and T and between v and T.

It has been already pointed out (Art. 40) that the curves of isothermal transformation are rectangular hyperbolas.

Fig. 66 exhibits two isothermals and two adiabatics. Thus, suppose the two rectangular lines Ov and Op to be taken as axes of volume and pressure, respectively; let the gas be contained in a cylinder the base alone of which is a conductor of heat—and we shall assume this base to be a *perfect* conductor, i. e., we can imagine it to be made of very thin copper; let the initial state of the gas be represented by the point A in the figure, i. e., its intensity of pressure, p_0, is represented by the ordinate Aa, and its volume by the abscissa Oa, while its *absolute* temperature in this state is T_0'.

248 *Hydrostatics and Elementary Hydrokinetics.*

Now let the base of the cylinder be placed on a perfectly nonconducting stand, and let work be done on the gas by forcing down the piston. The result will be a rise of temperature of the gas and, of course, an increase in its intensity of pressure, the relation between its volume and

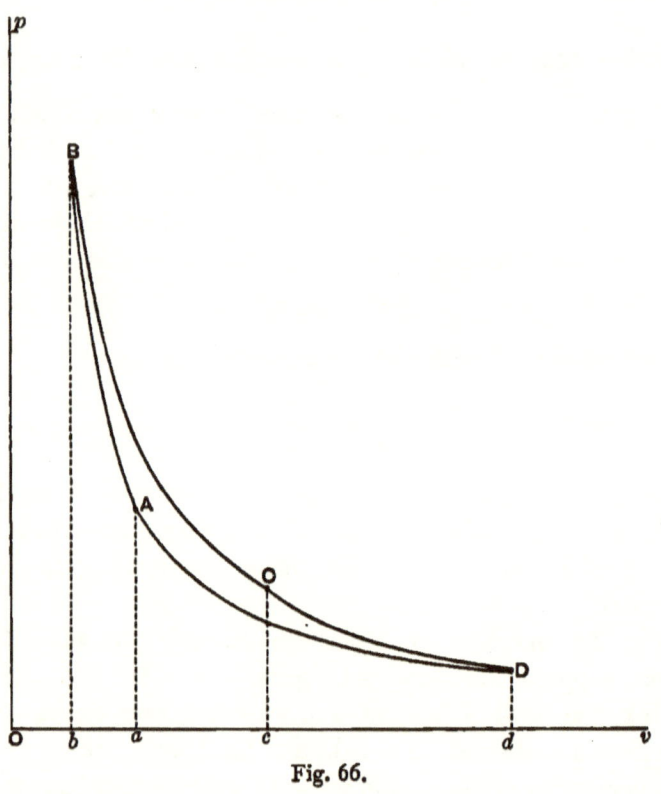

Fig. 66.

intensity of pressure being represented by the abscissae and ordinates of the adiabatic curve AB, whose equation is (5).

Let this adiabatic transformation be stopped when the absolute temperature becomes T_1 and the state of the gas is represented by the point B. Then the work

Gases. 249

done on the gas is represented by the area $AabB$, whose value is

$$-\int_{v_0}^{v_1} p\,dv, \quad \ldots \ldots \quad (7)$$

v_1 being the final volume Ob.

Substituting for p its value in terms of v from (5), we find this area to be

$$\frac{p_0 v_0^n}{n-1}\left(\frac{1}{v_1^{n-1}} - \frac{1}{v_0^{n-1}}\right),$$

or

$$\frac{p_1 v_1 - p_0 v_0}{n-1}. \quad \ldots \ldots \quad (8)$$

Now since for the given mass of gas, w, we have (2) of Art. 58, this expression is

$$\frac{Rw}{n-1}(T_1 - T_0) \quad \ldots \ldots \quad (9)$$

From (1) it appears that the work done on a gas when compressed adiabatically is also

$$cw(T_1 - T_0), \quad \ldots \ldots \quad (10)$$

where T_0 and T_1 are the initial and final absolute temperatures, and, as before stated, c is in work-units.

The co-ordinates, v_1, p_1, of the point B are easily found to be

$$v_1 = v_0 \left(\frac{T_0}{T_1}\right)^{\frac{1}{n-1}} \quad \ldots \ldots \quad (11)$$

$$p_1 = p_0 \left(\frac{T_1}{T_0}\right)^{\frac{n}{n-1}} \quad \ldots \ldots \quad (12)$$

When the gas has reached the higher temperature, T_1, let the base of the cylinder be removed from the nonconducting stand and placed in contact with a large reservoir of heat of the temperature T_1.

The external pressure on the piston being removed, the

250 *Hydrostatics and Elementary Hydrokinetics.*

piston will be driven outwards by the pressure of the gas, and this entails a fall of the temperature of the gas; but instantly heat flows into it from the reservoir, keeping the temperature constantly equal to T_1 as the gas continues to expand. The curve of expansion is a rectangular hyperbola, BC, whose equation is

$$pv = p_1 v_1 \text{ or } = RwT_1, \qquad \ldots \ldots (13)$$

and if the transformation is stopped at the point C, the work done *by* the gas on the piston is represented by the area $BCcb$, whose expression is $\int_{v_1}^{v_2} p\, dv$, where v_2 is the volume Oc.

This work is

$$RwT_1 \log_e \frac{v_2}{v_1}. \qquad \ldots \ldots (14)$$

Having reached the point C in this isothermal transformation, let the cylinder be removed from the reservoir and again placed with its base on the nonconducting stand. The pressure of the gas will continue to drive out the piston; but now the gas does this work at the expense of its own heat, and therefore its temperature falls, the relation between pressure and volume being given by the adiabatic curve CD. When the absolute temperature has fallen to the original value, T_0, let this transformation cease (the point D is supposed to represent the state of the gas when the temperature T_0 is reached), and let the cylinder be placed with its base in contact with a large reservoir of heat whose absolute temperature is T_0. Now let the piston be forcibly driven in, and therefore the gas compressed by work done on it, until the original volume, v_0, or Oa, is again reached; then the final curve DA is a hyperbola corresponding to the temperature T_0, and since there is work being continuously done on the gas during this isothermal compression, its temperature makes a continuous effort to rise; but, on account of the perfect conductivity of the base, the moment such a rise takes place, heat flows from the gas into the reservoir, and by this continuous flow of heat the temperature of the gas is steadily kept at T_0.

In the adiabatic expansion CD, the gas has done work of the amount
$$\frac{p_2 v_2 - p_3 v_3}{n-1}, \quad \ldots \ldots \quad (15)$$
if the co-ordinates of C and D are (v_2, p_2) and (v_3, p_3) respectively.

But since C belongs to the hyperbola corresponding to T_1, and D belongs to the hyperbola corresponding to T_0, we have $v_2 p_2 = RwT_1$ and $v_3 p_3 = RwT_0$; hence the work $CDdc$ is
$$\frac{Rw}{n-1}(T_1 - T_0), \quad \ldots \ldots \quad (16)$$
which has been proved to be also the work represented by $BAab$, so that the work done *on* the gas in the adiabatic compression AB and *by* it in the adiabatic expansion CD cancel each other.

The work done *on* the gas in the isothermal compression DA is represented by the area $DdaA$, and is equal to $p_0 v_0 \log_e \frac{v_3}{v_0}$, or $RwT_0 \log_e \frac{v_3}{v_0}$.

Now we can show that
$$\frac{v_2}{v_1} = \frac{v_3}{v_0}; \quad \ldots \ldots \quad (17)$$
for
$$p_2 v_2 = RwT_1,$$
$$p_2 v_2^n = p_3 v_3^n = RwT_0 \cdot v_3^{n-1};$$
$$\therefore v_2^{n-1} = v_3^{n-1} \frac{T_0}{T_1},$$

$$\therefore \quad v_2 = v_3 \left(\frac{T_0}{T_1}\right)^{\frac{1}{n-1}},$$

which, by (11), gives the result (17).

Hence the total work done *by* the gas in the whole series of transformations, which is represented by the area $ABCDA$, is

$$Rw(T_1 - T_0)\log_e \frac{v_2}{v_1}. \quad \ldots \quad \ldots \quad (18)$$

A series of transformations of a gas whereby, starting from a given state as regards volume, temperature, and intensity of pressure, submitting to changes of these by absorbing heat, doing work against external resistance, and having work done upon it by external agents, the gas is finally brought back exactly to its original state, is called a *cycle* of operations. When, as in the case just discussed, the changes consist of two isothermal and two adiabatic transformations, the operation is called a *Carnot cycle*, because such a simple cycle was first discussed by Carnot in the investigation of some of the fundamental principles of Thermodynamics.

Such a cycle it would be impossible to realise in practice, because perfect conductors and perfect nonconductors of heat do not exist, although isothermal and adiabatic transformations of a gas can be approximated to. For example, there is a common experiment which consists in igniting a piece of tinder, or cotton moistened with ether or bisulphide of carbon, at the bottom of a glass tube filled with air and tightly fitted with a piston, by very suddenly forcing the piston down the tube. In this case the transformation of the imprisoned air can be assumed to be adiabatic, because, during the time of the experiment, no heat can enter or leave the tube. This instrument is sometimes called a *pneumatic syringe*.

The order in which the changes in a Carnot cycle are made can be varied. Thus, we may start the operations from the point B and work round through C, D, and A to B again. If we call the reservoir at the higher absolute temperature, T_1, the *boiler*, and the reservoir at the lower temperature, T_0, the *condenser*, the two main features of the cycle are the abstraction of a certain amount of heat, represented by the area $BCcb$, in work-units, from the boiler, and the transference to the condenser of another amount of heat, represented by the area $DdaA$, the amounts of heat generated within the working gas itself in the abiabatic transformations neutralising each other.

Carnot, under the influence of the erroneous notion prevailing in his time, supposed that, since the working substance returns to exactly its original state in all respects, the quantity of heat which it receives from the boiler must be equal to that which it gives out to the condenser, *because heat is an indestructible substance*. But the fact remains that, although the gas has returned to its original state, a certain quantity of work, represented by the area $BCDAB$, has been done by the engine—if we use the term *engine* to denote the gas, cylinder, and piston. How, then, did Carnot explain the doing of this work, since (according to his view) the engine gave out, as heat, to the condenser all the heat that it received from the boiler?

Simply by saying that the letting down of the heat from the higher temperature, or heat level, T_1, to the lower level, T_0, constituted the doing of this work—just as the fall of a stone from a higher to a lower gravitation level constitutes the doing of work.

The view that heat is a substance (which used to be called *caloric*) is now abandoned, since the experiments of Joule and others prove that it is kinetic energy.

Of the quantity of heat which a Carnot engine receives

from the boiler, how much is converted into work done by the engine? If H_1 is the heat received from the boiler, and H_0 the heat which the engine transfers to the condenser, the amount $H_1 - H_0$ is converted into work.

Now
$$H_1 = RwT_1 \log_e \frac{v_2}{v_1},$$

$$H_0 = RwT_0 \log_e \frac{v}{v_1},$$

$$\therefore \frac{H_1 - H_0}{H_1} = 1 - \frac{T_0}{T_1}. \qquad \ldots \ldots \ldots (19)$$

The ratio of the quantity of heat converted into work to the quantity received from the boiler is called the *efficiency* of the engine, and we see that, unless $T_0 = 0$, i. e., the temperature of the condenser is absolute zero, the efficiency is always < 1. Thus, if the boiler is at the temperature of water boiling at the ordinary pressure, and the condenser at that of melting ice, $T_1 = 373$, $T_0 = 273$, and the efficiency is only $1 - \frac{273}{373}$, that is about ·268.

This is only a small amount, and yet such an imaginary engine (using a perfect gas as the working substance) has a much greater efficiency than any actual engine.

The cycle of operations with a Carnot engine is *reversible*, i. e., if we start from the point A in Fig. 66, we can work from A to D, then to C, then to B, and finally back to A.

In this way we should have to do work *on* the gas, this work being represented by the area $ADCBA$, and the result of this would be the transference of a certain amount of heat *from the condenser to the boiler*.

In the common steam-engine the various operations in a Carnot cycle with a perfect gas are roughly approximated to in the following way:

1. The isothermal expansion BC corresponds to the evaporation of the water in the boiler.

Gases.

2. The adiabatic expansion CD corresponds to the expansion of the steam in the cylinder.

3. The isothermal compression DA corresponds to condensation in the condenser.

4. The adiabatic compression AB corresponds to the forcing of the water into the boiler.

(See Cotterill's *The Steam Engine considered as a Heat Engine*, Chap. V.)

Examples.

1. The air column in a pneumatic syringe 10 inches long is suddenly compressed into a length of 1 inch; its original temperature was $15°C$; find its final temperature.

Ans. $463°\cdot 8$.

2. Air is contained in a vertical cylinder, closed at the lower end and open at the top, the area of whose cross-section is 2 square inches; the air is compressed, so that it occupies a length of 4 inches of the cylinder, by means of a piston whose mass is 1 pound, the intensity of pressure of the air being 150 pounds' weight per square inch; the intensity of atmospheric pressure is 15 pounds' weight per square inch. If the piston is suddenly released, find its velocity when it is 1 foot from the bottom of the cylinder, assuming the temperature of the air to be kept constant.

Ans. 75·55 feet per second.

3. In the above find the position of the piston when its velocity is a maximum.

Ans. 38·7 inches from the lower end.

4. Find how high the piston ascends before coming to rest for an instant.

Ans. About $142\frac{1}{2}$ inches from the end.

5. If the area of the cross-section of the cylinder is A square inches, the mass of the piston w pounds, the original intensity of pressure of the compressed air P pounds' weight per square

inch, that of the atmosphere p_0, and the piston is originally c inches from the bottom, find its velocity when it is x inches from the bottom, taking $g = 32$ feet per second per second.

Ans. $v^2 = \dfrac{16}{3w}\left\{APc\log_e\dfrac{x}{c} - (Ap_0+w)(x-c)\right\}$,

where v is feet per second.

6. If in example 2 the air in the cylinder expands without receiving or losing heat by conduction or radiation, find the velocity of the piston when it is 1 foot from the bottom.

Ans. 65·9 feet per second.

7. If in example 5 the air expands adiabatically, find the velocity of the piston.

Ans. $v^2 = \dfrac{16}{3w}\left\{\dfrac{AP}{n-1}\left[c - \left(\dfrac{c}{x}\right)^n \cdot x\right] - (Ap_0+w)(x-c)\right\}$,

where $n = 1·408$.

8. If a pound of air does 390·6 foot-pounds' weight of work without receiving heat or losing it by conduction or radiation, find its fall in temperature.

Ans. 3° F.

9. If a pound of air at 60° F and intensity of pressure 15 pounds' weight per square inch is compressed without gain or loss of heat by conduction or radiation until its temperature is 200° F, find its new pressure intensity.

Ans. About 34·1 pounds' weight per square inch.

✓ 10. The temperature of a given mass of air is observed to fall from 540° to 290° F when expanding to double its volume without gain or loss of heat by conduction or radiation; and at the same time the external work done is observed to be 32600 foot-pounds' weight per pound of air; find the specific heat of air at constant pressure.

Ans. We have from the data $c = 130·4$ work-units [see (10), Art. 59], and $2^{3-n} = 3$, where $n = \dfrac{C}{c}$; ∴ $C = 184·5$ work-units.

Gases. 257

11. Air contained in a cylinder at the atmospheric pressure is adiabatically compressed to an intensity of pressure of m atmospheres, and is then allowed to cool at constant volume to the temperature of the surrounding air; if it is then allowed to expand adiabatically until it reaches its original intensity of pressure, find the efficiency of this method of storing work, the work done on the air by the atmospheric pressure in the compression and *against* it in the expansion being deducted.

Ans. Efficiency $= \dfrac{m^{\frac{1}{n}} - 1 - n\left(m^{\frac{1}{n^2}} - 1\right)}{m - 1 - n\left(m^{\frac{1}{n}} - 1\right)}$, where $n = 1\cdot 408$.

60. Absolute Temperature. It has been already pointed out (Art. 41) that the notion of an *absolute zero* at a point 273 Centigrade degrees below the temperature of melting ice is not properly founded on the formula $vp = Rw\,(273 + t)$ which has been deduced for gases from experiments at ordinary temperatures, and from which it would follow that if $t = -273$, the volume of the gas would be zero. The position of the absolute zero—conceived to be such that at this temperature there would be no molecular motion in any body—is deduced from the principles of Thermodynamics as applied to the action of reversible engines (such as Carnot's) when we imagine the working substances in these engines to be any substances whatever that can undergo such a complete cycle of changes as that of Carnot.

Thus, we can imagine the cylinder to contain, as a working substance, a quantity of water and its vapour, by the expansion of which work is done on the piston.

It is not the aim of this work to discuss the details of Thermodynamics, which the student will find in treatises on this subject, such as Cotterill's *Steam Engine*, Clerk Maxwell's *Heat*, *The Mechanical Theory of Heat* by Clausius, Tait's *Heat*, Parker's *Thermodynamics*, and many other works; and therefore we shall assume the following proposition, which is fundamental in the subject, and the proof of which will be found in any of these works:—

if a reversible engine works between any two fixed temperatures, its working substance undergoing a complete cycle of operations indicated by two isothermals and two adiabatics, and if it receives a fixed quantity of heat from a reservoir at the higher temperature, it will convert the same fraction of this heat into external work, whatever be the nature of its working substance.

Of course the isothermals and adiabatics of the substance may be any curves whatever, and not the simple curves whose equations are $pv = $ const., and $pv^n = $ const., which belong to a perfect gas.

It may occur to the student as an objection that, inasmuch as we are now seeking for a means of measuring temperature, it is not permissible to speak of the working substance in Carnot's cylinder as receiving heat from a reservoir at the higher temperature and transferring some of it to the reservoir at the lower temperature. There is, however, nothing illogical in this, because, although we may not be able to measure temperature numerically, we are able to say when two bodies are at the *same* temperature. A mercurial or any other thermometer will tell us when two bodies are at the *same* temperature. We can say, for example, when the upper reservoir, or boiler, is at the temperature of water boiling at the normal pressure, and when the lower, or condenser, is at the temperature of melting ice.

Suppose, then, for definiteness, that the boiler is at the temperature of water boiling under the normal pressure, and that the condenser is at that of melting ice; and with any given substance in Carnot's cylinder let a quantity, H, of heat be taken isothermally from the boiler in a transformation such as that represented by the curve BC in Fig. 66, and let the quantity H_0 be transferred to the condenser in the transformation corresponding to DA. Then our assumption is that, whatever be the substance in the cylinder, the ratio

$$\frac{H_1}{H_0} \text{ is constant.}$$

When we speak of two *reservoirs* of heat each at a certain temperature, we mean two bodies so large in comparison with the dimensions of the Carnot cylinder that when the working substance in this cylinder takes heat from them or gives heat to them, their temperatures do not fall or rise. Let us imagine the cylinder to be small enough, and we can drop the term *reservoirs* and simply speak of two *bodies*, in connection with which the cylinder is supposed to work in a Carnot cycle. Denote the two bodies by B_0 and B_1, and imagine any number of Carnot cylinders $C, C', C'', \ldots$ containing different working substances.

Suppose, then, each of these cylinders successively to be put in connection with the body B_0, and each of them to be worked until it takes a quantity of heat, H_0, from it, this quantity being the same for all the cylinders; let each cylinder be worked in a Carnot cycle in which the second body (that to which heat is transferred) is B_1. Then all the cylinders transfer the same quantity of heat, H_1, to this body, and this quantity depends simply on the temperature of the body B_1. Hence H_1 can be taken as a measure of the temperature of the body. On the same scale, of course, H_0 would be the measure of the temperature of the body B_0. We may, if we please, imagine B_0 to be a body absolutely deprived of molecular motion, i. e., a body at the absolute zero of temperature. If B_0 is not such a body, let its temperature above the absolute zero be T_0, and let that of B_1 be T_1; then, on the scale adopted now, we have

$$\frac{H_1}{H_0} = \frac{T_1}{T_0}. \quad \ldots \ldots \ldots (1)$$

This definition gives only the *ratio* of the absolute temperatures of the two bodies, and makes it independent of the working substance in the Carnot cylinder, so that we may select the most convenient substance that we can find.

Now, supposing B_1 to be water boiling at the normal atmospheric pressure, and B_0 to be melting ice, and taking air as the most convenient substance, Thomson and Joule (Tait's *Heat*, chap. xxi) found that

$$\frac{H_1}{H_0} = 1\cdot365, \quad \ldots \ldots \ldots (2)$$

$$\therefore \frac{T_1}{T_0} = 1\cdot365. \quad \ldots \ldots \ldots (3)$$

The *magnitude* of each degree is still undefined. Let the magnitude be such that there are 100 degrees between these two temperatures; then $T_1 = T_0 + 100$, and (3) gives

$$T_0 = \frac{100}{\cdot365} = 274, \text{ very nearly.} \ldots \ldots (4)$$

Thus if the interval between the temperatures of melting ice and water boiling under the normal atmospheric pressure is divided into 100 equal parts, the freezing point is 274 of these degrees above the absolute zero of temperature. Hence the absolute zero coincides practically with that suggested by the air thermometer.

61. Total Heat of Steam. If a kilogramme of water is taken at the temperature $0°$ C and heated until it is completely converted into saturated vapour at the temperature $t°$ C, there will be two stages to note in the process:

1. The water has its temperature raised from 0° to the boiling point, showing a continuous rise of a thermometer placed in it.

2. At the boiling point the water is converted into saturated vapour, heat being continuously absorbed, but no rise of the thermometer being observed until all the water is converted.

M. Regnault found the total quantity of heat thus absorbed by 1 kilogramme of water, measured in metric thermal units, to be $606.5 + .305 t$. . . . . . . (1)

Of course the quantity of heat necessary to convert w kilogrammes is w times this.

Now the number of thermal units absorbed in heating the kilogramme of water from 0° to $t°$ is t; and as the quantity (1) exceeds t, it follows that the excess has been absorbed in overcoming external pressure and the molecular forces of the water, converting it into steam, which shows no rise of temperature above t although heat is being applied. This heat, which is not indicated by the thermometer and whose function is to perform internal and external work, is called the *latent heat* of the saturated steam. If we denote it by L, and denote the total heat used by H, per kilogramme of water, we have

$$H = 606.5 + .305 t, \quad . \quad . \quad . \quad . \quad (2)$$
$$L = 606.5 - .695 t. \quad . \quad . \quad . \quad . \quad (3)$$

If instead of a kilogramme of water at 0° we start with a kilogramme of ice at zero, and apply heat, we find that, though heat is being continuously absorbed by the body, a thermometer indicates no rise of temperature until the whole of the ice is melted. Hence during this process the absorbed heat is being used to perform the work of disintegrating the ice, and this heat is called the latent heat of water, or the latent heat of fusion of ice. Its amount per kilogramme is very nearly 80 thermal units.

If after the water has been completely converted into steam of maximum intensity of pressure (saturated steam) at the given temperature, t, heat is still applied, and the gas expands at constant pressure, its temperature will, of course, rise; and the further amount of heat absorbed when the temperature reaches $\theta°$ is $\cdot 4805\,(\theta - t)$, the number $\cdot 4805$ being the specific heat of aqueous vapour under constant pressure.

The expression analogous to (2) in English measures is deduced thus: the equation (2) can be expressed in the form

$$\frac{\text{total heat for any mass}}{\text{heat required to raise it } 1°\text{C}} = 606\cdot 5 + \cdot 305\,t$$

$$= 606\cdot 5 + \cdot 305 \times \tfrac{5}{9}(t' - 32),$$

where t' is the Fahrenheit temperature corresponding to $t°$ C.

$$\therefore \frac{\text{total heat for any mass}}{\text{heat required to raise it } \tfrac{9}{5}°\text{F}} = 606\cdot 5 + \cdot 305 \times \tfrac{5}{9}(t' - 32),$$

$$\therefore \frac{\text{total heat for any mass}}{\text{heat required to raise it } 1°\text{F}} = 606\cdot 5 \times \tfrac{9}{5} + \cdot 305\,(t' - 32).$$

If mass is measured in pounds, the left-hand side is the heat per pound, in the new units; and we have

$$H = 1081\cdot 94 + \cdot 305\,t, \quad \ldots \ldots \quad (4)$$

in which we have removed the accent from t'. This is the number of thermal units required to raise the water from the temperature of melting ice, i. e., from $32°$ F to t, and completely evaporating it at t. In this process $t - 32$ thermal units have been consumed in raising the tempera-

ture of the water to the evaporating point, and hence $H-(t-32)$ thermal units have been absorbed in doing the work of conversion; so that if L is the latent heat of the steam,
$$L = 1113\cdot 94 - \cdot 695\,t. \quad \ldots \quad \ldots \quad (5)$$

Thus, in evaporating 1 pound of water at the ordinary boiling temperature, 212° F, 966·6 thermal units are absorbed while the water is being changed into steam. The dynamical equivalent of this is about 746215 foot-pounds' weight. But this heat is employed in doing two things—

(a) disintegrating the water, i.e., converting it into vapour, or doing *internal* work;

(b) overcoming the resistance of the atmospheric pressure, i.e., doing *external* work.

It is easy to calculate each of these quantities in general. Let 1 pound of water be converted into steam at the temperature $t°$ F. Then, when the conversion into steam begins, the temperature and the intensity of pressure of the steam remain constant until the conversion is complete. Let the intensity of pressure be measured in pounds' weight per square foot and the volume of the steam in cubic feet; then from (4), Art. 49, we have

$$1 = \frac{1}{53\cdot 30222} \cdot \frac{vp \times \cdot 622}{T};$$
$$\therefore\; pv = 85\cdot 69\, T. \quad \ldots \quad \ldots \quad \ldots \quad (6)$$

Now since for a small expansion of a gas the element of work which it does against the external pressure is $p\,dv$, and since here p remains constant, the work done from volume v_0 to volume v is
$$p(v-v_0).$$

But v_0 is the volume occupied by the water, and v the volume of the steam when evaporation is complete, and

264 *Hydrostatics and Elementary Hydrokinetics.*

moreover v is vastly greater than v_0, so that v_0 may be neglected. Hence pv is the work done by the steam.

If this external work is denoted by W_e, we have, then,

$$W_e = 85 \cdot 69\, T. \quad \ldots \ldots \ldots (7)$$

The number of thermal units in this is obtained by dividing by 772; so that if H_e is the heat absorbed in this external work,

$$H_e = \cdot 111\, T = 51 \cdot 06 + \cdot 111\, t. \quad \ldots (8)$$

Now if H_i is the heat used in the internal work (that done in overcoming the molecular forces of the water),

$$L = H_i + H_e\,; \quad \ldots \ldots \ldots (9)$$

$$\therefore\; H_i = 1062 \cdot 88 - \cdot 806\, t \ldots \ldots (10)$$

Thus, then, of 966·6 thermal units constituting the latent heat of 1 pound of steam formed at 212°, there are 74·59 used in doing external work.

EXAMPLE.

A cylinder contains—

(*a*) water and its saturated vapour at (t, p) in contact with a hot body;

(*b*) a perfect gas at (t, p) in contact with a hot body:

the volumes of both being the same, v, or occupying the same length, h feet, of the cylinder; calculate the distances through which they must, respectively, drive the piston along the cylinder in order that each substance may take in the same quantity of heat, H thermal units, from the hot body.

Ans. If x is the distance for the steam and x' for the perfect gas,

$$x = h\,\frac{85 \cdot 69\,(460 + t)}{pv} \cdot \frac{H}{1113 \cdot 94 - \cdot 695\, t},$$

$$x' = h\,(e^{\frac{772\,H}{pv}} - 1)\,;$$

Gases.

or, taking the heat absorbed and the latent heat in work-units, and denoting by W_e the external work done in the process of evaporating the unit mass of water at t,

$$x = h\frac{W_e}{pv} \cdot \frac{H}{L}; \quad x' = h(e^{\frac{H}{pv}} - 1).$$

62. Hygrometry. Various methods have been adopted for determining the pressure intensity (and hence the quantity) of aqueous vapour present in the air. The hygrometers of Daniell and Regnault aim at lowering the temperature of the air so much that the vapour in it just saturates it, and therefore just begins to deposit as dew on the surface of the hygrometer. If at this instant the temperature of the hygrometer is read, the pressure intensity of *saturated* aqueous vapour corresponding to this temperature is *assumed to be* the pressure intensity of the aqueous vapour present in the air. The point of temperature at which the vapour is just thrown down on a surface as dew is called the *dew point.*

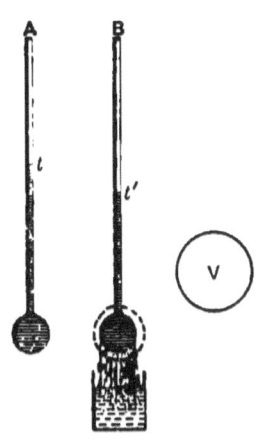

Fig. 67.

Another method is that of the *wet and dry bulb thermometers*, Fig. 67. A is a thermometer, fixed on a vertical stand, and indicating the temperature of the air; B is a thermometer, fixed on the same stand, whose bulb is covered with gauze which is kept moist by a bunch of threads which dip into a small vessel of water.

If the air is still, we may assume that the air surrounding the bulb of B is always saturated; but the temperature, t', indicated by B does not directly tell us the dew point or the pressure of vapour present in the air remote from the bulb of B.

Let t be the temperature of the air remote from this bulb, as indicated by A; f the intensity of pressure of the vapour present; p the total intensity of pressure indicated by a barometer ($\therefore p-f$ is that of the air alone); f' the maximum intensity of pressure of aqueous vapour corresponding to the temperature t'.

Consider a volume V of the air remote from the bulb of B to come to this bulb, and to fall in temperature from t to t'. Now in falling it will give out a certain quantity of heat which will be sufficient to evaporate a certain mass of the water at the wet bulb.

But V contains both air and vapour, and this heat will be contributed by each of them.

Without assuming specially either English or metric measures, let w and w' be the weights of the air and the vapour in V; then

$$w = k\frac{V(p-f)}{T}; \quad w' = k\frac{Vfs}{T}, \quad \ldots \quad (1)$$

$$\therefore w' = w\frac{fs}{p-f}, \quad \ldots \ldots \ldots \quad (2)$$

where $s = \cdot 622 = $ sp. gr. of vapour.

When V reaches the bulb and falls in temperature, it will be saturated by vapour, and fall to volume v; then if w'' is the weight of the vapour which it contains,

$$w = k\frac{v(p-f')}{T'}; \quad w'' = k\frac{vf's}{T'} \quad \ldots \ldots \quad (3)$$

$$\therefore w'' = w\frac{f's}{p-f'}; \quad \ldots \ldots \ldots \quad (4)$$

hence the weight of the vapour which has been formed at the bulb is

$$w\frac{(f'-f)ps}{(p-f')(p-f)} \quad \ldots \ldots \quad (5)$$

Now if L' is the latent heat of a unit weight of aqueous vapour at t', in thermal units (the unit weight being that involved in w), the heat required to produce the weight (5) is

$$L'w\,\frac{(f'-f)\,ps}{(p-f')(p-f)};\quad\ldots\ldots(6)$$

and it is assumed that this is the sum of the quantities of heat given out by the air w and the vapour $w\,\dfrac{fs}{p-f}$, in V in falling from t to t'. If c is the specific heat of air ($\cdot 237$) and c' that of aqueous vapour ($\cdot 48$), the sum of these quantities of heat is

$$w\left(c+\frac{c'fs}{p-f}\right)(t-t')\quad\ldots\ldots(7)$$

Equating (6) to (7), we have

$$L's\,\frac{p(f'-f)}{p-f'}=\{cp+(c's-c)f\}\,(t-t').$$

Now $c's-c = \cdot 061$ which we may neglect, so that

$$f=f'-\frac{c\,(p-f')}{sL'}(t-t'),\quad\ldots\ldots(8)$$

in which p is usually employed instead of $p-f'$, so that

$$f=f'-\frac{cp}{sL'}(t-t'),\quad\ldots\ldots(9)$$

from which f is found when f' is read from a table of pressures of saturated vapour. With English measures, we may take L' equal to 1080, and with metric units equal to 600.

Various objections have been urged to the assumptions involved in this (which Clerk Maxwell calls the *convection theory* of the wet and dry bulb thermometers); thus, it may happen that when the air V comes to the wet bulb it goes away from it without being saturated; and, indeed, this *must* happen if the air is in motion.

It is usual to assume a formula of the form (9), thus

$$f = f' - k \cdot p(t-t'), \quad \ldots \quad \ldots \quad (10)$$

where k is a constant which must be experimentally determined in each locality.

63. Liquefaction of Gases. All gases can be liquefied by compression *provided that the temperature is below a certain limit,* which limit is different for different gases.

Suppose that in a cylinder we have a volume of water vapour (steam) at the temperature 212°, and at an intensity of pressure of 10 pounds' weight per square inch. If now we gradually increase the pressure (keeping the temperature constant) and draw an indicator diagram, such as that in Fig. 55, representing the volumes assumed during the compression and corresponding to the various increasing pressures, when p reaches the value of about 15 pounds' weight per square inch, the steam begins to liquefy; and as we continue to diminish the volume, p remains at this value until the whole of the gas is liquefied; so that at the point where $p = 15$ the curve of pressures and volumes becomes a right line parallel to Ov. Hence the isothermal of steam for 212° F consists of a portion of an approximately hyperbolic curve and a right line parallel to the axis of volumes. If instead of taking the water vapour at 212°, we take it at 107° and an initial intensity of pressure of, say, ·1 pound weight per square inch, and trace the isothermal as we gradually increase the pressure, we shall obtain a similar result: the curve will at first be hyperbolic, and when p reaches the value of about 1 pound weight per square inch, the vapour will begin to liquefy, p remaining constantly equal to 1, until the whole becomes liquid; so that at the point at which $p = 1$ the curve will change to a right line.

The same result would be obtained with water vapour at all temperatures, *except for those above a certain very high value* (about 773° F). The mass of the steam in the cylinder being constant, the rectilinear portion in which every one of the isothermal curves ends becomes shorter as the temperature at which the steam is compressed is higher; but if the steam has a temperature higher than the above value, there is no rectilinear portion whatever: the curve is continuous, and, no matter how great a pressure we apply, we shall not see any distinction of vapour and liquid in the cylinder such as that which must occur at lower temperatures. The vapour of water is not a good gas, therefore, for observing *the absence* of an abrupt change from the gaseous to the liquid state, because the temperatures above which the change is absent are so high.

Carbonic acid gas is much more manageable for this purpose, because the temperature above which the change is absent is only about 30°·9 C, or 87·7 F.

The examination of the changes of this gas under increasing pressures and at various temperatures was made by Dr. Andrews about the year 1862; and by him *the temperature above which a given gas cannot be liquefied by any amount of pressure*—or, rather, above which the distinction between the liquid and the gaseous states is never emphasised—has been called *the critical temperature* for the gas. That there is such a temperature for each gas had previously been asserted by Cagniard de la Tour.

The changes of carbonic acid gas are represented in Fig. 68, which is copied, on a diminished scale, from the paper of Dr. Andrews in the Philosophical Transactions of the Royal Society, 1869, p. 583. The gas experimented upon was contained in a strong glass tube of very small bore; its

270 *Hydrostatics and Elementary Hydrokinetics.*

temperature was kept successively at the constant centigrade values 13°·1, 21°·5, 31°·1, &c., and at each temperature it was powerfully compressed.

In Fig. 68, Op is the axis of pressures, these being measured in atmospheres, and Ov the axis of volumes; the initial pressure marked at O is 47 atmospheres, and the final 110 atmospheres.

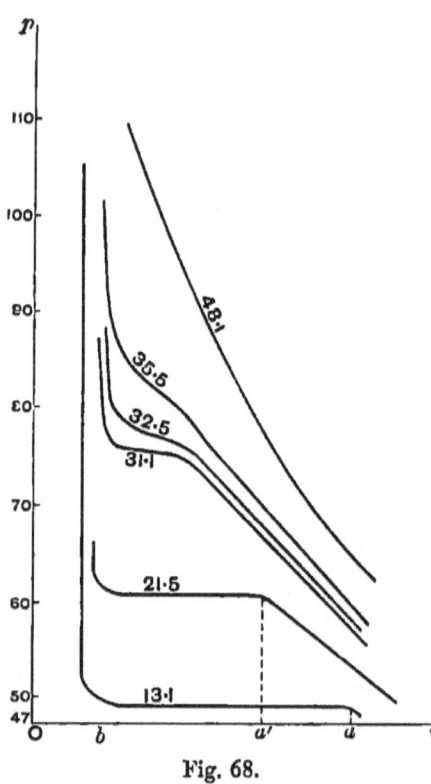

Fig. 68.

Now taking the gas at 13°·1 C, or 55°·6 F, when the pressure is below 47 atmospheres, the relation between p and v is shown by a curve, a small portion of which is shown to the right of the point a. At the pressure of 49 atmospheres liquefaction begins, and p remains equal to 49, from $v = Oa$ to $v = Ob$, the gas liquefying all the while, and the liquefaction being complete when $v = Ob$. When the body has become liquid completely, the further diminution of v requires enormous values of p, and the upward continuation of the curve is so nearly a right line parallel to Op, that the diagram will not show the difference.

Again, taking the gas at 21°·5 C, it remains gaseous

longer than in the previous case, liquefaction beginning when $v = Oa'$, and $p =$ about 60 atmospheres, so that the indicator diagram remains a right line parallel to Ov until the whole is liquid: the volume of this liquid is somewhat greater than Ob, as is seen in the figure.

In each of these experiments the eye could observe when liquefaction began, and until the process was complete the gas and the liquid could be clearly distinguished from each other in the tube.

Again, taking the gas at $31°\cdot1$ C, it was found that no part of the isothermal consisted of a right line during any part of the process of compression, and moreover the substance in the tube never exhibited the distinction of liquid and gas, as it did at the lower temperatures of the previous experiments: there was a continuity of state all through; but the isothermal had not its curvature at the same side all through; there was a part (where the mark 31·1 appears in the diagram) curved downwards, and this part took the place of the right line in the previous figures.

At the temperature $32°\cdot5$ C, the curve was of the same nature as that last described, but the suggestion of a rectilinear portion is less marked. It is still less marked at the temperature. 35·5; while the isothermal for $48°\cdot1$ C loses all trace of discontinuity, being curved the same way all through, and resembles the isothermals of a perfect gas, although it is not really one of these curves. Dr. Andrews found that the temperature at which if the gas was compressed it never exhibited the distinction of liquid and gas was $30°\cdot92$ C, or $87°\cdot7$ F.

The critical temperatures for ordinary gases are extremely low, and hence the great difficulty of liquefying them; because, until we get below this temperature for any gas, there is no use in endeavouring to liquefy it by compression; and, moreover, the pressure at which liquefaction begins is always extremely high.

The principle adopted by M. Cailletet, and independently by M. Pictet, for liquefying oxygen, hydrogen, and other gases which had previously escaped liquefaction, is that of enclosing the gas in a stout tube surrounded by a freezing mixture, then powerfully compressing the gas (500 atmospheres were required in the case of oxygen, 650 for hydrogen), and *suddenly liberating the gas* under the great pressure. The effect of this sudden expansion, (which may be regarded as an adiabatic transformation) whereby the gas (or a portion of it) is made to do work against enormous pressure, is (Art. 59) to lower its temperature still more and liquefy a part of it.

Thus for the liquefaction of any gas two distinct conditions must be satisfied—1°. the temperature must be below a certain point; 2°. the pressure must have a certain value.

For the vapours of ordinary liquids, on the contrary, the critical temperatures are very *high*. Thus, for water vapour, the critical temperature is 773° F; for alcohol vapour about 497° F, and at this point the intensity of pressure required is about 119 atmospheres; for bisulphide of carbon vapour the temperature must be below 505° F; at this temperature liquefaction would require about 67 atmospheres. Of course, if these vapours are at ordinary low temperatures, they can be liquefied by very small pressures: the theory of critical temperature merely asserts that, if they are *above* these individual high temperatures, no amount of compression will liquefy them.

CHAPTER VII.

HYDRAULIC AND PNEUMATIC MACHINES.

64. Water Pumps. *The Common Pump.* Let DB, Fig. 69, represent a vertical section of an iron cylinder terminating in a much narrower cylinder or pipe, BA, which dips into a well from which water is to be raised. In the cylinder DB, or barrel, works a piston having a valve, v, which opens upwards, the piston rod, r, being connected at c with a lever, fH, working about the fulcrum f.

At the top, B, of the suction pipe is a valve which also opens upwards.

Let the piston be at the bottom of the barrel, the level of the water in the well being A, and the pipe and barrel both completely filled with air.

When the piston is raised by means of the handle, H, of the lever, the valve v remains closed, and the pressure of the air in the pipe opens the valve at B, the air in AB rising

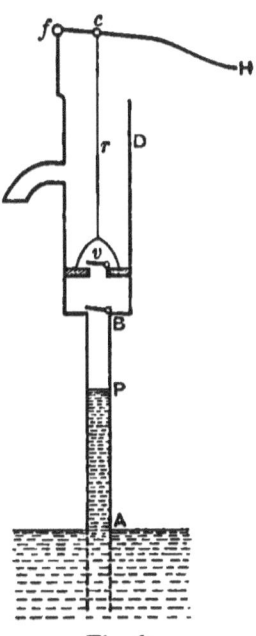

Fig. 69.

and partly filling the barrel, its intensity of pressure also diminishing. As a result of this diminution of pressure below its original (atmospheric) value, the atmospheric pressure on the water forces some of the liquid into the pipe. Let P be the level of the water in the pipe at the end of the upward stroke of the piston. On the downward stroke of the piston the valve B closes and v opens allowing the air in the barrel—which the downward motion of the piston tends to compress—to escape through the piston into the atmosphere, until the piston again reaches the bottom of the barrel. On again raising the piston, the valve v closes and that at B opens, thus allowing the air in BP to expand and to diminish its intensity of pressure; and, in consequence, more water is forced up into the pipe, and perhaps into the barrel. This process being continued, the water ultimately reaches the level of the spout and is thus driven out.

To find the height to which the water is raised in the pipe by the first stroke of the piston, let A = area of cross-section of barrel, a = area of cross-section of pipe, l = length of stroke of piston, c = length BA, h = height of a water barometer, and $AP = x$. The volume of air in the suction-pipe before the stroke is ac, and its intensity of pressure is represented by h. At the end of the stroke the volume of this mass is $a(c-x)+lA$, and its intensity of pressure is represented by $h-x$. Hence

$$(h-x)\{a(c-x)+lA\} = ach,$$
$$\therefore ax^2 - (ac+ah+lA)x + lhA = 0,$$

which determines x.

When the water is flowing out of the spout, there will be a tension in the piston rod on an upward stroke, which is found thus. Let z be the vertical height of the piston above the level A, and let z' be the height of the column of water (which we may suppose to reach to any point, D, of the barrel) above the piston. The valve at B being open, there is a continuous water communication between A and the bottom of the piston. Hence if w is the weight of a unit volume of water, the intensity of the upward pressure exerted by the water on the bottom of the piston is $w(h-z)$; and the intensity of downward pressure exerted on the top of the piston is $w(h+z')$; therefore the total downward pressure on the piston is $w(z+z')A$. This is equal to T, the tension of the rod, if we neglect any acceleration of the piston. Hence, approximately,

$$T = w \cdot A \cdot DA,$$

which shows that the tension of the rod is the weight of a vertical column of water having the piston for base, and for height the difference of level of the water in the barrel and that in the well.

On the downward stroke there is a *pressure* in the rod, which is approximately equal to the weight of the column of water above the piston.

When the water is flowing out, the force required at H to work the piston on the upward stroke is $T \times \dfrac{fc}{fH}$, where T has the above value.

It is obvious that, for the working of the pump, the length AB of the suction-pipe above the well must be less than the height of a water barometer, i.e., about 34 feet; and, owing to imperfect fittings, AB must be considerably less than this— say about 25 feet.

In the middle ages a curious modification of the common pump, called the *bellows pump*, was employed in Europe. Instead of a piston worked by a lever, fH, (Fig. 69) a large bellows was attached firmly to the top of the barrel, and the nozzle of the bellows was the spout through which the water was forced.

The top of the barrel fitted into the interior of the bellows through a hole in the lower board of the bellows; there was no valve in the top board, but there was one opening outwards fixed in the nozzle. The action was, of course, the same as that in our modern pump.

The Forcing Pump. This is an instrument for raising water to a great height. It differs from the previous pump in having a completely solid piston.

To the barrel of the pump is attached the pipe through which the water is to be raised. This delivery pipe is provided with a valve at D, Fig. 70, opening upwards, and this pipe is represented as being provided with a spout and stop-cock by means of which the machine can be made to act as a Common pump.

The action is the same as in the previous case.

On the downward stroke of the piston, the valve at B closes and that at D opens, and through this latter the water is forced out of the barrel into the delivery pipe, DV.

There is then a pressure in the piston rod, r, equal to the weight of a column of water having the piston for base, and for height the difference of level between the piston and the water, V, in the delivery tube.

On the upward stroke there is a tension in the rod, whose value is the same as in the previous pump.

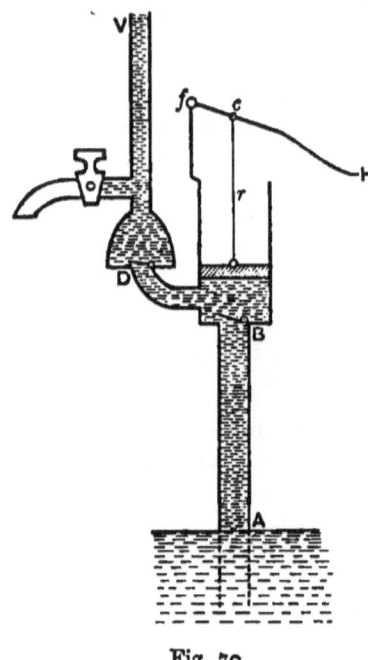

Fig. 70.

The Fire Engine. This is simply a double forcing pump. The figure (Fig. 71) represents the two barrels, P and Q, as immersed in a tank, DET, full of water; and from this tank the pumps, which are both worked by the lever AB, force the water through a hose connected with the chamber C at h. The water is forced through this hose to the place where the fire is to be extinguished. The action of the valves is obvious in the figure. Such would be the arrangement in a small fire engine, the tank DEI being filled by buckets of water brought by hand.

When large fires have to be dealt with, a supply thus

derived would be of no use, and the water which is pumped through the hose must be derived from a well or other reservoir by means of a suction pipe. The figure represents at *c* the place where such a pipe can be attached to the engine.

The chamber *C* is partly filled with water and partly with air, and is called an *air chamber*. Such a chamber may be, and often is, fitted to forcing and other pumps, the object being to render the stream of water from

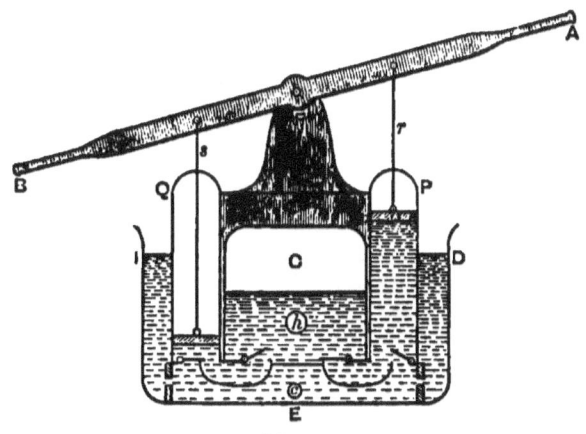

Fig. 71.

the delivery pipe at *h* continuous instead of intermittent; and this result is evidently secured by means of the compressed air at the top of the chamber; for, since this air was originally at the atmospheric pressure (when it filled the whole of the chamber) its intensity of pressure after the chamber is partly filled with water is greater than this value. This increased pressure is therefore continuously exerted on the top of the water in the chamber and helps to drive the stream through the hose.

The Hydraulic Screw. This is one of the most ancient machines for raising water, and is still employed in some

278 *Hydrostatics and Elementary Hydrokinetics.*

countries. It is often called the *Screw of Archimedes*, because its invention is supposed to be due to the philosopher of Syracuse. There is, however, reason to think that it was first employed in Egypt.

As represented in Fig. 72, it consists of a pipe wound spirally on an axis which is fixed in a position inclined to the vertical, its extremities fitting into two solid supports,

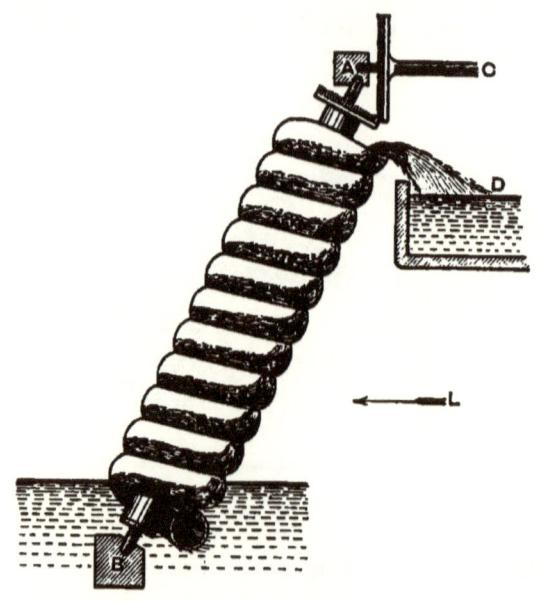

Fig. 72.

A, B, the axis (and, with it, the screw) being able to revolve freely. Both ends of the spiral pipe are open, and the lower is immersed in the water which is to be raised to the level D.

The revolution of the screw can be effected in various ways: in the figure it is represented as produced by the revolution of a shaft C fitted with a toothed wheel which gears with another fitted to the top of the axis of the screw.

Hydraulic and Pneumatic Machines. 279

When the screw is made to revolve so that the lower end comes towards us in the figure, water entering at this end continually drops to the lowest point of each part of the spiral, and is thus carried continuously up to the top where it is discharged.

There is a certain condition that must be satisfied by the inclination of the axis of the screw to the vertical and the angle of the spiral in order that the machine may be able to raise the water. The condition is this: *the inclination of the axis of the screw to the horizon must be less than the inclination of the tangent line of the spiral to the axis of the screw.* To prove this, we may put the matter thus: *the axis of the screw must be so much inclined to the vertical that it is possible to draw a horizontal tangent to the spiral.* This is obvious, because if we imagine a single particle (suppose a small marble) entering the lower end of the pipe, it would not drop down farther through the opening unless there were in the pipe a place in which the particle could rest under the action of its own weight and the reaction of the pipe on it; and at such a place the tangent to the spiral must be horizontal.

Let i be the inclination of the axis of the screw to the vertical, a the angle of the spiral, i.e., the inclination of the spiral to the axis (the spiral signifying the central helical axis of the pipe), r the radius of the cylinder on which this helix lies, and let m be used for tan a.

Then if we take as axis of z the axis AB of the screw, and as plane of xz the plane of the vertical line at B and the axis BA, the axis of x being the line at B perpendicular, in this plane, to BA, and the axis of y the perpendicular to this plane at B, the direction-cosines of the vertical at B with reference to these axes of co-ordinates are

$$-\sin i, \quad 0, \quad \cos i. \qquad \qquad (1)$$

Also if x, y, z are the co-ordinates of any point, P, on the helix,

$$\left. \begin{array}{l} x = r \cos \dfrac{mz}{r}, \\[4pt] y = r \sin \dfrac{mz}{r}. \end{array} \right\} \qquad (2)$$

If ds is an element of length of the helix at P, we have $ds = \sec a \, . \, dz$, and

$$\left.\begin{aligned}\frac{dx}{ds} &= -\sin a \sin \frac{mz}{r}, \\ \frac{dx}{ds} &= \sin a \cos \frac{mz}{r}, \\ \frac{dz}{ds} &= \cos a,\end{aligned}\right\} \quad \ldots \quad (3)$$

and the direction-cosines of the tangent to the helix at P are the expressions (3).

Supposing this tangent to be at right angles to the vertical (1), we have
$$\sin i \sin a \sin \frac{mz}{r} + \cos i \cos a = 0,$$

$$\therefore \sin \frac{mz}{r} = -\cot i \cot a \ . \ \ldots \ \ldots \ (4)$$

which shows that $\cot i \cot a$ must be < 1, i.e., $\cot i < \tan a$, or $\frac{\pi}{2} - i < a$, that is, the inclination of the axis, BA, of the screw to the horizon must be less than the inclination of the helix to BA.

The point P (which is now the lowest point in one turn of the pipe) does not lie in the vertical plane through BA, as might at first be supposed; for (2) gives

$$y = -r \cot i \cot a \ \ldots \ \ldots \ (5)$$

for this point.

To find the force necessary to turn the screw when a particle of weight W is to be raised, let us consider the screw to be either in equilibrium or in uniform motion. The forces acting upon it are W, the reactions at the supports B, A, its own weight, and the force applied to turn it. Of these the reactions at B and A and the weight of the screw intersect BA. We have, then, the result that the moment of W about BA is equal to the moment of the applied turning force. (Hence an obvious reason why the equilibrium position, P, of the ascending particle cannot be in the vertical plane through BA.)

Now the moment round the axis of z of a force whose components are X, Y, Z acting at the point x, y, z is $xY - yX$; and the force at P is W, acting vertically down, whose components are $W \sin i$, 0, $-W \cos i$; hence the moment of W about BA is $-Wy \sin i$, i.e., by (4)

$$W r \cos i \cot a.$$

If a is the radius of the toothed wheel attached to the axis of the screw, and f the force on it, we have then

$$f = W \frac{r}{a} \cos i \cot a.$$

If b is the radius of the toothed wheel attached to the driving shaft C, and if this shaft is turned by an effort F applied at a distance R from the axis C, we have $f.b = F.R$, and

$$F = W \frac{rb}{aR} \cos i \cot a.$$

If instead of a single particle which, always occupying the lowest point of the convolution in which it lies, ascends to the top, we consider the whole tube full of water, the moment of the weight of this water about the axis is $w \sin i \int_0^l y \, ds$, where w is the weight of water in the tube per unit length, and l is the length of the tube. This is

$$r^2 w \sin i \, \text{cosec} \, a \left(1 - \cos \frac{l \sin a}{r} \right).$$

The Hydraulic Screw is capable of a *differential* form. Suppose the screw in Fig. 72 not to dip into water at its lower extremity, but to receive into the upper end of the pipe at A a stream of water from any source. Then, the screw being fixed exactly as represented, would be driven by this stream in the direction opposite to that in which it was caused to rotate under the previous circumstances.

Now suppose, that it is, as before, desired to raise water from a well at the extremity B to a position D, and suppose that there is available a stream of water at some lower level, represented by the arrow L. Let a second pipe be spirally wound round the axis BA outside the tube represented in the figure *and in the reverse sense*, its upper end terminating at the level L. If

282 *Hydrostatics and Elementary Hydrokinetics.*

the stream L is led into the upper end of this second pipe, it will cause the whole machine to rotate in the sense required to raise the water from the well by means of the internal spiral pipe.

The idea of the Differential Hydraulic Screw appears to be due to the ingenious Marquis of Worcester, who published his notions on this machine and on many others in a work called *A Century of Inventions*, in the reign of Charles II.

The Hydraulic Ram. This is a machine in which the momentum of a stream of water is suddenly stopped, with

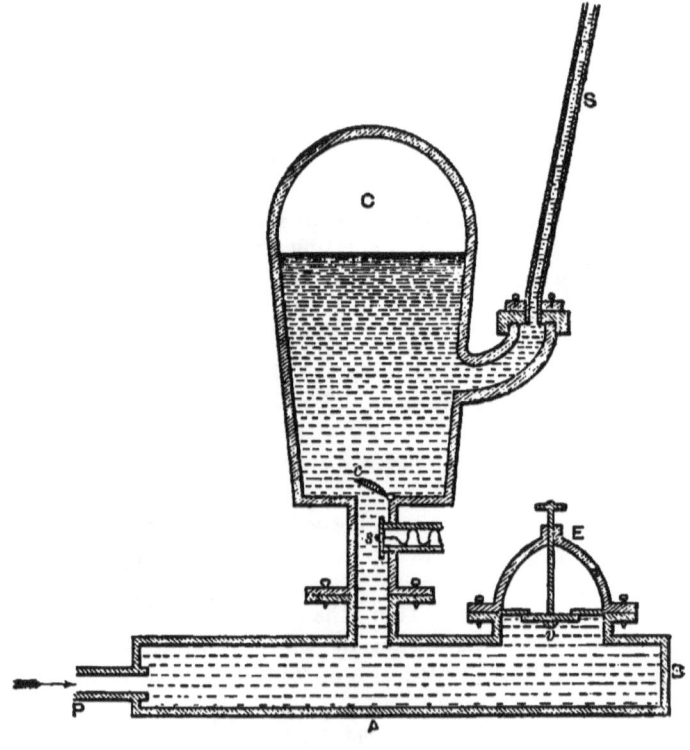

Fig. 73.

the result that a portion of the water is forced up to a considerable height.

Let AB, Fig. 73, be a stout iron chamber to which

is attached a pipe P which admits a flow of water from a stream or reservoir the level of which may be only slightly higher than that of AB. The vessel AB is fitted with a support, E, for a valve, v, which can move freely up and down. When this valve falls, there is a free communication between the interior of AB and the atmosphere, and if AB is full of water, some of this water flows away through the opening at v, and is wasted.

To the top of the vessel AB is screwed a chamber, C, which has a valve c opening upwards, and also a valve s opening inwards. This latter valve is attached to a rather weak spring fixed to a side pipe opening in to the chamber C. Finally, a supply pipe, S, is attached to the chamber C.

Imagine the whole machine free of water, so that the valves c and v are down; and then let the stream flow in at P. At first the water will rush through the opening at v; but soon the rush of water will close this valve, and at this instant, the water being suddenly checked, some will be forced through the opening at c. This valve will then close and v will drop, allowing an outflow again from the vessel AB. The same process will be again and again repeated until the water forced into C rises in the pipe S. The upper part of the chamber C contains imprisoned air, the pressure of which serves to keep the flow up the pipe S continuous. The valve v falling and rising thus regularly, the machine is self-acting. After a long time the air in the air chamber C would be absorbed by the water, and thus the advantage of an air chamber C would be lost. The object of the *snifting valve* s is to prevent this, and the air is renewed as follows. When the valve v drops, the intensity of pressure immediately under it is that of the atmosphere, and therefore the intensity of pressure of the water at s is less than that of the outside air, so that (as the spring which closes s is a weak one) this valve s is forced inwards, thus allowing some air to enter the water; and this air when the valve c is next opened will rise to the top of C and replace the air absorbed by the water.

(It must not be supposed that the forcing of water by this self-acting machine to a height vastly greater than that of the source whence the water was derived involves any contradiction of the principle of Work and Energy; for, it is by means of the kinetic energy generated by gravity in the very large mass of water which flows into and out of the ram that the comparatively small mass of water is raised in the pipe.)

The Hydraulic Ram was invented in 1772 by Whitehurst of Derby, his machine, however, not being self-acting. With him, instead of the self-acting valve v, there was a stop-cock through which the water flowed; and it was on the sudden closing of this stop-cock that the water was forced through the valve c.

65. Air Pumps. *The Common Air Pump.* In Fig. 74, B is a cylinder or barrel in which works a piston with a valve, c, opening upwards. The barrel is screwed, or otherwise firmly attached, to a plate, D, through which runs a groove which communicates with the interior of the barrel through an opening which can be closed by a valve a; the other end, n, of this groove opens up through a large plate PQ, the upper surface of which is perfectly flat. On this plate is placed a large glass vessel, A, called the *receiver*, the mouth of which rests on the plate PQ; the rim of the receiver is ground, and it fits the plate so accurately that the junction is air-tight, especially as a layer of grease is rubbed on the rim before it is placed on the plate.

The object is to remove the air, partially or completely, from the receiver, and therefore from any body or vessel that may be placed under it.

Hydraulic and Pneumatic Machines.   285

The manner in which this is effected is obvious from the figure. Suppose the piston in the lowest position in the barrel; then when it is raised, a vacuum tends to form above a, so that the air of the receiver raises this valve and fills the barrel at the end of the stroke. On the descent of the piston, a closes and c opens, and the air in the barrel is thus expelled into the atmosphere. This process is repeated many times, with the result that the air in A is greatly diminished in mass.

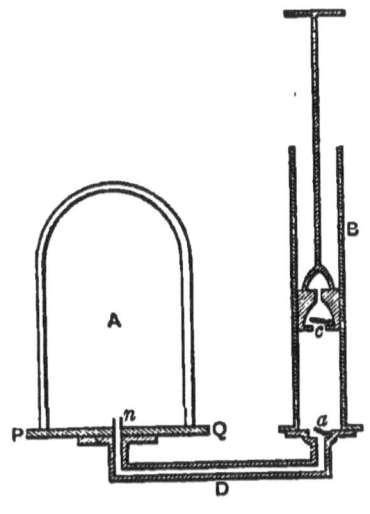

Fig. 74.

To calculate the degree of exhaustion after n strokes of the piston, let the volumes of the receiver and barrel be A and B; let the original intensity of pressure of the air in A be p_0, and let p_1, p_2, p_3 ... be the intensities after 1, 2, 3,... strokes.

Then after the piston has been raised the first time the mass of air whose volume and intensity of pressure were (A, p_0) becomes $(A+B, p_1)$; hence

$$(A+B)p_1 = Ap_0. \quad \ldots \quad \ldots \quad (1)$$

When the down stroke is ended there is a different mass of air in A, and it is denoted by (A, p_1); and this same mass of air is denoted by $(A+B, p_2)$ at the end of the second upward stroke; hence

$$(A+B)p_2 = Ap_1. \quad \ldots \quad \ldots \quad (2)$$

Similarly
$$(A+B)p_3 = Ap_2, \quad \ldots \quad \ldots \quad (3)$$

. . . . .
. . . . .

286 *Hydrostatics and Elementary Hydrokinetics.*

Hence, by multiplication,
$$(A+B)^n . p_n = A^n . p_0,$$
$$\therefore \; p_n = p_0 \left(\frac{A}{A+B}\right)^n, \quad \ldots \ldots \quad (a)$$

which gives the final intensity of pressure. If W_n is the weight of the air finally left, and W_0 the original weight, we have from (a) of Art. 47,
$$W_n = W_0 \left(\frac{A}{A+B}\right)^n,$$

and a similar relation between the final density, ρ_n, and the original ρ_0.

For the very high exhaustions required in the globes of incandescent electric lamps, and in the interior of vacuum tubes, an air pump of this kind would be quite insufficient, because, after the exhaustion has reached a certain limit, the pressure of the gas is insufficient to raise the valve *a*.

Condensing Air Pump. When it is desired to fill a vessel, *A*, Fig. 75, with air or any other gas at a given intensity of pressure, a condensing pump is employed. This machine consists of a cylinder or barrel, fitted with a solid piston, and having a valve, *c*, opening downwards. At the side of the barrel there is attached a pipe having a valve, *a*, opening inwards. When any other gas than air is to be forced into *A*, the vessel supplying this gas is attached to the pipe at *a*. The valve at *a* opens while the piston is raised, and the barrel is filled with the gas. When the piston is lowered, *a* closes, *c* is forced open, and if the stopcock, *s*, fitted to the vessel *A* is properly turned, the gas enters *A*. On the upward stroke of the piston, *c* closes, *a* opens, and the process is repeated.

Let *A* and *B* denote the volumes of the chamber *A* and the barrel, and consider the gas which fills *A* after *n* strokes of the piston. Let p be its intensity of pressure, and let p_0 be the intensity of pressure of the gas which fills the barrel: if A is being filled with atmospheric air, p_0 is the atmospheric intensity.

Then the gas whose volume and intensity of pressure are (A, p) was once represented by

$$(A + nB, p_0),$$

supposing that A contained the gas originally. Then

$$A \cdot p = (A + nB) p_0,$$

$$\therefore p = \left(1 + n\frac{B}{A}\right) p_0.$$

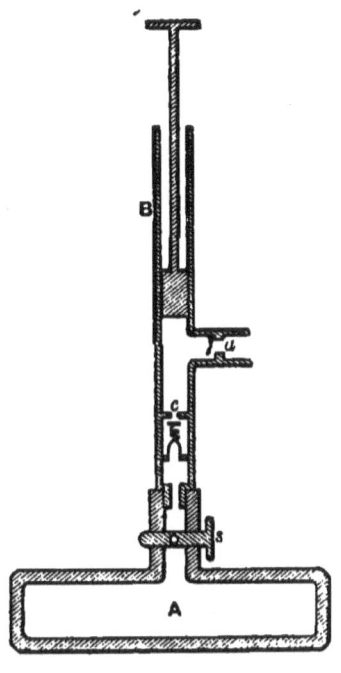

Fig. 75.

The Geissler Pump. When very high exhaustions are required, air pumps with valves and pistons are replaced by pumps in which a column of mercury plays the part of a piston. Of the latter kind is that represented in Fig. 76.

To a certain extent, all air pumps are identical in principle. In each of them a given mass of gas occupying a volume V is made to occupy a larger volume, $V + U$, and then the portion occupying U is mechanically expelled. If this process could be repeated indefinitely, an exhaustion of any degree desired could be obtained; but we have seen that the raising of valves in the common air pumps puts a limit to the process.

288 *Hydrostatics and Elementary Hydrokinetics.*

The mercury pumps of Geissler and Sprengel are free from this drawback.

AB is a glass tube of greater length than the height of the mercury barometer, having part of the Torricellian

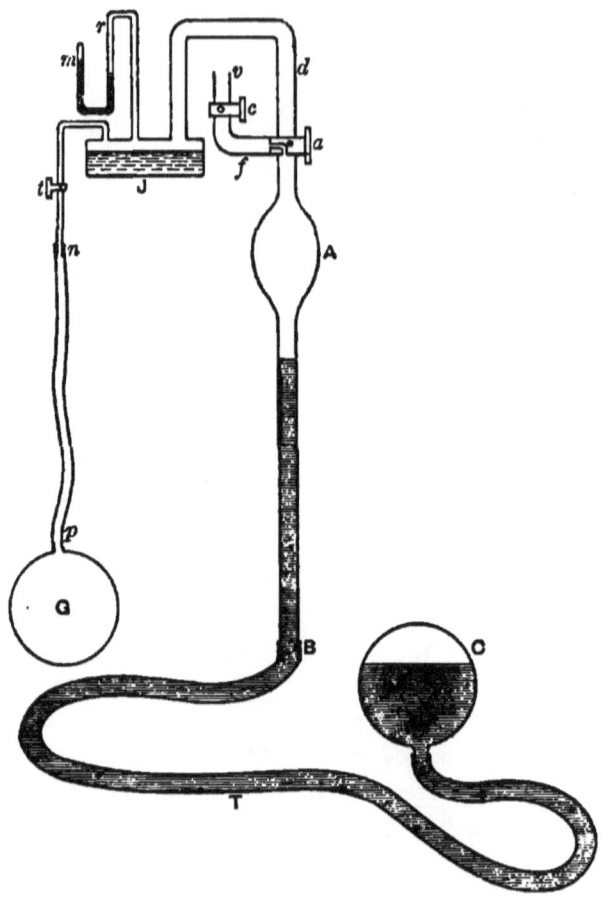

Fig. 76.

vacuum enlarged into a chamber, A, of large capacity. Above A is inserted a two-way stop-cock, a, which in the position represented establishes a communication

Hydraulic and Pneumatic Machines. 289

between the chamber A and a side tube, f, fitted to the tube BAd. This tube f has a stop-cock, c, which in the position represented closes f; but if c is turned through a right angle it will establish a communication between f and the external atmosphere at v. If a is turned through a right angle from its present position, it will close the communication of A with f, and open one between A and a vessel J to which the portion d is joined as represented. The vessel J contains some sulphuric acid the object of which is to remove aqueous vapour from air which may pass over it; and, by means of a stop-cock, t, J communicates with a very stout indiarubber tube, np, which is connected with the vessel G which is to be exhausted.

To the vessel J is connected a truncated manometer, that is, a bent glass tube, mr, containing mercury which when the air in J is at atmospheric pressure quite fills the leg m. If the air in J is completely removed, the columns of mercury in the legs m and r will assume exactly the same level.

To the end B of the tube BA is attached a stout flexible tube T, which is also fixed to a large reservoir, C, of mercury.

Suppose now that c is turned so as to establish communication between f and the atmosphere at v, and that a is in the position represented (i. e., closing communication of A with d); and let C be raised until the surface of the mercury in BA reaches the stop-cock a, thus expelling all the air from A through fv. Then turn a and t so as to admit air from G through J and d, and lower C to its original position. The air of G now occupies the volume $G+A$ together with the volumes of the communicating tubes.

Turn t so as to break communication with the rarefied air in G, and then turn a so as to establish communication between A and the atmosphere at v. Again, raise C until

U

the mercury in BA drives out the gas from A into the atmosphere; and repeat the process of establishing communication with G, &c. In this way, by repeated operations, the air in G is exhausted almost completely. By this pump the air left in G can be reduced to an intensity of pressure represented by only $\frac{1}{10}$ of a millimetre of mercury.

The Sprengel Pump. Fig. 77 represents this pump, in which, as in the Geissler, the vessel, G, to be exhausted is made part of the Torricellian vacuum of a barometer tube, H.

A funnel, F, prolonged into a narrow tube fitted with a stop-cock, f, is supported in a vertical position (support not represented in figure) and dips into a wide tube, B, also supported. B is connected by indiarubber tubing with a narrow vertical tube, C, above which is a large chamber, A, open at the top, and fitted with a stopper, s, the tube D being, like C, connected with the chamber. D is connected by indiarubber tubing with the vertical tube I, which communicates freely with the very narrow tube H, the top of which is connected with the vessel G, and the bottom of which, curved up a little, dips into a vessel V full of mercury. There is an overflow from V to a trough T, as represented; and there is a clamp, c, by means of which communication between D and I can be established or broken. The vessel G is provided with a stop-cock, g.

The order of operations is as follows. The tubes being all completely occupied by air, fix the clamp c, remove the stopper s, lower the system of tubes D, C, and pour mercury into F and through its tube into B, until this mercury completely fills the tubes C, D and the chamber A. Close A with the stopper s, and raise the system C, D to its original position; open the clamp c, and let the mercury run from B, C, D into I up to the top expelling air from H

Hydraulic and Pneumatic Machines. 291

through the mercury in V. The mercury coming over at a into H will occupy a certain portion of H. Now let the stop-cock g, be turned so that G is connected with the upper part of H. Mercury may be poured into F to keep up the flow from a through H, and the rate of supply of this mercury can be regulated by turning the stop-cock f more or less.

Now as each drop of mercury falls down through H, it forces the air in front of it down through the end of H; and hence the gas of G which keeps flowing into the upper part of H is perpetually driven down and out by the successive drops of mercury which fall over from a. If the mercury in A has fallen down through D into I, II, and V, the chamber A is a vacuum.

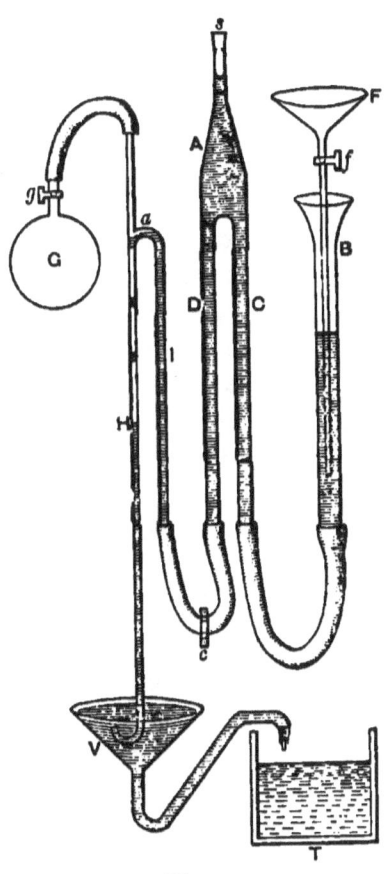

Fig. 77.

When the exhaustion of G has not been carried very far, the successive threads of mercury falling down H (and represented in the figure) succeed each other comparatively slowly, and they can be seen forcing the gas which reacts against their fall; but when the exhaustion is nearly complete, these drops fall much more freely through the now exhausted

U 2

space; and when there is only a very small quantity of gas left in G, the drops falling from a on the top of the mercury surface, H, produce a sharp metallic sound, like that of a water hammer. This sound is an indication of a high degree of exhaustion.

When the exhaustion is complete the surface of the mercury in H will be at the barometric height above the level in V, and the difference of the level of the mercury in B and C will also be the barometric height.

The object of allowing the tube from F to dip into a much wider tube, B, is, partly, to let any air that may be carried down with the mercury from F escape into the atmosphere through the mercury in the wide tube, and partly to avoid filling F a very great number of times; this incessant filling will not be necessary if the tube B is very much wider than the other tubes.

The object of turning up the end of the tube H in the mercury in V is to allow the gas (whatever it may be) that is expelled from G through this end to be collected in another vessel, a tube from which is led to the point at the end of H.

The object of having the tubes C, D and the chamber A is (when A is vacant of mercury) to catch in this chamber any air that may have been carried over by the mercury from F, so that the exhaustion in the tube H shall be performed by mercury devoid of air. Hence the chamber A is called an *air trap*. So far as the principle of the Sprengel pump is concerned, we might dispense with the tubes C, D and the air trap, A, and connect B directly with I—and the Sprengel pump is, in fact, usually so represented.

66. Manometers. A manometer is an instrument for measuring the intensity of pressure of a condensed or an exhausted gas. The instrument has various forms, and the

Hydraulic and Pneumatic Machines. 293

principle of all will be easily understood from that represented in Fig. 78.

Let $HFED$ be a vertical bent glass tube, having a portion, ED, of one leg enlarged into a capacious reservoir, and let two necks C, D, project from this reservoir so that vessels may be connected with the reservoir by means of indiarubber tubes fitting on C and D. The leg HF is closed at the top.

Suppose the cross-section of this leg to be uniform, as also that of the reservoir except near its ends. Let mercury be poured into the instrument, and when the air thus imprisoned in HF assumes the temperature and pressure of the surrounding atmosphere (which enters at C and D) let AB be the level of the mercury in both legs, and let the number 1 be marked on the leg HF where the surface of the mercury stands, this point being, of course, in the prolongation of BA. The number indicates that when the air in HF occupies the length HI, at a given temperature, its intensity of pressure is 1 atmosphere. Let c be the length HI.

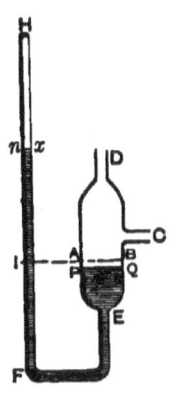

Fig. 78.

Now suppose that it is desired to fill a globe, or other vessel, with air at a great pressure and to measure the intensity of the pressure. Let the vessel be connected with the neck C, and let D be connected with a condensing pump. When this pump forces air in through D, and therefore through C into the vessel, the surface of the mercury falls in the reservoir ED and rises in the tube HF. After the pump has been working for any time, let n atmospheres be the intensity of pressure of the air in the reservoir (and therefore in the vessel which was to be filled); let PQ be the depressed surface in the reservoir, and let the surface of the mercury in HF be at x, which is x inches or millimètres above the level AB. If the intensity of pressure of the air in the reservoir is n atmospheres, the number n is to be marked at the level x on the tube HF. Let the depth of PQ below AB be y inches or millimètres; a = area of cross-section of HF, A = that of the reservoir; and let h inches or millimètres represent an intensity of pressure of 1 atmosphere. Then the difference of level between the point x and the surface PQ being $x+y$, and the volume of the air in HF being now $a(c-x)$ with an intensity of pressure denoted by $nh-x-y$, we have by Boyle's law

$$a(c-x)(nh-x-y) = ach.$$

But evidently $Ay = ax$; hence

$$a(c-x)\left\{nh-\left(1+\frac{a}{A}\right)x\right\}-ach = 0,$$

which determines the number, n, to be marked on any part of the tube.

67. Hydrometers. These are instruments for the determination of the specific gravities of liquids and solids. We shall describe two only.

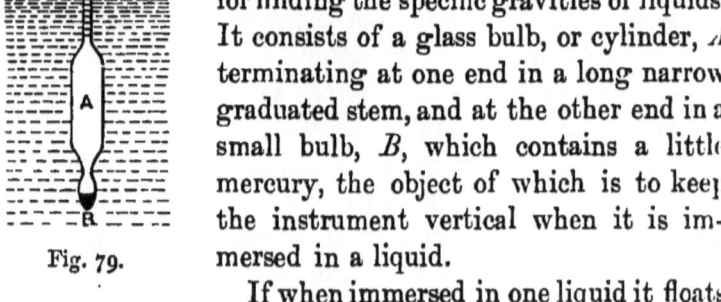

Fig. 79.

The Common Hydrometer, Fig. 79, is used for finding the specific gravities of liquids. It consists of a glass bulb, or cylinder, A terminating at one end in a long narrow graduated stem, and at the other end in a small bulb, B, which contains a little mercury, the object of which is to keep the instrument vertical when it is immersed in a liquid.

If when immersed in one liquid it floats with a volume v immersed, and in another with a volume

v', the specific weights of these liquids being w and w', respectively; and if W is the weight of the instrument itself, we have

$$v \cdot w = W; \quad v' \cdot w' = W,$$

$$\therefore \frac{w}{w'} = \frac{v'}{v}, \quad \ldots \ldots \ldots (1)$$

so that the specific weights are inversely as the volumes immersed.

The volume of the portion AB irrespective of the stem can be found by graduating the stem (supposed of uniform bore) into any number of equal parts, and then observing the weight, W, of the instrument. Let masses p, p' be successively attached to the top of the stem, and with these let the instrument float in water up to the nth and n'th division, respectively. Then if V is the volume AB, and a the volume per division of the stem,

$$(V+na)w = W+p; \quad (V+n'a) = W+p',$$

which determine V and a.

If when the hydrometer is immersed in two different liquids the readings on the stem are n and n', we have from (1)

$$\frac{w}{w'} = \frac{V+n'a}{V+n\,a},$$

which shows that if a is very small, n and n' must be very widely different, i. e., the instrument is exceedingly sensitive to small differences of specific weight.

Sikes's Hydrometer is a form of the above in which the stem is a very thin flat strip of metal, for which, of course, a is very small.

Nicholson's Hydrometer. This hydrometer is employed to measure the specific gravities of both solids and liquids.

It consists of a hollow metallic cylinder, A, Fig. 80, having a very fine stem on which there is a fixed mark, P; the lower end of the cylinder is connected by a wire with a closed cone, D, which is heavy enough to keep the stem vertical; the base, C, of this cone serves as a platform on which a solid body can be placed; the stem terminates in cup, B, in which solids can be placed.

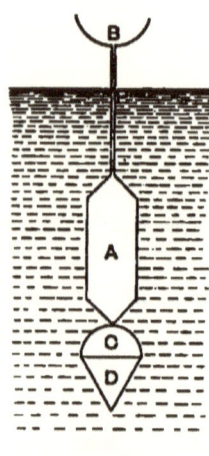

Fig. 80.

To find the specific gravity of a solid, place masses in B until the mark P is just sunk to the surface of the water; then place the given body in B: this will cause P to sink lower; remove weight from B until P again reaches the surface; if the weight removed is W, then W is the weight of the given body. Now remove the body from B to the platform C, and add weight, W', to B until P sinks to the surface; then W' is the weight of a volume of water equal to the volume of the body; and $\dfrac{W}{W'}$ is the required specific gravity.

To find the specific gravity of a liquid, let H be the weight of the hydrometer itself; let the instrument be immersed in the given liquid; add weight, p, to B until P sinks to the surface; let p_0 be the weight which must be added to B to cause P to sink to the surface when the instrument is immersed in water; then evidently the specific gravity of the given liquid is $\dfrac{H+p}{H+p_0}$.

Hydraulic and Pneumatic Machines. 297

MISCELLANEOUS EXAMPLES.

1. A semicircular area is immersed vertically in water, with its diameter in the surface; show how to divide it into n sectors all equally pressed.

Ans. Divide the diameter into n equal parts, and draw ordinates at the points of division; then the points in which these ordinates cut the circumference determine the sectors.

2. If w_1, w_2, w_3 are the apparent weights of a given body in fluids whose specific gravities are s_1, s_2, s_3, show that

$$w_1(s_2-s_3) + w_2(s_3-s_1) + w_3(s_1-s_2) = 0.$$

3. A ship sailing from the sea into a river sinks m inches, and, on discharging P tons of her cargo, she rises n inches; calculate the mass of the ship assuming her sides vertical at the water line, and that s is the specific gravity of sea water with respect to fresh (about 1·025).

Ans. $\dfrac{ms}{n(s-1)} \cdot P$ tons.

4. A uniform rod, AB, of length $2a$ is moveable round a horizontal axis fixed at A which is in a liquid of specific weight w_2; the end B projects into a liquid of specific weight w_1 which rests on top of the other liquid; find the positions of equilibrium, and determine whether they are stable or unstable.

If $2h$ is the depth of A below the surface of the lower liquid, $\theta =$ inclination of AB to the vertical, $w =$ specific weight of rod, the oblique position is given by the equation

$$\sec^2 \theta = \frac{a^2}{h^2} \cdot \frac{w-w_1}{w_2-w_1},$$

and if it exists, it is stable.

CHAPTER VIII.

MOLECULAR FORCES AND CAPILLARITY.

(*This Chapter may be omitted on first reading.*)

68. Molecular Forces. Common observations on the resistance which solid bodies oppose to any effort to elongate or twist them have compelled physicists to assume the existence of forces between the molecules of such bodies other than the ordinary action of Newtonian gravitation.

Thus, let us fix our attention on any one molecule, m, inside a body. It is surrounded by a group of molecules, and if we take all those molecules which lie within a sphere of extremely small radius whose centre is m, there is a special action exerted on m by each molecule within this sphere, those molecules nearest to m exerting a more powerful action than those near the surface of the sphere. This holds, whatever be the sizes, the shapes, or the distances between the molecules.

Beyond a certain distance, ϵ, from m these special actions are assumed to be insensible; this length ϵ is the radius of the aforesaid sphere, called the *sphere of molecular activity*.

Now if dm and dm' are two elements of mass, *the linear dimensions of each being infinitely smaller than the length of any line from the surface of the one to that of the other*, it is assumed that these elements exert on each other a force whose magnitude is $f(r) \, . \, dm \, dm'$, ... (1)

where r is the distance between the elements—i. e., the length of a line drawn from any point on one to any point on the other—and this force acts in the line joining them.

If the elements dm and dm' were homogeneous spheres, such a law of force as (1) could be assumed to hold, though their dimensions were even large compared with the distance between their centres, which distance would be the value of r in (1); but if they are not spherical, such a law could not be admitted (because it would be utterly devoid of meaning) if the elements are so close together that their linear dimensions are of the same order of magnitude as lines drawn from points on the surface of one to points on the surface of the other.

Now there are several suppositions that may be made with regard to the arrangement of matter in a body, such as the following:—

1. The matter is absolutely continuous within the volume of the body, there being no vacant spaces, however small.

2. The matter consists of molecules (in the chemical sense) which are packed very closely together, their linear dimensions being great compared with the distances between their surfaces.

3. The matter consists of molecules (in the chemical sense) which are very distant from each other, so that the space surrounding any molecule is comparatively void of matter.

If the third supposition is made, it is clear that the application of mathematical calculation becomes exceedingly difficult, if not impossible. It is true that Lamé in his *Élasticité des Corps Solides* objects strongly to the method applied by Navier and others in the theory of Elasticity, because, in applying the Integral Calculus to the determination of the action produced on a molecule of a body by the neighbouring molecules, they thus assume the continuity of matter, an assumption which Lamé describes as a 'hypothése absurde et complétement inadmissible.' His own method is a molecular one in which the existence of vacant spaces between the molecules is admitted; and the process of *integration* round a molecule is replaced by a process of mere algebraic summation—which, no doubt, is a much safer process, and should be adopted if it could be legitimately applied. It is not, however, satisfactorily applied by Lamé, since he has no hesitation in assuming a molecule to be wherever he wants one, and this assumption is not essentially different from integration.

If the second of the above suppositions is adopted, the matter surrounding a molecule, although not continuously filling space with mathematical strictness, may be assumed to be practically continuous, and the method of integration round a point becomes permissible as a very close approximation to the truth. The shapes of the molecules may possibly be such as to allow of their filling space much more effectively than if they were spheres.

But in adopting this supposition when calculating the forces produced on any molecule, m, by those within the range, ϵ, of molecular force, it will be necessary to imagine m and any *very* close neighbour, m', as *both* divided into infinitely smaller elements, of which dm is the type for the first and dm' that for the second, each of these elements being now infinitely smaller than the distance between them, and then assuming the force between them to be given by the expression (1). Thus for a pair of molecules so close that it is logically impossible to define anything that could be called the 'distance between them' we must imagine a special process of integration performed before we proceed to calculate the action of the more distant molecules within the sphere of molecular activity.

Such a process it is, of course, quite impossible to follow in detail because the form of $f(r)$ and the shapes of molecules are unknown; nevertheless, on account of the symmetry of arrangement of molecules round all points in a homogeneous body, it is possible to represent the result of such a process by a mathematical expression and to base further calculation thereon.

Various forms for $f(r)$ have been suggested, such as $A - \dfrac{B}{r}$ and e^{-ar}: these are, of course, merely conjectural; but it is conceivable that the observation of certain phenomena measurable *in the total* might afford a clue to, if not a necessary demonstration of, the law of this assumed molecular force.

If, then, we admit the second supposition, with the above notions, the first of our three suppositions becomes unnecessary, and Lamé's objection to the integration method loses its force.

In the study of the forms assumed by the surfaces of liquids in contact with each other and with solid bodies it is with these molecular forces that we have chiefly to deal. Indeed, the curious forms of such surfaces become explicable on no other hypothesis than that of the existence of very intense molecular forces having an extremely small range of action.

Supposing that the force between two elements of matter is given by the expression (1), its component along any fixed line (axis of x) is

$$\frac{x'-x}{r}f(r)\,dm\,dm',$$

if the co-ordinates of dm and dm' are x and x', so that the total component force acting on dm has for expression

$$dm \int_0^\epsilon \frac{x'-x}{r} f(r)\, dm',$$

if the integration is performed with reference to r, the limits of r being 0 and ϵ. Now, since the forces are zero beyond the distance ϵ, no error is introduced by assuming r to extend to ∞, so that such an expression is often written in the form

$$dm \int_0^\infty \frac{x'-x}{r} f(r)\, dm'.$$

Some notion of the magnitude of ϵ may be obtained from experiments such as the following. Quincke covered surfaces of glass with extremely thin layers of different bodies, and on these layers then deposited drops of mercury and other liquids. Now it will be seen presently that there is a definite angle between the tangent plane to the free surface of a liquid and the tangent plane to a solid with which it is in contact; this angle is constant all round the curve in which the two surfaces intersect; and it matters not whether the solid is a millimètre or 100 millimètres thick, the value of the angle does not alter. But if the solid is, say, the millionth of a millimètre thick, the angle alters. Covering the surface of glass with a layer of sulphide of silver, Quincke found that there was no change in the angle between the surface of the drop of mercury and the plate until the thickness of the silver layer was reduced to $\frac{46}{10^6}$ mm.; and when the glass was coated with a layer of iodide of silver, no change was observed until the thickness of the layer was reduced to $\frac{59}{10^6}$ mm., or, say, the

Molecular Forces and Capillarity. 303

wave length of yellow light. These thicknesses, then, indicate the order of magnitude of the distances at which molecular attractions are sensible.

Granting the existence of these molecular forces, it follows very obviously that within a layer of a fluid just at the surface, and of the extremely small thickness ϵ, there is a special intensity of pressure which increases in magnitude as we travel from any point P (Fig. 81) along the normal Pb to the surface, AB, of the fluid, towards the interior of the fluid.

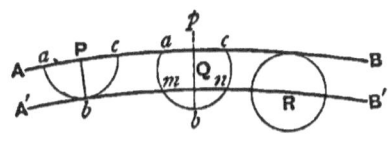

Fig. 81.

For, consider a molecule of the fluid at P; round P as centre describe the sphere of molecular activity; of this only a hemisphere, abc, exists within the fluid, so that the molecular forces acting on the particle P come from the molecules of this hemisphere. Now it is obvious that the symmetrical grouping of these molecules about the line Pb results in their producing a resultant force on P inwards along Pb.

Describe a surface, $A'B'$, parallel to AB at the distance ϵ, or, Pb, from AB.

Consider now the molecular actions on a molecule Q anywhere within this layer. Describe round Q the sphere, $ambnc$, of molecular activity. Of this sphere the portion apc does not contain any molecules of the fluid L, so that the action at Q is due to the portion $ambnc$, and the resultant force will obviously be directed along the normal Qb and will be less than the force at P, since there is some component of force in the sense Qp. Finally, consider a molecule at any point R on the inner surface $A'B'$, and we see that since this molecule is completely surrounded by attracting molecules, there is no resultant force whatever.

Now if F is the force exerted at Q per unit mass, and dn denotes an element of length of the normal Qb at Q measured towards b, while ϖ denotes the pressure intensity at Q due to the forces under consideration, we have

$$\frac{d\varpi}{dn} = wF, \quad \ldots \quad \ldots \quad (2)$$

w being the density of the fluid.

Since as we travel along the normal Pb from P towards b, or from Q towards b, the force F constantly preserves the sense Pb, although with diminishing value, we see that $\frac{d\varpi}{dn}$ is constantly positive, that is, ϖ continuously increases inwards until the surface $A'B'$ is reached, when F vanishes and $\frac{d\varpi}{dn} = 0$, i.e., ϖ becomes constant when we pass inwards through $A'B'$.

Hence the intensity of pressure due to molecular forces is constant throughout the interior of the fluid below $A'B'$, but it varies within the layer between AB and $A'B'$.

It is a matter of doubt with physicists whether we are or are not entitled to assume in the case of a liquid that the density within the layer contained between the surfaces AB and $A'B'$ is constant and equal to the density within the main body of the liquid. M. Mathieu, following Poisson, denies this constancy (*Théorie de la Capillarité*), but arrives, by the method of Virtual Work, at results of the same form as those obtained on the supposition of constant density.

Whatever may be the nature of the molecular forces, at any point close to the surface of separation of a liquid

from another medium, we can represent the magnitude of the resultant molecular force of the liquid on a molecule m by the expression
$$m \cdot F(\omega),$$

where ω is the area of that part of the surface of the sphere of molecular activity which exists round m within the liquid; and this force vanishes when $\omega = 4\pi\epsilon^2$. Or we might represent this resultant force (along the normal to the surface) by
$$m \cdot F(z),$$

where z is the distance of the molecule m from the surface; and the force $= 0$ when $z = \epsilon$. Of course the form of the function F is unknown, but it is the same at all points which are at the same normal distance from the surface, and this fact is sufficient for the purpose of calculation.

69. Calculation of Molecular Pressure. Let AB (Fig. 82) be the bounding surface of a liquid, P any point on the surface, TM the tangent plane at P, $A'B'$ the surface parallel to AB within the liquid at the depth ϵ; take an infinitely small element, σ, of area at P, and on the contour of this area describe a right cylinder, PR, extending indefinitely into the liquid. Consider now the action of molecular forces only on the liquid contained within this cylinder PR.

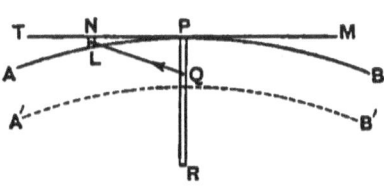

Fig. 82.

If ϖ is the intensity of molecular pressure at R, and the cylinder is terminated at R by a normal section, $\varpi\sigma$ is the pressure exerted on this section; then, for the equilibrium of the fluid in the cylinder we see that $\varpi\sigma$ must be equal to the integral of the molecular attraction exerted by the whole mass of the fluid on the portion of fluid contained in this slender canal. Now below the point P' there is no change in the molecular pressure, and there is no molecular force exerted on the elements of liquid; hence we might have taken the slender canal as reaching only to P'.

Let Q be any point in the canal between P and P' and let dm be an element of mass at Q. We shall calculate the attraction produced by the whole body of liquid below AB in the direction QR, which we know to be the direction of the resultant molecular attraction.

Now the total action on the canal PP' (or PR) can be calculated on the supposition that the liquid extends up to the tangent plane TM, and then deducting the attraction, in the sense QR, which is due to the meniscus, $ABMT$, of liquid thus added. The attraction of this added meniscus is obviously in the sense QP, so that this must be added to the attraction of the liquid terminated by TM. Let the attraction of this fictitiously completed liquid on the canal PP' be denoted by $K\sigma$ where K is obviously the same at all points on the bounding surface AB.

Let the plane of the figure be a normal section of the surface AB making the angle θ with the principal section at P whose radius of curvature is R_1, and let the radius of curvature of the other principal section at P be R_2.

Let any point, N, be taken on the tangent TM; let $NP = x$, and on the element of area $x\,dx\,d\theta$ at N construct the small cylinder NL terminated by AB.

If w is the mass per unit volume of the liquid, the mass of this cylinder is $NL . w x\,dx\,d\theta$; and if its distance, NQ, from Q is r, the molecular force which it produces on dm at Q is
$$w\,dm\,x\,dx\,d\theta f(r) . NL. \quad . \quad . \quad . \quad . \quad (1)$$

Molecular Forces and Capillarity. 307

Now if ρ is the radius of curvature of the section AB, we have
$$NL = \frac{x^2}{2\rho}, \text{ nearly};$$

and if $PQ = z$, the component of the force (1) along QP is obtained by multiplying it by $\frac{z}{r}$; hence this component is

$$w\,dm\,\frac{zx^3\,dx\,d\theta}{2\,r\rho}f(r). \quad \cdots \quad (2)$$

Integrate this with respect to θ, keeping x and r constant.

Now (Salmon's *Geometry of Three Dimensions*, Chap. XI) we know that
$$\frac{1}{\rho} = \frac{\cos^2\theta}{R_1} + \frac{\sin^2\theta}{R_2},$$

and the integration in θ from 0 to 2π makes (2) become

$$\frac{\pi}{2}w\,dm\left(\frac{1}{R_1} + \frac{1}{R_2}\right)\frac{zx^3\,dx}{r}f(r). \quad \cdots \quad (3)$$

But since $r^2 = x^2 + z^2$, $x\,dx = r\,dr$, and (3) becomes

$$\frac{\pi}{2}w\left(\frac{1}{R_1} + \frac{1}{R_2}\right)dm\,z\,(r^2 - z^2)f(r)\,dr. \quad \cdots \quad (4)$$

Let
$$f(r)\,dr = -d\phi(r), \quad \cdots \quad (5)$$

then since $f(r)$ rapidly diminishes with an increase of r, $\phi(r)$ is a positive quantity.

The resultant action of the meniscus on dm is obtained by integrating (4) from $r = z$ to $r = \epsilon$, or to $r = \infty$. Hence the resultant is

$$\frac{\pi}{2}w\left(\frac{1}{R_1} + \frac{1}{R_2}\right)z\,dm\int_z^\infty (-r^2 + z^2)\,d\phi(r). \quad \cdots \quad (6)$$

x 2

Now $\phi(\infty) = 0$, and

$$\int -r^2 d\phi(r) = -r^2\phi(r) + 2\int r\phi(r)\,dr,$$

$$\therefore \int_z^\infty -r^2 d\phi(r) = z^2\phi(z) + 2\int_z^\infty r\phi(r)\,dr,$$

therefore (6) becomes

$$\pi w\left(\frac{1}{R_1} + \frac{1}{R_2}\right) z\,dm\int_z^\infty r\phi(r)\,dr. \quad \ldots \quad (7)$$

Again, let

$$r\phi(r)\,dr = -d\psi(r), \quad \ldots \quad (8)$$

where $\psi(r)$ is obviously positive; therefore, since $\psi(\infty)$ is evidently zero, (7) becomes

$$\pi w\left(\frac{1}{R_1} + \frac{1}{R_2}\right) z\,\psi(z)\,.\,dm \quad \ldots \quad (9)$$

Now put dm at Q equal to $w\sigma dz$, and (9) becomes

$$\pi w^2\sigma\left(\frac{1}{R_1} + \frac{1}{R_2}\right)\,.\,z\,\psi(z)\,dz, \quad \ldots \quad (10)$$

and the integral of this from $z = 0$ to $z = PP'$, or $= \infty$, is the total action of the meniscus on the canal PR. Denoting by H the integral $\int_0^\infty z\,\psi(z)\,dz$, this action is

$$\pi w^2\sigma H\left(\frac{1}{R_1} + \frac{1}{R_2}\right), \quad \ldots \quad (11)$$

and hence equating $\varpi\sigma$ to the sum of (11) and the force $K\sigma$, we have

$$\varpi = K + \pi w^2 H\left(\frac{1}{R_1} + \frac{1}{R_2}\right), \quad \ldots \quad (12)$$

which is the form obtained by Laplace for the molecular

pressure intensity at all points in the liquid below P'. (See the *Mécanique Céleste*, Supplement to book X.)

We have supposed the surface of the liquid at P to be concave towards the liquid. If it is convex,

$$\varpi = K - \pi w^2 H \left(\frac{1}{R_1} + \frac{1}{R_2}\right). \quad \ldots \quad (13)$$

Hence we have the following obvious consequences.

1. If a liquid is acted upon by molecular forces only (no *external* forces) the quantity $\frac{1}{R_1} + \frac{1}{R_2}$ must be constant at all points of its bounding surface; for, otherwise we should obtain conflicting values for the intensity of molecular pressure at one and the same point in the body of the liquid.

2. The molecular pressure at a point strictly *on* its bounding surface is zero; for on the portion of liquid contained within the canal PP' and included between P and a point infinitely close to P the resultant force exerted by the fluid is infinitely small (since the mass contained is infinitely small).

3. The value of the intensity of molecular pressure at a point within the body of a liquid is not a constant related solely to its substance; it depends on the curvature of its bounding surface. If this surface is plane, the intensity is K.

4. If (owing, as we shall see, to the action of *external* forces) it happens that some parts of the bounding surface are plane, others are curved and have their concave sides turned towards the liquid, and others again have their convex sides towards the liquid, the intensity of molecular pressure just below the second kind of points is *greater* and below the third *less* than it is at the plane portions.

310   *Hydrostatics and Elementary Hydrokinetics.*

(We shall presently see how this is verified in the rise or fall of liquid in capillary tubes when gravity is the external force.)

The constant K can be easily expressed in terms of the function ψ thus: let CD, Fig. 83, represent the plane surface of a liquid, the liquid lying below CD; at any point A on the surface take an infinitely small element of area, σ, and describe on it a normal cylinder or canal, AMR, extending into the liquid indefinitely; then $K\sigma$ is equal to the whole force produced on the liquid within this canal by the whole body of liquid below CD.

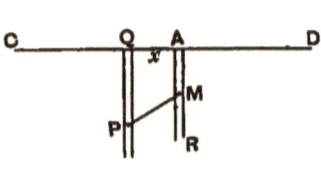

Fig. 83.

Take any point Q on CD; let $AQ = x$, and take the circular strip $2\pi x dx$ round A; along this strip describe normal canals (represented by QP) extending indefinitely into the liquid; take any point, P, in one of these canals; let $PQ = y$; take any point, M, in the canal at A, and let $AM = z$. Then if s = area of normal section of the canal QP, the element of mass at P is $swdy$, and its action on a mass m at M is $mswdy f(r)$, where $r = MP$; the component of this along MR is $mswdy f(r) \cdot \dfrac{y-z}{r}$, or $mswf(r)dr$; therefore taking the points P at a constant distance y from CD all round the strip $2\pi x dx$, we see that their action on m is
$$2\pi wmx dx \cdot f(r) dr.$$

Integrating this from $r = MQ$ to $r = \infty$, we have (putting $MQ = r_1$)
$$2\pi wmx dx \cdot \phi(r_1).$$

Now we must integrate with respect to x from 0 to ∞,

Molecular Forces and Capillarity. 311

and observe that $x\,dx = r_1\,dr_1$, so that the limits of r_1 are MA (or z) and ∞. Thus we get

$$2\pi w m \int_z^\infty r_1\,\phi(r_1)\,dr_1,$$

i. e., $2\pi w m\,\psi(z)$,

which is therefore the action of the whole mass of liquid on the particle at M.

Also $m = w\sigma\,dz$; therefore the action on the canal AR is

$$2\pi w^2 \sigma \int_0^\infty \psi(z)\,dz,$$

$$\therefore\ K = 2\pi w^2 \int_0^\infty \psi(z)\,dz.$$

This constant K is called by Lord Rayleigh the *intrinsic pressure* of the liquid, *Philosophical Magazine*, October, 1890.

70. Pressure on Immersed Area. Suppose a vessel to contain a heavy homogeneous liquid of specific weight w, and let P be a point in the liquid at a depth z below the plane portion of the free surface. Then, as has been shown in the earlier portion of this work, the intensity of pressure at P due to gravitation is wz; and, as has just been proved, the intensity of pressure at P due to molecular forces is K, the intrinsic pressure. Hence the total intensity of pressure is
$$wz + K.$$

Now it is known that K is enormously great: Young estimated its value for water at 23000 atmospheres, while Lord Rayleigh (*Phil. Mag.*, Dec., 1890) mentions, with approval, a hypothesis of Dupré which leads to the value 25000 atmospheres for water. The hypothesis is that the value of K is deducible from the dynamical equivalent

312 *Hydrostatics and Elementary Hydrokinetics.*

of the latent heat of water; that evaporation may be regarded as a process in which the cohesive forces of the liquid are overcome. Now the heat rendered latent in the evaporation of one gramme of water = 600 gramme-degrees (about), or $600 \times 42 \times 10^6$ ergs, and 1 atmosphere = 10^6 dynes per square centimètre; hence $K = 25000$ atmospheres (about).

If this is so, the question must naturally present itself to the student: what becomes of our ordinary expressions for the liquid pressure exerted on one side of an immersed plane area? Instead of being merely $A\bar{z}w$, must it not be very vastly greater—in fact $A(\bar{z}w+K)$? And moreover it should always act practically at the centre of gravity of the area.

We shall see, however, by considering closely the nature of molecular forces, that this large pressure does not influence in any way the value of the pressure exerted by a liquid on the surface of an immersed body. Let us revert to Fig. 82 and consider the result arrived at in Art. 69. This result may be stated thus: at all points on the surface which terminates a liquid—whether this be a free surface or a surface of contact with any foreign body—there is a resultant force intensity due to molecular actions; this diminishes rapidly as we travel inwards along the normal to the surface, and vanishes after a certain depth has been reached.

If we consider a slender normal canal of any length, PR (Fig. 82) at the boundary of a liquid, this canal will experience *from the surrounding liquid itself* a resultant force acting inwards along its axis PR; this force is due to molecular actions and the imperfect surrounding of points near the liquid boundary (as explained in Art. 69), and its effects are felt along only a very small length PP' (or ϵ) of the canal. If the boundary of the liquid at P is plane, and

Molecular Forces and Capillarity. 313

the canal has a cross-section σ, this resultant inward molecular force on the canal is $K\sigma$.

Now let Fig. 84 represent a vertical plane, AB, immersed in a liquid having a portion, at least, of its surface, LM, horizontal, and let us consider the pressure exerted per unit area on this plane AB at P at the left-hand side. At P take an infinitesimal element of area, σ, and on it describe a horizontal canal of any length PQ, closed by a vertical area at Q. Consider the equilibrium of the liquid in this canal.

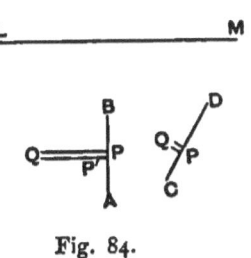

Fig. 84.

Now since AB is a foreign body, there is a termination to the liquid along the surface AB, and hence there will be resultant molecular force exerted by the liquid at points on and very near AB. Hence if along the canal PQ we take the length $PP' = \epsilon$, the liquid in PQ experiences a resultant molecular force from the surrounding liquid, of magnitude $K \cdot \sigma$, this force acting from P towards Q and being confined to the length PP'. In addition, the solid plane AB exerts a certain attraction, $a \cdot \sigma$, on the liquid in the canal, together with a certain pressure, $q \cdot \sigma$, on it. Finally, at Q the canal experiences the pressure $(wz + K)\sigma$ from the liquid. Hence we have

$$(wz + K)\sigma = K\sigma + (q - a)\sigma,$$
$$\therefore q - a = wz,$$

which shows that K disappears. Now the action of the canal on the plane at P is exactly $(q - a)\sigma$ in the sense QP, and this action is that which is described in ordinary language as the pressure on the plane at P.

If the immersed plane is inclined, as at CD, the resultant action of the liquid on the plane at P on the element of surface σ is seen in the same way to be $wz\sigma$, by considering the equilibrium of a canal PQ normal to the plane, PQ being equal to ϵ, the radius of molecular activity.

Laplace is somewhat obscure on the subject of the action between a liquid and an immersed plane (see Supplement to Book X, *Mécanique Céleste*, p. 41). Thus he says: the action experienced by the liquid in the canal PQ is equal, 1°, to the action of the fluid on this canal, and this action is equal to K; 2°, to the action of the plane on the canal; 'but this action is destroyed by the attraction of the fluid on the plane, and there cannot result from it in the plane any tendency to move; for, in considering only reciprocal attractions, the fluid and the plane would be at rest, action being equal and opposite to reaction; these attractions can produce only an adherence of the plane to the fluid, and we can here make abstraction of them.' He is considering the action experienced by the canal at the extremity P where it touches the plane. But, in considering the forces exerted *on the fluid* by the plane, it does not seem allowable to balance any force exerted by the plane on the canal by an opposite force produced *on the plane* by the fluid.

According to the view which we have taken, the action which is commonly called the fluid pressure on the plane is, in reality, a difference action—the difference between a pressure proper and a molecular attraction between the fluid and the plane.

71. Liquid in contact with a Solid. Admitting the existence of molecular forces operative within infinitesimal distances, the surface of a liquid near its place of contact with a solid body must, in general, be curved, even when gravity is the only external force acting throughout the mass of the liquid.

For, let PAB, Fig. 85, represent the surface of a liquid in contact at P with the surface PQ of a solid body.

Consider the forces acting on a molecule at P. We have gravity in the vertical direction Pw; also the molecular

Molecular Forces and Capillarity. 315

forces exerted by the solid evidently produce a resultant along Pn, the normal to the solid at P; and the molecular forces of the fluid molecules adjacent to P produce a resultant, Pf, acting somewhere between the tangent plane to the liquid surface at P and the surface of the solid.

Now in all cases the resultant force, due to all causes, exerted on a molecule of a perfect fluid at its free surface must be normal to the surface. Hence the resultant of the forces Pn, Pw, and Pf will determine the normal to the fluid surface at P; and, in general, this resultant will not act along Pw, so that the surface of the fluid at P is not, in general, horizontal.

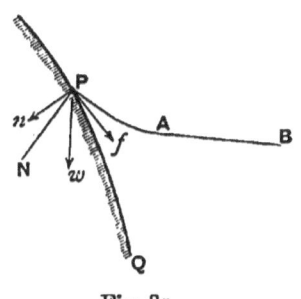

Fig. 85.

The form of the surface remote from the solid body is, of course, that of a horizontal plane; because at such points as A and B there are only two forces acting, viz., gravity and the molecular attraction, the latter of which is normal to the surface, and if the resultant of it and gravity is also normal, the force of gravity must act in the normal, i. e., the surface must be horizontal.

72. Application of Virtual Work. When, under the action of any forces whatever, a system of particles assumes a configuration of equilibrium, this configuration is signalised by the fact that if it receives, or is imagined to receive, any small disturbance whatever, the total amount of work done by all the forces acting on the various particles is zero.

We shall now apply this principle to the equilibrium of a liquid contained within an envelope ACB (Fig. 86), the

316 *Hydrostatics and Elementary Hydrokinetics.*

surface of the liquid being APB, and the forces acting being—

1°. molecular forces between particle and particle of the liquid,

2°. molecular forces between the envelope and the liquid,

3°. Any assigned system of external force.

Let m, m' denote indefinitely small elements of mass of the liquid at a distance r, and assume the force between them to be

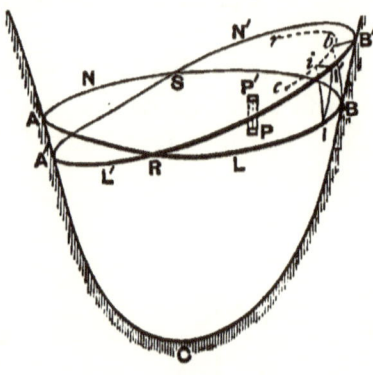

Fig. 86.

$$mm'f(r). \quad \ldots \quad \ldots \quad \ldots \quad (1)$$

Let μ denote an element of mass of the envelope, and m any element of mass of the liquid very close to μ, and assume the force between m and μ to be

$$m\mu F(r). \quad \ldots \quad \ldots \quad \ldots \quad (2)$$

The value of r in (1) must be $< \epsilon$, otherwise the force between the elements of mass would be zero; and r in (2) must be $< \epsilon'$, the radius of molecular activity for the solid and the fluid.

The virtual work of the force (1) is $-mm'f(r)dr$. Now if, as in Art. 69, we put

$$\int_r^\epsilon f(r)dr = \phi(r),$$

$$\int_r^{\epsilon'} F(r)dr = \psi(r),$$

Molecular Forces and Capillarity. 317

the total work done by the molecular forces for any system of small displacements is

$$\tfrac{1}{2}\delta \Sigma mm' \phi(r) + \delta \Sigma m\mu \psi(r), \quad \ldots \quad (3)$$

the summations extending to all pairs of elements between which the molecular force is exerted, and one-half of the result of the summation relating to pairs of liquid elements being taken, because this summation will bring in each term twice.

If V is the potential, per unit mass, at any point of the liquid where the element of mass dm is taken, the virtual work of the external forces is

$$\delta \int V dm. \quad \ldots \ldots \quad (4)$$

Hence the equation of virtual work for any system of displacements of the liquid elements is

$$\delta \left[\int V dm + \tfrac{1}{2}\Sigma mm' \phi(r) + \Sigma m\mu \psi(r) \right] = 0. \quad \ldots \quad (5)$$

It will be necessary, therefore, to calculate the functions

$$\Sigma mm' \phi(r) \text{ and } \Sigma m\mu \psi(r).$$

Since if we take any one element m of the liquid and perform the summation $\Sigma m' \phi(r)$ round it, the process is confined within the sphere of radius ϵ having m for centre, we may obviously put

$$\Sigma m' \phi(r) = L, \quad \ldots \ldots \quad (6)$$

L being a constant throughout the whole of the fluid contained in the vessel and bounded—not by the surface APB but—by the parallel surface $A'B'$ (see Fig. 82) which is at the constant distance ϵ below APB, and also by a surface inside the fluid parallel to that of the vessel at a distance ϵ from the surface of the vessel; for each element m within this space is completely surrounded by a liquid sphere of radius ϵ, while the liquid elements between the surfaces AB and $A'B'$, and between the vessel and the second surface named are not completely surrounded. Or, if we please, we may imagine the summation L to extend up to the bounding surface AB and to that of the vessel, and subtract a summation relating to a fictitious layer, $A''B''$, above AB of constant thickness ϵ, included between AB and $A''B''$ (Fig. 87), and a fictitious layer outside the surface of contact with the vessel, also of thickness ϵ.

Hence if M is the whole mass of the liquid, the summation can be expressed in the form

$$M \cdot L - \sigma\, mm'\, \phi(r). \quad \ldots \quad (7)$$

in which σ denotes a summation confined within the superficial layer, which is everywhere of the constant thickness ϵ, and which embraces the free surface and the surface of contact of the liquid with the vessel.

As regards the summation $\Sigma m \mu\, \psi(r)$, it is obviously confined between two surfaces each of which is parallel to the surface, ACB, one inside the liquid and the other inside the solid envelope, the distance between these surfaces being $2\epsilon'$.

Hence equation (5) becomes

$$\delta \left[\int V dm - \tfrac{1}{2} \sigma\, mm'\, \phi(r) + \Sigma m \mu\, \psi(r) \right] = 0. \quad . \quad (8)$$

Now we can easily see that the summation σ is proportional to the sum of the areas of the surface AB and that of contact with the vessel; for, if we draw any surface, XY, Fig. 87, parallel to AB and $A''B''$ within the fictitious surface layer above indicated, and at any point Q on XY take the element dm of mass, describing round Q a sphere of radius ϵ, the summation $dm \cdot \sigma m'\, \phi(r)$ will extend to the volume of the sphere included between AB and $A''B''$; and if z is the normal distance, Qp, of Q from AB, the summa-

Molecular Forces and Capillarity. 319

tion $\sigma m' \phi(r)$ can obviously be written $w \Pi(z)$; also if dS is a small element of area of XY at Q, we can take

$$dm = w\, dS\, dz,$$

where w is the mass per unit volume of the fluid; hence we have for this element of the fluid the term $w^2 dS . \Pi(z) dz$.

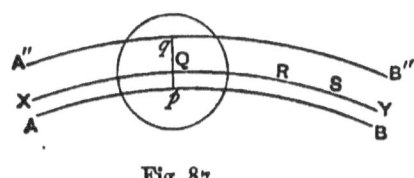

Fig. 87.

Now we can make a summation from p to q along a cylinder whose cross-section is everywhere dS if the radii of curvature of the surfaces AB, XY, $A''B''$ are infinitely greater than ϵ. The result of this summation is

$$w^2 dS \int_0^\epsilon \Pi(z)\, dz,$$

the definite integral being the same at all points, p, of AB. If the definite integral is denoted by A, and we then sum the results all over AB, we have AwS, where S is the area of AB; and similarly for the part of σ which extends over the surface of the vessel.

In the same way it is obvious that the term $\Sigma m\mu\psi(r)$ is proportional to the product of the densities of the envelope and the liquid and to the area, Ω, of their surface of contact. We may therefore write (5) in the form

$$\delta\left[\int V dm - \frac{k}{2}.(S+\Omega) + \lambda.\Omega\right] = 0, \quad . \quad . \quad . \quad (9)$$

where k and λ are constants which depend on the densities of the liquid and solid and on the laws of intermolecular action.

A result of the same form will obviously hold if the

320 *Hydrostatics and Elementary Hydrokinetics.*

density of the liquid varies both at the surface AB and at the surface of contact with the envelope, provided that the thickness of the stratum of variable density near AB is everywhere the same, and the same at all points along each surface, XY, parallel to AB; and similarly for the stratum near the envelope.

Hence, then, the work done by the molecular forces for any imagined displacement is entirely superficial, and its two parts are proportional to the small increments of the area of the surface AB of the liquid and the surface of contact with the envelope.

We shall now calculate δS and $\delta \Omega$.

In Fig. 86 the new surface, $A'L'I'B'N'$, of the fluid (resulting from the imagined small disturbance of the fluid) can be considered as consisting of two parts: firstly, the portion bounded by the curve $cibr...$ which is formed by the feet of the normals to this new surface or to the old one (since they differ infinitely little in position) drawn at all the points $A, L, I, B, ...$ of the contour $ALIBN$; and secondly, the small strip included between the contour $A'L'B'N'$ and the curve $cibr...$; so that δS is the area of this strip plus the excess of the first of these portions over the area of the old surface of the fluid. Also $\delta \Omega$ is represented in the figure by the surface $BILRI'B'N'SB$ minus the surface $ARL'A'SNA$, each of these lying on the interior of the vessel.

A simple geometrical investigation of the first part of δS is as follows: at any point P (Fig. 86) on the old surface of the fluid draw the two *principal* normal sections, PQ, PJ (Fig. 88) of the surface, and the normal, $PC_1 C_2$, to the surface at P; take the element of area, $PQFJ$, determined by the elements $PJ = ds_1$ and $PQ = ds_2$; this is an infinitesimal rectangle since ds_1 and ds_2 are at right angles.

Let C_1 and C_2 be the centres of curvature of the principal sections, and $PC_1 = R_1, PC_2 = R_2$, the radii of curvature of these sections.

Produce the normals to the surface at the points $PQFJ$ to meet the new surface of the fluid in $P'Q'F'J'$, and denote PP' by δn. Then we determine the small rectangular area $P'Q'F'J'$ on the new surface, and the excess of this above $PQFJ$ when integrated over the whole of the old surface is the first part of δS. The figure assumes that the *concave* side of the surface is turned towards the liquid.

Fig. 88.

Now
$$P'J' = \left(1 + \frac{\delta n}{R_1}\right) PJ,$$

$$P'Q' = \left(1 + \frac{\delta n}{R_2}\right) PQ,$$

therefore if $dS =$ area $PQFJ$, and $dS + \delta dS =$ area $P'Q'F'J'$, we have
$$\delta dS = \left(\frac{1}{R_1} + \frac{1}{R_2}\right) \delta n\, dS.$$

Hence the first part of δS is

$$\int \left(\frac{1}{R_1} + \frac{1}{R_2}\right) \delta n\, dS. \quad \ldots \quad (10)$$

To find the second part of δS, i. e., the sum of all such elements as $iI'B'b$, let $d\omega$ be the element of area $II'B'B$ of the interior of the vessel, and let θ be the angle between the tangent plane to the surface of the liquid at B and the tangent plane to the surface of the vessel at B, i. e., the angle between the plane $iI'B'b$ and the plane $II'B'B$; then

$$\text{area } iI'B'b = \cos\theta \cdot d\omega. \quad \ldots \quad (11)$$

322 *Hydrostatics and Elementary Hydrokinetics.*

Hence the second part of δS is

$$\int \cos \theta \, d\omega \quad \ldots \ldots \quad (12)$$

taken all round the curve of contact of the fluid and the vessel.

Then, we have

$$\delta S = \int \left(\frac{1}{R_1} + \frac{1}{R_2}\right) \delta n \, dS + \int \cos \theta \, d\omega ; \quad . \quad (13)$$

and (9) becomes

$$\delta \int V dm - \frac{k}{2} \int \left(\frac{1}{R_1} + \frac{1}{R_2}\right) \delta n \, dS$$
$$+ \frac{1}{2} \int (2\lambda - k - k \cos \theta) \, d\omega, \quad (14)$$

since $\delta \Omega$ is obviously the integral of all such elements as $II'B'B$, i.e., $\int d\omega$.

Now observe, however, that (14) is the equation of Virtual Work irrespective of the condition that the volume of the fluid remains the same after displacement as before. The excess of the new volume over the old is obviously the sum of such prismatic elements of volume as that contained between the areas $PQFJ$ and $P'Q'F'J'$ (Fig. 88) whose expression is $\delta n \, dS$, added to the sum all round the curve $ARLIBNA$ of such wedge elements as $IibBB'I'$. The area of this wedge is $\frac{1}{2} \delta n_0 . \cos \theta \, d\omega$, if δn_0 denotes Ii, the normal distance between the new and the old surface of the fluid at any contour point, I; and hence the sum of the wedges will add nothing to the contour integral

$$\int (2\lambda - k - k \cos \theta) \, d\omega$$

in (14), since each element of this sum is an infinitesimal compared with $d\omega$.

Molecular Forces and Capillarity. 323

Hence, the whole volume of fluid being constant,
$$\int \delta n\, dS = 0. \quad \ldots \quad \ldots \quad (15)$$

We know from the principles of the Lagrangian method (*Statics*, vol. ii., chap. xv.) that the condition of unchanged volume combined with the principle of Virtual Work is expressed by multiplying the left-hand side of (15) by an arbitrary constant, h, and adding it to the left-hand side of (14). Hence, then, the complete equation is

$$\delta \int V dm + \int \left\{ h - \frac{k}{2} \left(\frac{1}{R_1} + \frac{1}{R_2} \right) \right\} \delta n\, dS$$

$$+ \frac{1}{2} \int (2\lambda - k - k \cos \theta)\, d\omega = 0. \quad (16)$$

We may finally simplify the term $\delta \int V dm$. It means simply the variation of the potential of the external forces due to changed configuration of the liquid; and this variation is due merely to the two wedges $BILRI'B'N'SB$ and $ARL'A'SNA$, being positive for one and negative for the other. The type of the variation is the variation for the element of mass contained in the small prism PP', Fig. 86, that is $w \delta n\, dS$; so that if V is the potential of the external forces (per unit mass) at any point P on the surface of the liquid, the work of these forces for any small change of configuration is
$$w \int V \delta n\, dS, \quad \ldots \quad \ldots \quad (17)$$

and therefore (16) finally becomes

$$\int \left\{ wV + h - \frac{k}{2} \left(\frac{1}{R_1} + \frac{1}{R_2} \right) \right\} \delta n\, dS$$

$$+ \frac{1}{2} \int (2\lambda - k - k \cos \theta)\, d\omega = 0. \quad (18)$$

The first integral is one extended over the surface of the

liquid, and the second is one relating only to its contour, i. e., its bounding curve $ALIBSNA$.

Now, owing to the perfectly arbitrary displacement of every point on the surface, each element of the first integral must vanish, and hence at every point of the surface we have

$$wV + h - \frac{k}{2}\left(\frac{1}{R_1} + \frac{1}{R_2}\right) = 0, \quad \ldots \quad (19)$$

which is the equation of the surface.

Every element, also, of the contour integral must vanish, and hence at all points of contact of the surface of the liquid with the vessel

$$\cos\theta = \frac{2\lambda - k}{k}, \quad \ldots \quad (20)$$

which shows that *the liquid surface is inclined at the same angle to the surface of the vessel all round.* The angle θ is called the *angle of contact* of the liquid and the solid, which we shall definitely suppose to be the angle contained between the normal to the liquid surface drawn into the substance of the liquid and the normal to the solid drawn into the substance of the solid.

If $\lambda > k$, the angle of contact is imaginary, and equilibrium of the liquid in the vessel is impossible.

If the *convex* side of the surface is turned towards the liquid, we shall have

$$P'J' = \left(1 - \frac{\delta n}{R_1}\right) PJ,$$

$$P'Q' = \left(1 - \frac{\delta n}{R_2}\right) PQ,$$

$$\therefore \delta dS = -\left(\frac{1}{R_1} + \frac{1}{R_2}\right) dS,$$

and (19) is replaced by

$$wV + h + \frac{k}{2}\left(\frac{1}{R_1} + \frac{1}{R_2}\right) = 0. \quad \ldots \quad (21)$$

If the density of the liquid is not constant (owing to the variable molecular pressure) in the layers near the surface, it will be the same at all points on a surface parallel to AB (Fig. 82) at a distance $< \epsilon$ from AB; and hence it is obvious that the virtual work of the molecular forces for any small displacement will still be proportional to the variation, δS, of area of the surface, but the value of the constant k, will not be the same as on the supposition of constant density. The equation of Virtual Work, will, then, be still of the form (9), and the results (19) and (20) will still hold.

If above the surface AB there is another fluid, virtual work of its external and molecular forces will give terms of the same form as before, as will be shown in a subsequent article.

If at each point of the free surface of a liquid there is an external pressure whose intensity at a typical point on the surface is p_0, the virtual work of this pressure must be brought into equation (9) or (16). This virtual work is obviously $-\int p_0 \delta n \, dS$, so that the equation (19) of the free surface becomes

$$wV - p_0 + h - \frac{k}{2}\left(\frac{1}{R_1} + \frac{1}{R_2}\right) = 0. \quad \ldots \quad (22)$$

From (20) we see that the angle of contact of a liquid with a solid will be $< \frac{\pi}{2}$ if $\lambda > \frac{k}{2}$, i. e., the surface of the liquid will be convex towards the liquid at the place of contact. If the law of attraction between the liquid elements themselves is the same as that of attraction between a liquid element and an element of the solid in contact with it, ϕ and ψ in (5) differ only by constant multipliers; and we can state the last result thus: if the attraction of the solid on the fluid is greater than half the attraction of the fluid on itself, the surface of the fluid will at the curve of contact be convex towards the fluid; and if $\lambda < \dfrac{k}{2}$, θ will be $> \dfrac{\pi}{2}$, i.e., the surface of the fluid will be concave towards the fluid. We shall subsequently see that in the case in which a capillary tube dips into a liquid which is under the action of gravity, the liquid must rise in the tube in the first case, and fall in the second. These results were first enunciated by Clairaut.

The experimental determination of the angle of contact of a liquid and a solid has been made by means of the measurement of a large drop, $OCyD$, Fig. 89, of the liquid placed on a horizontal plane, AB, made of the solid. If the drop is a very large one, it is virtually a plane surface at its highest point O. Suppose the figure to represent a vertical section. Then at any point P the two principal sections are the meridian curve PO and the section made by a plane perpendicular to the plane of the paper through the normal to the curve at P. The curvature of this section may be neglected in the case of a large drop; and if ρ is the radius of curvature of the meridian at P, we have from (12) of Art. 69 the intensity of molecular pressure at P equal to $K + \dfrac{T}{\rho}$, where T is a constant. At O the intensity of molecular pressure is K, and if the depth, Pm, of P below

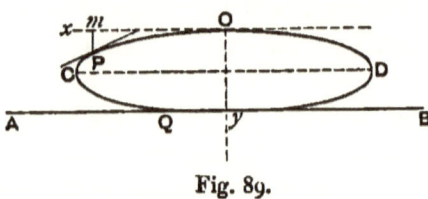

Fig. 89.

the tangent Ox is y, the intensity of pressure at P is also $K + wy$, by transmission from O. Hence

$$wy = \frac{T}{\rho}. \qquad (23)$$

If the arc $OP = s$ and θ is the angle made by the tangent at P with Ox, $\frac{1}{\rho} = \frac{d\theta}{ds}$, and $\sin\theta = \frac{dy}{ds}$; hence (23) becomes

$$wy\,dy = T\sin\theta\,d\theta,$$

$$\therefore wy^2 = 2T(1 - \cos\theta). \qquad (24)$$

Let the angle of contact at Q be i, let $Oy = a$; then

$$wa^2 = 2T(1 + \cos i). \qquad (25)$$

Let CD be the equatorial section of the drop; then at C we have $\theta = \frac{\pi}{2}$, and if the depth, b, of C below O is measured, we have

$$wb^2 = 2T. \qquad (26)$$

This last gives T, which is called (see Art. 76) the surface tension of the liquid in contact with air; and then (25) gives

$$\cos i = \frac{a^2}{b^2} - 1, \qquad (27)$$

which determines the angle of contact.

The above arrangement is suitable in the case of a drop of mercury.

To find the angle of contact between water and any solid body, a somewhat similar method has been employed. Imagine Fig. 89 to be inverted, and suppose AB to be the horizontal surface of a mass of water (which then occupies the lower part of the figure). Along this surface fix a plane

328 *Hydrostatics and Elementary Hydrokinetics.*

of the given substance, and under this plane insert a large bubble of air, $QCOD$, the lowest point of the bubble being O.

Then, exactly as before, by measuring the thickness, Oy, of the bubble, and the depth of C below AB, we obtain the angle of contact at Q between the water surface and that of the solid (the water surface being bounded by air). If $a = Oy$, and $b =$ the vertical height of C above O, we have, as before

$$\tfrac{1}{2} w b^2 = T,$$

$$\tfrac{1}{2} w a^2 = T(1 + \cos i),$$

i being the angle of contact CQA.

It is very difficult, if not impossible, to find a definite value of the angle of contact between a given liquid and a given solid, because any contamination or alteration of either surface during the experiment will affect the result. Thus, the angle of contact between water and glass is often said to be zero, while some experimenters quote it at 26°. Again; it is known that in the case of mercury and glass the angle varies with the time during which they are in contact: at the beginning of an experiment the angle was found to be 143° and some hours afterwards 129°.

73. Analytical investigation of general case. The expression (13) of Art. 72 can be analytically deduced from the general theory of the displacements of points on any unclosed surface. Thus if, as in *Statics*, vol. ii., Art. 291, we denote the components of displacement of any point (x, y, z) by u, v, w, and if at any point, P, Fig. 86, on the surface we put $\dfrac{dz}{dx} = p$, $\dfrac{dz}{dy} = q$, $\epsilon = \sqrt{1 + p^2 + q^2}$, we know that the change, δdS, of the infinitesimal area dS at P is given by the equation

Molecular Forces and Capillarity.

$$\delta dS = \left\{\epsilon\left(\frac{du}{dx} + \frac{dv}{dy}\right) + \frac{p\delta p + q\delta q}{\epsilon}\right\} dx\, dy. \quad . \quad . \quad (1)$$

Also (*Statics*, Art. 283)

$$\delta p = \frac{dw}{dx} - p\frac{du}{dx} - q\frac{dv}{dx}, \quad . \quad . \quad . \quad . \quad (2)$$

$$\delta q = \frac{dw}{dy} - p\frac{du}{dy} - q\frac{dv}{dy}. \quad . \quad . \quad . \quad . \quad (3)$$

Hence

$$\delta S = \iint \frac{1}{\epsilon}\left\{(1+q^2)\frac{du}{dx} + (1+p^2)\frac{dv}{dy} - pq\left(\frac{dv}{dx} + \frac{du}{dy}\right)\right.$$
$$\left. + p\frac{dw}{dx} + q\frac{dw}{dy}\right\} dx\, dy \quad (4)$$

Now, by the method of integration by parts, we have the double integral equal to

$$\int \frac{1}{\epsilon}\{(1+q^2)u\,dy - pq\,u\,dx + (1+p^2)v\,dx - pq\,v\,dy$$
$$+ pw\,dy + qw\,dx\}$$
$$- \iint\left\{u\left(\frac{d}{dx}\frac{1+q^2}{\epsilon} - \frac{d}{dy}\frac{pq}{\epsilon}\right) + v\left(\frac{d}{dy}\frac{1+p^2}{\epsilon} - \frac{d}{dx}\frac{pq}{\epsilon}\right)\right.$$
$$\left. + w\left(\frac{d}{dx}\frac{p}{\epsilon} + \frac{d}{dy}\frac{q}{\epsilon}\right)\right\} dx\, dy, \quad 5)$$

in which, of course, the single integral is one carried along the bounding curve, $ALBN$ (Fig. 86) of the surface, while the double integral is one carried over the surface itself.

Dealing with the double integral in (5) first, we easily find that the coefficient of u is

$$-p\frac{(1+p^2)t + (1+q^2)r - 2pqs}{\epsilon^3}, \quad . \quad . \quad (6)$$

where $r = \frac{dp}{dx}$, $t = \frac{dq}{dy}$, $s = \frac{dp}{dy}$.

But (Salmon's *Geometry of Three Dimensions*, Chap. XI) the multiplier of $-p$ in (6) is $\frac{1}{R_1} + \frac{1}{R_2}$. Similarly the coefficient of v in the double integral is $-q\left(\frac{1}{R_1} + \frac{1}{R_2}\right)$, and that of w is $\frac{1}{R_1} + \frac{1}{R_2}$; so that the double integral in (5) is

$$\iint \left(\frac{1}{R_1} + \frac{1}{R_2}\right)(pu + qv - w)\, dx\, dy. \quad . \quad . \quad . \quad (7)$$

But, u, v, w being the components of displacement of the point P, and $\dfrac{p,\, q,\, -1}{\epsilon}$ the direction-cosines of the normal to the surface at P, the projection of the displacement of P along the normal, which we have called δn, is given by

$$\delta n = \frac{pu + qv - w}{\epsilon}\, dx\, dy$$
$$= (pu + qv - m)\, dS.$$

Hence (7) is

$$\iint \left(\frac{1}{R_1} + \frac{1}{R_2}\right) \delta n\, dS,$$

as before found.

Dealing now with the single integral in (5), and carrying it *continuously* round the bounding curve, we see that the sign of every term in dx must be changed, as is fully explained in *Statics*, Art. 316 a.

Of course this integral is one which we may consider as carried round the projection on the plane xy of the bounding curve. Hence the correct form of the single integral is

$$\int \frac{1}{\epsilon} \{(1 + q^2)\, u\, dy + pq\, u\, dx - (1 + p^2)\, v\, dx - pq\, v\, dy$$
$$+ pw\, dy - qw\, dx\}, \quad (8)$$

Molecular Forces and Capillarity.

in which u, v, w are components of displacement of the points on the surface, Ω, of the solid body where it is intersected by the surface of the liquid; so that if $\dfrac{p', q', -1}{\epsilon'}$ are the direction-cosines of the normal to Ω at any point, I, we have now
$$w = p'u + q'v.$$

Also, the projection of the element $d\omega$ of area of the surface of the vessel at I on the plane of xy is $u\,dy - v\,dx$, so that
$$d\omega = \epsilon'(u\,dy - v\,dx).$$

And since dx, dy, dz are proportional to the direction-cosines of the element IB of the bounding curve which is at right angles both to the normal to the liquid and to the normal to the solid, we have $p\,dx + q\,dy - dz = 0$ and $p'dx + q'dy - dz = 0$, from which
$$(p-p')\,dx + (q-q')\,dy = 0.$$

Hence the terms multiplying $\dfrac{1}{\epsilon}$ in (8) are equivalent to $(pp' + qq' + 1)(u\,dy - v\,dx)$, i.e., to $\epsilon \cos\theta\, d\omega$; so that (8) is simply
$$\int \cos\theta\, d\omega,$$

as was found in (12), Art. 72.

(The alteration of signs in the terms of the single integral in (5) which is rendered necessary by the carrying of the integral by continuous motion in the same sense round the bounding curve $ALBN$, or its projection on the plane xy, is a circumstance which, perhaps, the student would be very likely to overlook.)

74. Liquid under the action of Gravity. In the particular case in which gravity is the only external force acting on a liquid which has air or any other gas above its surface, if z is the height of any point on the surface above a fixed horizontal plane, $V = -z$, and as w' is negligible compared with w, the equation (19) of the surface of the liquid is

$$wz + \frac{k}{2}\left(\frac{1}{R_1} + \frac{1}{R_2}\right) = \text{Constant}, \quad \ldots \quad (1)$$

from which it follows that if there are any points on the surface at which the concavity is turned towards the liquid, i.e., $\frac{1}{R_1} + \frac{1}{R_2}$ is positive, those points must be at a lower level than the points at which the surface is plane; and points where the surface is convex towards the liquid must be at a higher level than the plane portions.

Thus, supposing that two cylindrical capillary tubes, BC, FE, immersed vertically in a given liquid, and of such different materials that the surface of the liquid in one is concave, and in the other convex, towards the liquid, if L is plane from which z is measured and if R, R' are the radii of curvature of the surfaces of the liquid within the tubes at C and F, and ζ, ζ' the heights of these points above L, we must have

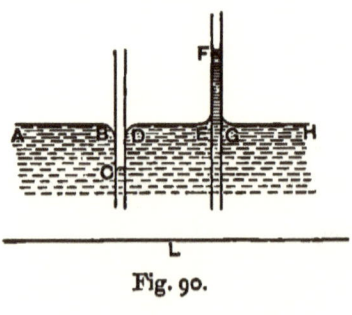

Fig. 90.

$$w\zeta + \frac{k}{R} = w\zeta' - \frac{k}{R'} = wz,$$

where z is the height of the plane portion, AB, DE, GH, of the liquid above L.

A simple experiment with water serves to illustrate the result that if in a continuous body of this liquid there is a part of the surface plane and another convex towards the liquid, this latter must be at a higher level than the former.

Molecular Forces and Capillarity. 333

Let a large glass vessel be connected with a capillary tube, t, Fig. 91, and let water be poured continuously into the large branch. A stage will be reached at which the water in t will just reach the top of this tube, and then the surface of the water is acb. If the glass is quite clean, this surface will have its tangent planes vertical round the rim ab; and the level of the water in the large branch will be at C, which is lower than ab. As water continues to be poured in, the surface in t will become quite horizontal and represented by the right line ab, and the surface in the other branch will be AB, which is at the same level as ab. Continuing to pour in water, the surface at the top of t will become *concave* downwards, as represented by adb, and then the level in the other branch is at D, which is higher than d.

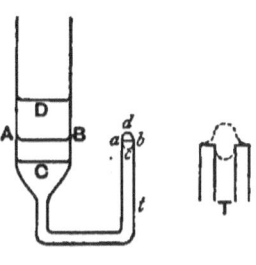

Fig. 91.

A side-figure at T shows how the surface of the liquid can be adb. The horizontal edges at the top of t are of appreciable breadth, and when the water rises above the line ab, the surface of the glass is the horizontal rim of the tube t (and the angle of contact being 0°) the surface of the water at the rim lies horizontally.

75. Rise or fall of Liquids in Capillary Tubes. It is a well known fact that if a tube of very small diameter is plunged into a mass of liquid contained in a vessel, the level of the liquid in the vessel will not, in general, be the same as its level in the narrow tube. What is the cause of this? To say that it is 'capillary attraction' is to use an expression which is at once inaccurate and vague. To say that it is molecular attraction is to use an expression which is true but vague. This was evidently in the mind of

Laplace when he said (Supplement to Book X, p. 5) that the attraction of capillary tubes has no influence on the elevation or depression of the fluids which they contain, except in determining the inclination of the surface of the fluid to the surface of the tube along the curve of intersection of the free surface with the tube, and thereby determining the curvature of the free surface. That the angle of contact does determine the curvature of the surface when this surface is inside a very narrow tube is obvious.

For, suppose that the angle of contact for glass and a certain liquid is 45°, and that the liquid is contained in a vertical glass tube one-tenth of a millimètre in diameter; then it is evident that the free surface of the liquid within the tube must be very much curved, because its tangent planes where it meets the tube must all be inclined at 45° to the vertical, while its tangent plane at its vertex must be horizontal; and in order that such a great amount of change in the direction of the tangent plane should be possible, the surface must be very much curved.

Now, great curvature of surface means great intensity of molecular pressure, if the surface is concave towards the liquid, and small intensity if the surface is convex towards the liquid (Art. 69).

Hence, owing merely to the fact that within a very narrow tube, the free surface of a liquid is curved—and not to any special action due to the *narrowness* of the tube—this liquid must rise or fall within the tube below the level of the plane portions in any vessel into which the narrow tube dips.

Let FE, Fig. 90, be a capillary tube dipping into a vessel containing a liquid such that the angle of contact (as defined in Art. 72) for the liquid and the tube is $< \dfrac{\pi}{2}$.

Molecular Forces and Capillarity. 335

In this case the surface is concave upwards, and therefore the intensity of molecular pressure at F is $K - \frac{k}{R}$, where R is the radius of curvature of the liquid surface at the lowest point of the surface at F (where the two radii R_1, R_2 are evidently equal), and the liquid must rise in the tube until the intensity of pressure due to the weight of the column FE added to this molecular intensity produces the intensity of pressure which exists along DE. If p_0 is the intensity of atmospheric pressure, and $EF = z$, the intensity of pressure inside the tube at the level E is $p_0 + wz + K - \frac{k}{R}$; and the intensity of pressure along the plane surface DE is $p_0 + K$; hence

$$z = \frac{k}{wR} \quad \quad \quad \quad (1)$$

determines the height to which the liquid rises in the tube.

Let i be the angle of contact of the liquid with the tube and $r =$ internal radius of tube; then $R = r \sec i$, very nearly; hence

$$z = \frac{k \cos i}{wr}, \quad \quad \quad \quad (2)$$

and the weight of the liquid raised in the tube above E is

$$\pi r k \cos i. \quad \quad \quad \quad (3)$$

Equation (2) shows that the heights to which the same liquid rises in capillary tubes of the same substance are inversely proportional to the diameters of the tubes.

If the tube is such that the angle of contact is $> \frac{\pi}{2}$, the free surface within the tube is concave towards the liquid, and therefore the intensity of molecular pressure is greater than at the plane surface DE. Hence the liquid must be depressed in the tube, as represented in BC, and the amount of depression is calculated as above.

76. Surface Tension of a Liquid. The amount of rise or fall of a liquid, under the action of gravity, in a capillary tube is usually calculated by means of the introduction of the notion of *surface tension*. The free surface of the liquid is considered to be in a state of tension resembling that of a stretched surface of indiarubber—with, however, this important difference, that, whereas the tension of the indiarubber surface increases if the surface is further increased, the tension of the liquid surface remains absolutely constant whether the surface expands or contracts.

Let $ABCD$, Fig. 92, represent a part of the bounding surface of a liquid; let any line QPR be traced on it; and along this line draw small lengths Qq, Pp, Rr into the liquid and normal to the surface. Consider now the action exerted over the area $RrpqQP$ by the liquid at the right side on that at the left.

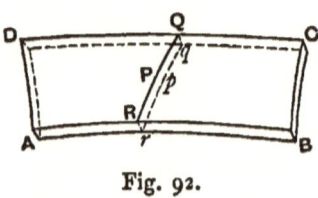

Fig. 92.

One part of this action will consist of molecular attraction, towards the right; and if the depth of the line qpr below QPR is nearly equal to ϵ (Art. 68) or greater than ϵ, another part of the action will consist of pressure, towards the left, in the lower parts of the area $QqrR$. If qpr instead of being at an infinitely small depth is at any finite depth, the molecular attraction exerted across any of the lower portions of the area $QqrR$ is exactly balanced by the molecular pressure on such portions. But if qpr is at a depth *very* much less than ϵ, the molecular pressure (towards the left) on any part of the area $QqrR$ is negligible,

Molecular Forces and Capillarity. 337

and we may consider the portion of liquid at the right of the line QR as simply exerting a pull or tension on the portion at the left. Observe that on the line QPR itself the intensity of molecular pressure is zero, and that if qpr is at a distance infinitely less than ϵ from QPR, although the intensity of pressure on any portion of the area $QqrR$ is infinitely small, the force of molecular attraction exerted across this area may be large. The value of the pressure depends on the force by a *differential* relation $\dfrac{d\varpi}{dn} = wF$ (see (2) of Art. 68), where F is the intensity of molecular force in the direction of the normal to the surface of the liquid at any point, P; and we know that at P, where F is greatest, ϖ is zero.

Hence at points of the imagined surface $QqrR$ of separation which are infinitely near to the surface we are to imagine the stress to be merely tension; at points whose distances from the surface become comparable with ϵ we are to imagine this molecular pull, or attraction, as accompanied by a contrary pressure; and at points which are at and beyond the distance ϵ from the surface, the molecular pull is balanced by the molecular pressure.

Hence, however far the imagined surface $QqrR$ extends into the liquid, the whole stress exerted on the liquid at the left by that at the right is confined to an action which terminates at a curve, qpr, at the depth ϵ, this action being a mixed one consisting of molecular attraction and an opposing molecular pressure which latter grows in intensity from zero at the surface to a maximum value at the depth ϵ.

At any point P on the surface the amount of this stress (which is, on the whole, a tension) *per unit length of the curve PQ is called the surface tension exerted across the curve PQ by the liquid at the right on that at the left of PQ.*

z

It is obvious that the amount of the stress per unit length is the same across *all* curves drawn at P on the surface, and is normal to these curves. This is another point of difference between the surface tension of a liquid and the stress of an elastic membrane in general; for, in the latter the stress exerted across any curve PQ at P is not, in general, normal to PQ, nor is it of constant magnitude for all curves drawn in the membrane at P.

Now, although it is obvious that, if we grant the existence of molecular forces, we must admit the existence of this mixed surface stress (i. e., within a layer of thickness ϵ at the surface) no such stress has explicitly presented itself in our investigation, by means of Virtual Work, of the conditions of equilibrium of the liquid. This fact, however, involves no difficulty or contradiction; for, in taking the molecular actions exerted between all possible pairs of elements of mass, we are sure of having omitted no forces that act; but in this way surface tension (which is obviously a resultant, and not a simple, action) could not have specially presented itself.

Knowing now of the existence of this stress, we can see why the terms in the expression for the Virtual Work of the molecular forces, (9), Art. 72, consist of constants multiplied by the changes of the areas $S+\Omega$, and Ω.

For, if a surface of area A is subject to a tension T which is the same at all points and of constant intensity in all directions round a point, the work done by the stress in an increase δA of the area is

$$-T \cdot \delta A.$$

A liquid contained in a vessel, or resting as a drop on a table, is sometimes spoken of as having a 'skin' within which the liquid proper is contained.

A drop of water hanging from the end of a tube and

Molecular Forces and Capillarity. 339

ready to fall is spoken of as being contained in an 'elastic bag.' Of course, if the surface of the liquid is oxidised, contaminated by foreign particles of any kind, or in any way rendered different from the liquid below the surface, we may, if we please, say that the pure liquid is contained within a surface which is not pure liquid; but even such a contaminated surface is radically different from an elastic bag, for the magnitude of the tension in a stretched bag increases with the stretch of the bag, whereas the tension of the bounding layer of the liquid does not. In no case—either that of a perfectly pure liquid or that of a liquid with an oxidised or contaminated surface—is there any skin or bag. In the case of a liquid with a pure surface there is no material thing at the surface which there is not everywhere else in the liquid; and we must not imagine that, because we see a drop of water hanging, a globule of mercury lying on a table, or a column of water with a concave surface standing in a capillary tube above the level of the water outside, such conditions require bags for enclosing the liquids or skins by means of which to catch hold of them. We can assure ourselves that molecular forces, with special circumstances near the surface (owing to incomplete surrounding of molecules, &c.) will amply account for all such forms of equilibrium.

The height to which a liquid rises in a capillary tube may be calculated by the introduction of surface tension.

For, in Fig. 93, let the tube $ABB'A'$ have any form (not necessarily cylindrical); let l be the length of the curve of contact of the liquid surface at BB' with the surface of the tube, let T be the surface tension of the liquid, and i the angle of contact with the tube. Then consider the equilibrium of the column in the tube above the level, OCx, of the plane portion outside.

We may suppose the layer of particles round the tube at

B which are in actual contact with the tube as exerting the tension T per unit length of the curve l on the particles just outside them; hence these supply an upward vertical force $Tl \cos i$ on the column $BOO'B'$. If σ is the area of the cross-section of the tube, and $z =$ height of B above O, the weight of the liquid is $w\sigma z$. There is the downward atmospheric pressure, $p_0 \sigma$, at B, and an upward pressure at O consisting of $p_0 \sigma$ and of a molecular part, $K\sigma$, and there is finally a downward molecular attraction exerted on those particles in the tube which stand on the area σ at OO' and are contained within the distance ϵ from σ.

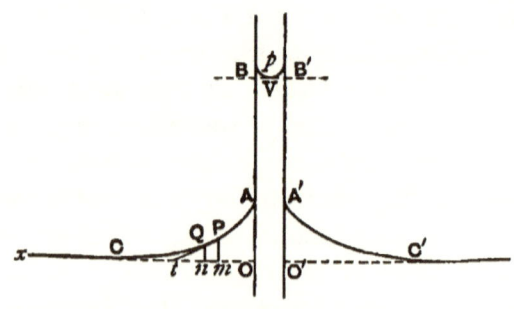

Fig. 93.

Now this downward attraction is precisely equal to $K\sigma$ (Art. 69), so that this force balances the upward pressure $K\sigma$, and we have for the equilibrium of the contained column
$$Tl \cos i = w\sigma z.$$

If the tube is cylindrical,
$$l = 2\pi r,\ \sigma = \pi r^2,$$
and we have
$$z = \frac{2 T \cos i}{wr},$$
as in (2), Art. 75.

It is obvious that $\frac{k}{2}$ in the general equation (9), Art. 72, is T, the surface tension.

By taking any element of area of the curved surface BVB', the principal radii of curvature of this element being R_1 and R_2, and considering the equilibrium of the vertical cylinder described round the contour of this element, we at once deduce (1) of Art. 74; for, if dS is the area of the element, the component, along its normal, of the surface tension all round dS is $T\left(\frac{1}{R_1} + \frac{1}{R_2}\right)dS$, and the vertical component of this is $T\left(\frac{1}{R_1} + \frac{1}{R_2}\right)d\sigma$, if $d\sigma$ is the horizontal projection of dS; also the weight of the column is $wzd\sigma$; therefore

$$wz - T\left(\frac{1}{R_1} + \frac{1}{R_2}\right) = 0.$$

Let the capillary tube be replaced by two very close parallel vertical plates, AB, $A'B'$.

Then, considering the equilibrium of the column $BOO'B'$ of unit thickness perpendicular to the plane of the figure, we have
$$2T\cos i = wzd,$$

where d is the distance OO' between the plates; hence

$$z = \frac{2T\cos i}{wd},$$

which shows that the liquid rises twice as high in a cylindrical tube as between two parallel plates whose distance is equal to the diameter of the tube.

The existence of surface tension in a liquid may be shown experimentally in many ways, of which we select two. Take a rectangle formed of brass strips or wires, AB, BC, CD, and EF (Fig. 94) of which the first three form one rigid piece, while the last, *EF*, is capable of sliding up and down on the bars *AB* and *DC*. The space, *abcd*, enclosed by the bars being vacant, dip the system into a solution of soap in water, thus forming a film (represented by the shading) in this space. This film attaches itself to all the bars; and if the moveable bar *EF* is not restrained by the hand, it will be drawn along the others by the film until it

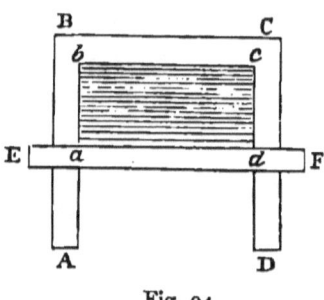

Fig. 94.

reaches *BC*. If the system is held in a vertical plane, *EF* being below *BC*, the former will be raised, in opposition to gravity, if it is not too heavy.

As a second example, take a circular brass wire, A, Fig. 95, dip it into the soap solution, thus covering its area on withdrawal with a thin film (represented by theshading); then form a loop of a piece of thread and place it gently on the surface of the film. This loop is represented by *ab*. Now per-

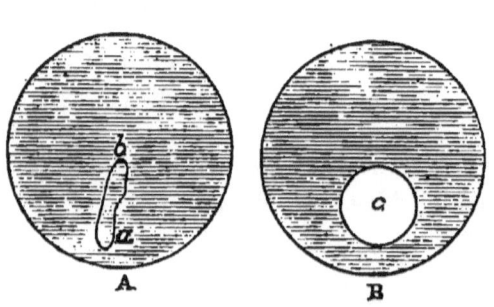

Fig. 95.

forate the film inside the loop by a pin, and we shall see the loop of thread instantly drawn out into a circle, *c*, by the contracting film. By the formation of this circle the area of the remaining film is the least that it can be—a result which is necessary because, the virtual work of the tension being $T\delta A$, the static energy of the film

Molecular Forces and Capillarity. 343

is proportional to its area, and every material system which is subject to given conditions assumes as a configuration of stable equilibrium one in which the static energy of its forces is least.

The following table, taken from Everett's *Units and Physical Constants*, gives the values of a few surface tensions in dynes per linear centimètre at the temperature 20° C:

	Tension of surface separating the liquid from		
	Air	Water	Mercury
Water	81	0	418
Mercury	540	418	0
Bisulphide of Carbon	32·1	41·75	37·5
Alcohol	25·5	.....	399
Olive Oil	36·9	20·56	335
Petroleum	31·7	27·8	284

from which it is seen how large the surface tension is for mercury as compared with other liquids.

77. Two Liquids and a Solid. To illustrate further the application of the principle of Virtual Work, take the case in which two liquids, w, w' (Fig. 96), are in contact with each other over a surface Λ and with a solid body. The liquid w is contained within the space represented by $ABDC$, and the second within $CDB'A'$. Let S be the area of the free surface of the first, and Ω the area of the solid in contact with this liquid; and S' and Ω' be the corresponding things for the second liquid.

Then, exactly as in Art. 72, the Virtual Work of all

the forces acting will reduce to the sum of terms relating to the bounding surfaces alone; and since the whole bounding surface of the liquid w is $S+\Lambda+\Omega$, the virtual work of its own molecular forces will give the term

$$-\frac{k}{2}\delta(S+\Lambda+\Omega);$$

and we see that the equation of virtual work is

$$\delta\int Vdm + \delta\int V'dm' - \frac{k}{2}\delta(S+\Lambda+\Omega) - \frac{k'}{2}\delta(S'+\Lambda+\Omega')$$
$$+ l\delta\Lambda + \mu\delta\Omega + \mu'\delta\Omega' = 0, \quad (1)$$

where the term $l\delta\Lambda$ relates to the molecular forces exerted at the surface Λ between particles of w and particles of w'; and $\mu\delta\Omega$ relates to the forces between the particles of the liquid w and the solid.

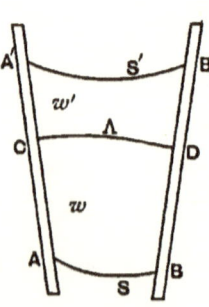

Fig. 96.

Now, denoting by δn, $\delta n'$ elements of normals at points on S, S' drawn outwards from the liquids; $\delta\nu$ an element of normal of Λ drawn outwards from the liquid w; $d\omega$ an element of Ω at the intersection of S and Ω (as in Art. 72), $d\psi$ an element of Ω at the intersection of Ω and Λ; $d\omega'$ an element of Ω' at the intersection of Ω' and S', and by θ, χ, θ' the angles of contact with the solid at S, Λ, S', we have, exactly as shown in Art. 72,

$$\delta\Omega = \int d\omega + \int d\psi,$$
$$\delta\Omega' = \int d\omega' - \int d\psi,$$

$$\delta S = \int \left(\frac{1}{R_1} + \frac{1}{R_2}\right) \delta n\, dS + \int \cos\theta\, d\omega,$$

$$\delta S' = \int \left(\frac{1}{R'_1} + \frac{1}{R'_2}\right) \delta n'\, dS' + \int \cos\theta'\, d\omega',$$

$$\delta \Lambda = \int \left(\frac{1}{\rho_1} + \frac{1}{\rho_2}\right) \delta v\, d\Lambda + \int \cos\chi\, d\psi,$$

$$\delta \int V\, dm = \int V \delta n\, dS + \int V \delta v\, d\Lambda,$$

$$\delta \int V'\, dm' = \int V' \delta n'\, dS' - \int V' \delta v\, d\Lambda,$$

where ρ_1, ρ_2 are the principal radii of curvature at any point on Λ.

To the left-hand side of (1) must be added the terms

$$h \int \delta n\, dS + h' \int \delta n'\, dS' + h'' \int \delta v\, d\Lambda,$$

which are rendered necessary by the constancy of the volumes of the liquids in the supposed displacement of the system.

As before, the coefficients of δn, δv, $\delta n'$ must each be zero, as also the coefficients of the terms relating to $d\omega$, $d\omega'$, and $d\psi$. Hence, for example, we have the equation

$$V + h - \frac{k}{2}\left(\frac{1}{R_1} + \frac{1}{R_2}\right) = 0$$

at all points of S, and similar equations at all points of S' and Λ; and at all points of meeting of S and Ω

$$\frac{k}{2}(1 + \cos\theta) - \mu = 0,$$

which proves the constancy of the angle of contact at such points; a similar result holding for S' and Ω', while the terms relating to $d\psi$ give

$$(k - k' - 2l)\cos\chi = 2(\mu - \mu') - k - k',$$

346 *Hydrostatics and Elementary Hydrokinetics.*

which proves the constancy of the angle of contact between the solid and the common surface of the liquids.

78. Drop of Liquid on another Liquid. Let Fig. 97 represent a drop of one liquid resting on the surface of another, the area of contact being Λ, the free surface of the drop being S' and that of the supporting liquid S.

If the sides of the vessel in which this liquid is contained are very distant from the drop, in considering a small deformation of the system and applying the equation (1), Art. 77, of Virtual Work, we may neglect terms relating to Ω; so that the equation is

$$\delta \int V dm + \delta \int V' dm' - \frac{k}{2}\delta(S+\Lambda) - \frac{k'}{2}\delta(S'+\Lambda) + l\delta\Lambda = 0. \quad (1)$$

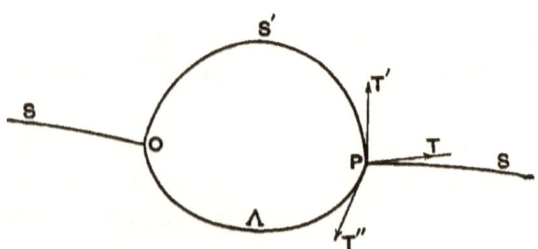

Fig. 97.

This equation will, as has already been seen, give equations satisfied at all points of S, S', Λ, as well as equations relating to their common bounding curve. Considering merely the latter for the present, we may take

$$-\frac{k}{2}\delta(S+\Lambda) - \frac{k'}{2}\delta(S'+\Lambda) + l\delta\Lambda = 0, \quad . \quad . \quad (2)$$

in which the variations of the surfaces are only those portions at their common boundary curve; or

$$\frac{k}{2}\delta S + \frac{k'}{2}\delta S' + \left(\frac{k+k'}{2} - l\right)\delta \Lambda = 0, \quad \ldots \quad (3)$$

or again
$$T\delta S + T'\delta S' + T''\delta \Lambda = 0, \quad \ldots \quad (4)$$

where T, T', T'' are the surface tensions in the surfaces S, S', Λ.

Now when any surface Σ having for bounding edge a curve C receives a very small deformation whereby it becomes a surface Σ' having for bounding edge a curve C', the Calculus of Variations leads (see Arts. 72 and 73) to the result that the variation $\Sigma' - \Sigma$ is obtained by drawing normals to Σ all round the contour C, these normals being terminated by Σ' and enclosing a surface Ω on Σ', and then adding to $\Omega - \Sigma$ a linear integral taken all along the curve C, the elements of this linear integral involving the displacements of points on C to their new positions on C'. The term $\Omega - \Sigma$ is

$$\int \left(\frac{1}{R_1} + \frac{1}{R_2}\right) \delta n \, dS,$$

while the term given by the line-integral along C is (8) of Art. 73.

Now take the case in which the figure of the drop and that of its submerged part are surfaces of revolution round the axis of z, and suppose Fig. 97 to represent the section of it made by the plane xz. Also let the displacement of the point P be confined to the plane xz, and let its components be u, w. Hence in (8) of Art. 73 we are to put $q = 0$, $v = 0$, and the terms of the linear integral which relate to the displacement of P are

$$\frac{1}{\epsilon}(u + pw)dy,$$

where dy relates to P and a point on the curve (a circle) which is the common bounding edge of the surfaces S, S', Λ, this circle being represented in projection on the plane of the figure by the right line OP. We may, then, omit dy; and the terms in (4) given by the three surface tensions are

$$\left(\frac{T}{\epsilon} + \frac{T'}{\epsilon'} + \frac{T''}{\epsilon''}\right)u + \left(\frac{Tp}{\epsilon} + \frac{T'p'}{\epsilon'} + \frac{T''p''}{\epsilon''}\right)w.$$

Now u and w are quite arbitrary and independent, and the coefficient of every independent displacement for each point involved in the line-integral must be zero. Hence

$$\frac{T}{\epsilon} + \frac{T'}{\epsilon'} + \frac{T''}{\epsilon''} = 0,$$

$$\frac{Tp}{\epsilon} + \frac{T'p'}{\epsilon'} + \frac{T''p''}{\epsilon''} = 0.$$

But if the tangent line to S in the plane of the figure makes the angle θ with the axis of x, we have

$$-\frac{1}{\epsilon} = \sin\theta \text{ and } \frac{p}{\epsilon} = \cos\theta;$$

similarly for the tangent lines to S' and Λ; so that these become

$$T\sin\theta + T'\sin\theta' + T''\sin\theta'' = 0,$$

$$T\cos\theta + T'\cos\theta' + T''\cos\theta'' = 0,$$

which plainly assert that three forces, T, T', T'', supposed acting along the tangents in the senses represented have no resultant; in other words, if a plane triangle is formed by three lines proportional to the surface tensions, the directions of the distinct surfaces of the two liquids and that of their common surface of contact are parallel to the sides of this triangle.

Molecular Forces and Capillarity. 349

Hence equilibrium of the drop is impossible unless each surface tension is less than the sum of the other two.

79. Liquid under no external forces. When a mass of liquid is in equilibrium under its own molecular forces only, its surface can assume several forms. In this case (19) of Art. 72 gives as the equation of the surface

$$\frac{1}{R_1} + \frac{1}{R_2} = \frac{1}{a}, \quad \ldots \ldots (1)$$

where a is a constant length

We shall confine our attention to surfaces which are of revolution, and we shall suppose the axis of revolution to be taken as axis of x.

Now if at any point, P, of the revolving curve (Fig. 98) or meridian, PDE, which by revolution round the axis AB generates the surface, ρ is the radius of curvature and n is the length, Pn, of the normal terminated by the

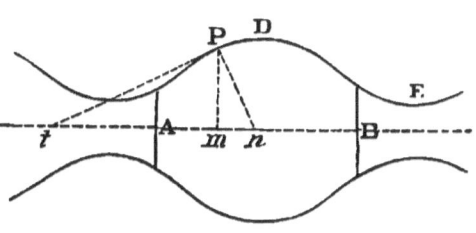

Fig. 98.

axis of revolution, the principal radii of curvature of the surface generated are ρ and n, so that (1) becomes

$$\frac{1}{\rho} + \frac{1}{n} = \frac{1}{a}. \quad \ldots \ldots (2)$$

Now let the tangent at P make the angle θ with the axis of x, and let ds be an element of arc measured along the curve from P towards D; then $\rho = -\dfrac{ds}{d\theta}$, and $n = y \sec\theta$; hence (2) becomes

$$-\frac{d\theta}{ds} + \frac{\cos\theta}{y} = \frac{1}{a}; \quad \ldots \ldots (3)$$

350 *Hydrostatics and Elementary Hydrokinetics.*

or, since $\dfrac{d\theta}{ds} = \sin\theta \dfrac{d\theta}{dy}$,

$$\dfrac{1}{y} \cdot \dfrac{d}{dy}(y\cos\theta) = \dfrac{1}{a}, \quad \ldots \ldots (4)$$

$$\therefore y\cos\theta = \dfrac{y^2}{2a} + h, \quad \ldots \ldots (5)$$

where h is a constant.

We may observe that if the constant $\dfrac{1}{a}$ is zero, (1) gives as the property of the surface of the fluid that at every point the two principal radii are equal and opposite; the two principal sections have their concavities turned in opposite directions. If the surface is one of revolution, this property at once identifies it with the surface generated by the revolution of a catenary round its directrix, and the surface is called a *catenoid*.

Before proceeding to integrate (5), we can show that all the curves satisfying it are generated by causing conic sections to roll, without sliding, along the axis AB: the curves satisfying (5) are the loci traced out by the foci of these rolling conics. For, if $Pn = n$, (5) gives

$$y^2 \left(\dfrac{1}{n} - \dfrac{1}{2a}\right) = h. \quad \ldots \ldots (6)$$

Now if p is the perpendicular from the focus of an ellipse on the tangent at any point, and r the distance of this point from the focus, we have

$$p^2 \left(\dfrac{1}{r} - \dfrac{1}{2a}\right) = \dfrac{b^2}{2a}, \quad \ldots \ldots (7)$$

a and b being the semiaxes. Comparing (7) with (6), we see that P in Fig. 98 is the focus of an ellipse touching AB at n, the semiaxes being a and $\sqrt{2ah}$, and this ellipse is therefore invariable whatever be the position of P on the meridian. The locus of P when the rolling conic is an ellipse is called the *unduloid*, and is the locus PDE actually represented in the figure.

If the rolling conic is a parabola, the locus of the focus is a catenary, which gives by revolution the catenoid.

If the rolling conic is a hyperbola, the locus is a curve having a series of loops, and the surface which it generates is called a *nodoid*.

Since $\tan \theta = \dfrac{dy}{dx}$, we have from (5)

$$dx = \pm \frac{y^2 + 2ah}{\sqrt{4a^2 y^2 - (y^2 + 2ah)^2}} \cdot dy \quad \ldots \quad (8)$$

$$= \pm \frac{y^2 + 2ah}{\sqrt{-y^4 + 4a(a-h)y^2 - 4a^2 h^2}} \cdot dy \quad \ldots \quad (9)$$

$$= \pm \frac{y^2 \pm 2a\beta}{\sqrt{(a^2 - y^2)(y^2 - \beta^2)}} \cdot dy, \quad \ldots \quad (10)$$

by putting $a^2 + \beta^2 = 4a^2 - 4ah$, and $a^2 \beta^2 = 4a^2 h^2$; so that a and β are the greatest and least values of the ordinate.

Equation (10) is best integrated by expressing y, and therefore x, in terms of a variable angle ϕ; thus, let

$$y^2 = a^2 \cos^2 \phi + \beta^2 \sin^2 \phi, \quad \ldots \quad (11)$$

$$\therefore \quad y = a \sqrt{1 - k^2 \sin^2 \phi} \quad \ldots \quad (12)$$

where $k^2 = \dfrac{a^2 - \beta^2}{a^2}$. This gives, if $\Delta \phi \equiv \sqrt{1 - k^2 \sin^2 \phi}$,

$$dx = \left\{ a \Delta(\phi) \pm \frac{\beta}{\Delta(\phi)} \right\} d\phi. \quad \ldots \quad (13)$$

When $\phi = 0$, $y = a$, and when $\phi = \dfrac{\pi}{2}$, $y = \beta$; so that if

D and *E* are the points of maximum and minimum ordinate, all points between them on the curve are given by values of ϕ between 0 and $\frac{\pi}{2}$.

In the common notation of elliptic integrals

$$\int_0^\phi \Delta(\phi)\, d\phi = E(\phi), \quad \int_0^\phi \frac{d\phi}{\Delta(\phi)} = F(\phi), \quad . \quad . \quad (14)$$

therefore we have the abscissa and ordinate of every point on the curve expressed in terms of the variable ϕ by the equations

$$x = a E(\phi) \pm \beta F(\phi), \quad . \quad . \quad . \quad . \quad . \quad (15)$$

$$y = a \Delta(\phi). \quad . \quad . \quad . \quad . \quad . \quad . \quad . \quad (16)$$

In the unduloid the tangent can never be parallel to the axis of y, i. e., $\frac{dx}{dy}$ can never be zero, so that in (10) the sign + in the numerator belongs to this curve, and therefore in (15) the signs ± belong, respectively, to the unduloid and the nodoid.

In the unduloid $\frac{d^2x}{dy^2} = 0$ when $\tan \phi = \left(\frac{a}{\beta}\right)^{\frac{1}{2}}$, or $y = \sqrt{a\beta}$, and this gives the point of inflexion on the curve.

If *s* is the length of the arc between *D* and any point *P*, we have $s = (a + \beta)\phi$.

When $a = \beta$, the surface generated becomes a cylinder.

When a is very slightly greater than β, the generating curve becomes, approximately, a *curve of sines*.

The case of a liquid unacted upon by any external forces was realised by M. Plateau by inserting a drop of olive oil in a mixture of water and alcohol arranged so as to have the same specific gravity as the oil. By seizing this drop between two wires in the shapes of any closed curves, or by allowing it to form round a solid of any shape held in the water-alcohol mixture, we can obtain a large number of liquid surfaces each satisfying the common equation (1) of such surfaces.

We shall subsequently see that the same surfaces can be produced by means of soap-bubbles instead of large masses of liquids.

A full account of all such experiments will be found in Plateau's celebrated work, *Statique Expérimentale et Théorique des Liquides soumis aux seules Forces Moléculaires*.

80. Liquid under action of Gravity. Taking now the case in which the only external force is gravity, the equation of its surface is of the form

$$\frac{1}{R_1} + \frac{1}{R_2} = \frac{z-h}{a^2}, \quad \ldots \ldots \quad (1)$$

where h and a' are constant lengths, and z is the height of any point on the surface above a fixed horizontal plane; also $a^2 = \dfrac{T}{w}$, where T is the surface tension.

We shall begin by investigating the form of the surface of a liquid in contact with a broad vertical plane, or wall.

Let this plane be supposed normal to the plane of the paper, and let Fig. 93 represent the section of the plane and the liquid surface made by the plane of the paper (supposed also vertical), this section being far removed from the edges of the immersed vertical plane BAO. Of the two principal radii of curvature of the liquid surface at any point P one will be infinite, since one principal section at P is the right line through P perpendicular to the plane of the paper, and the other will be ρ, the radius of curvature of the curve APC.

Taking the axis of x horizontal and the axis of y vertical, we replace z in (1) by y, and the equation becomes

$$\frac{1}{\rho} = \frac{y-h}{a^2}, \quad \ldots \ldots \quad (2)$$

A a

354 *Hydrostatics and Elementary Hydrokinetics.*

which shows that the curve APC belongs to the class of *elastic curves*, i. e., those formed by a thin elastic rod which when free from strain was straight, but under the action of terminal pressures is bent. (See *Statics*, vol. ii., Art. 306.)

Since at a considerable distance from the wall the surface is plane and $\rho = \infty$, we see that if we measure y from the level the plane portion, the equation is

$$\rho y = a^2. \qquad \qquad (3)$$

Let Ox be the plane level, which is now taken as axis of x.

Putting $p = \dfrac{dy}{dx}$, which in the figure is $-$ the tangent of the inclination of the tangent at P to the axis of x, we have

$$\frac{1}{\rho} = \frac{p\dfrac{dp}{dy}}{(1+p^2)^{\frac{3}{2}}}$$

and a first integral of (3) gives

$$\frac{-2}{\sqrt{1+p^2}} = \frac{y^2}{a^2} + C, \qquad \qquad (4)$$

where C is a constant. Since $p = 0$ when $y = 0$, $C = -2$, and then we have

$$\frac{2a^2 - y^2}{y\sqrt{4a^2 - y^2}} \cdot dy = \pm dx. \qquad \qquad (5)$$

Putting $\qquad\qquad y = 2a \sin \phi, \qquad \qquad (6)$

we have $\qquad \left(\dfrac{1}{\sin \phi} - 2 \sin \phi\right) d\phi = \pm \dfrac{dx}{a}. \qquad (7)$

Now if the angle of contact is acute (as represented

in the figure), $\dfrac{dx}{dy}$ can never vanish, and $\dfrac{dy}{dx}$ is continuously negative. Hence y is always $< a\sqrt{2}$, and the negative sign in (5) is the one to be used. In this case the integral of (7) is

$$-\frac{x}{a} = 2\cos\phi + \log_e \tan\frac{\phi}{2}. \quad \ldots \quad (8)$$

This shows that the plane surface of the liquid is in reality asymptotic to the curve APC, because when $\phi = 0$, $x = \infty$.

If the angle of contact is zero, we have $\dfrac{dy}{dx} = \infty$ at A, $\therefore y = a\sqrt{2} = OA$.

If i is the angle of contact, $\dfrac{dy}{dx} = -\cot i$ at A, and we have from (4)
$$OA = a\sqrt{2(1-\sin i)}, \quad \ldots \quad (9)$$
$$= \sqrt{\frac{2T}{w}(1-\sin i)}, \quad \ldots \quad (10)$$

which determines the height to which the liquid rises against the plate; and, if i is known, by measuring this height the surface tension T can be found.

The equation (3) can be immediately deduced by elementary principles from the notion of surface tension. For, let Q be a point on the curve indefinitely near P; draw the ordinates Pm, Qn, and consider the equilibrium of the prism of liquid $PmnQ$ of unit breadth perpendicular to the plane of the paper. We may consider this prism as kept in equilibrium by the surface tensions, each equal to T, at P and Q, and its weight, the atmospheric pressure cancelling at the top and bottom. Now the vertical upward component of T at P is $T\sin\theta$, and the vertical component at Q is $T\sin\theta + ds \cdot \dfrac{d}{ds}(T\sin\theta)$; hence

$$-T\,ds \cdot \frac{d}{ds}(\sin\theta) = wy\,dx,$$

$$\therefore\ -T\cos\theta\,\frac{d\theta}{ds} = wy\,\frac{dx}{ds},$$

$$\therefore\ \frac{T}{\rho} = wy,$$

since $\rho = -\dfrac{ds}{d\theta}$, and $\cos\theta = \dfrac{dx}{ds}$. This equation is the same as (3), since $\dfrac{T}{w} = a^2$.

The integration of (3) may be effected in another way which gives the intrinsic equation of the curve. It can be written

$$-\frac{d\theta}{ds} = \frac{y}{a^2}, \qquad \ldots \qquad (11)$$

$$\therefore\ -\frac{d^2\theta}{ds^2} = \frac{1}{a^2}\frac{dy}{ds} = -\frac{1}{a^2}\sin\theta \ . \ . \ . \ (12)$$

$$\therefore\ \left(\frac{d\theta}{ds}\right)^2 = C - \frac{2}{a^2}\cos\theta. \qquad \ldots \qquad (13)$$

Now $-\dfrac{1}{\rho} = \dfrac{d\theta}{ds} = 0$ when $\theta = 0$, therefore $C = \dfrac{2}{a^2}$, and we have

$$\frac{d\theta}{ds} = -\frac{2}{a}\sin\frac{\theta}{2}, \qquad \ldots \qquad (14)$$

$$\therefore\ \log_e \tan\frac{\theta}{4} = -\frac{s}{a} + C,$$

where C is a constant. Now at A we have $\theta = \dfrac{\pi}{2} - i$, therefore

$$\tan\frac{\theta}{4} = e^{-\frac{s}{a}} \tan\left(\frac{\pi}{8} - \frac{i}{4}\right) \qquad \ldots \qquad (15)$$

is the intrinsic equation of the curve.

Pass now to the case in which two large parallel vertical plates, BO, $B'O'$ (Fig. 93), are immersed, very close together in a liquid. Let BVB' be the curve in which the liquid surface between them is intersected by the plane of the figure, V being the lowest point, or vertex, of the curve.

One of the principal radii of curvature at every point, p, of this surface is still ∞, and the other is ρ, the radius of curvature of the curve BVB' at the point. Hence, measuring the height of p above the plane surface Ox, we have still.
$$\rho y = a^2;$$
and if θ is the inclination of the tangent at p to the horizon and $s = Vp$,
$$\rho = \frac{ds}{d\theta}, \quad \sin\theta = \frac{dy}{ds};$$
therefore
$$\frac{d^2\theta}{ds^2} = \frac{\sin\theta}{a^2},$$
$$\therefore \frac{d\theta}{ds} = \frac{\sqrt{2}}{a}\sqrt{C-\cos\theta}, \quad \ldots \quad (16)$$
where C is a constant. Hence
$$y = a\sqrt{2}\sqrt{C-\cos\theta}; \quad \ldots \quad (17)$$
and if h is the height of V above OO',
$$h = a\sqrt{2}\sqrt{C-1}. \quad \ldots \quad (18)$$
An approximate value of h has been already found (p. 341). If the abscissa of p with reference to V as origin is x, we have $\dfrac{dx}{ds} = \cos\theta$; therefore
$$dx = \frac{a}{\sqrt{2}}\frac{\cos\theta\, d\theta}{\sqrt{C-\cos\theta}} \quad \ldots \quad (19)$$

358 *Hydrostatics and Elementary Hydrokinetics.*

Substituting for C in terms of h from (18), we have

$$y = 2a\left(\frac{h^2}{4a^2} + \sin^2\frac{\theta}{2}\right)^{\frac{1}{2}} \quad \ldots \quad (20)$$

$$dx = \frac{a}{2} \cdot \frac{\cos\theta\, d\theta}{\left(\frac{h^2}{4a^2} + \sin^2\frac{\theta}{2}\right)^{\frac{1}{2}}} \quad \ldots \quad (21)$$

The value of x can be expressed in terms of the ordinary elliptic integrals by putting $\theta = \pi - 2\phi$. If then we put $k^2 = \frac{4a^2}{4a^2 + h^2}$, so that k is < 1, we have

$$dx = \frac{2a}{k}\left\{\frac{k^2-2}{2}\frac{d\phi}{\Delta(\phi)} + \Delta(\phi)\, d\phi\right\}, \quad \ldots \quad (22)$$

where, as usual, $\Delta(\phi) \equiv \sqrt{1 - k^2 \sin^2\phi}$. The limits of θ being 0 and $\frac{\pi}{2} - i$, where i is the angle of contact, those of ϕ are $\frac{\pi}{2}$ and $\frac{\pi}{4} + \frac{i}{2}$. The figure supposes the angle of contact to be acute, as when water rises between two glass plates; if it is obtuse, as when mercury is depressed between two glass plates, the discussion proceeds in the same manner.

Two plates close together in a liquid move towards each other, as if by attraction, whether the liquid rises or is depressed between them—as was first explained by Laplace. In all such cases of approach between bodies floating on a liquid the result is due to an excess of pressure on their backs, or remote faces, over the pressure on their adjacent faces. Thus, on the plate OB above B the intensity of pressure is the same on both sides, being that of the atmosphere; also below O the intensities are the same, and

Molecular Forces and Capillarity. 359

again at both sides of OA; but between A and B the intensity of pressure on the left side is that of the atmosphere while on the right it is less. For if on the surface AB between A and B we take any point R, at a height z above Ox, the intensity of pressure exerted by the liquid on the plane is $p_0 - wz$, where $p_0 =$ atmospheric intensity, since it has been shown in Art. 70 that the intrinsic pressure K disappears. Hence the total pressure on the plane AB from left to right is less than that from right to left; and similarly for $A'B'$; therefore the planes approach. The same result follows if (as in Fig. 90) the liquid is depressed between both planes.

But if the liquid rises in contact with one plane and is depressed in contact with the other, the two plates are urged away from each other.

Suppose the liquid to rise in contact with the plate AB (Fig. 99) and to fall in contact with the plate $A'B'$; then the level of the liquid at the left of the first must be higher than that at the right; and the depression of the liquid at the right of the second plate is greater than at the left. Evidently, then, the portion AB of the first plate experiences an excess of pressure towards the left, while, the pressure at the left of $B'A'$ being greater than that of the atmosphere, the second plate experiences an excess of pressure towards the right. Thus the plates move away from each other.

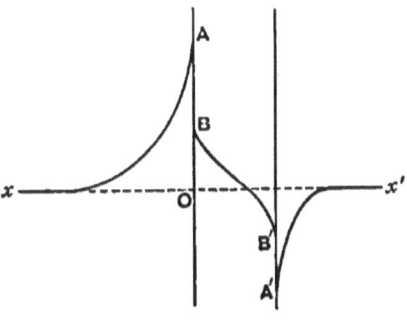

Fig. 99.

This can be shown experimentally by placing on the

surface of water two small pieces of cork, one of which is clean and the other greased.

Next, taking the case in which a liquid rises inside a narrow vertical cylindrical tube, the free surface of the liquid inside the tube will be one of revolution; and if y is measured from the line Ox, which is the level of the plane surface, the equation of the surface of the liquid is

$$\frac{1}{\rho} + \frac{1}{n} = \frac{y}{a^2}, \quad \ldots \ldots (23)$$

where ρ is the radius of curvature of the meridian at any point, and n the length of the normal between this point and the axis of the tube.

If the horizontal line through the lowest point, V, of the meridian (Fig. 93) is taken as axis of x, the equation becomes

$$\frac{1}{\rho} + \frac{1}{n} = \frac{h+y}{a^2}; \quad \ldots \ldots (24)$$

and if we put $\dfrac{d\theta}{ds}$ and $\dfrac{\sin\theta}{x}$ for $\dfrac{1}{\rho}$ and $\dfrac{1}{n}$, this becomes

$$\frac{1}{x}\frac{d}{dx}(x\sin\theta) = \frac{h+y}{a^2}. \quad \ldots \ldots (25)$$

This equation cannot be integrated accurately; but an approximate solution can be obtained by the following method, which is, in principle, the same as that employed by Laplace (Supplement to Book X of the *Mécanique Céleste*).

Take a circle having its centre on the vertical through V and having a radius c; and let us determine this circle in such a way that its ordinate (for any abscissa) differs very little from the ordinate of the point, p, on the curve of the liquid which has the same abscissa. The ordinate of the circle is given by the equation

$$y = b - \sqrt{c^2 - x^2},$$

so that for any point on the curve we have

$$y = b - \sqrt{c^2 - x^2} + \zeta, \quad \ldots \quad (26)$$

where ζ is a very small quantity. This gives

$$\frac{dy}{dx} = \frac{x}{\sqrt{c^2 - x^2}} + \frac{d\zeta}{dx}, \quad \ldots \quad (27)$$

$$\therefore x \sin \theta = \frac{x^2}{c} + \frac{x(c^2 - x^2)^{\frac{3}{2}}}{c^3} \frac{d\zeta}{dx}, \quad \ldots \quad (28)$$

neglecting the square of $\dfrac{d\zeta}{dx}$. Hence (25) becomes

$$\frac{d}{dx} \left\{ \frac{x^2}{c} + \frac{x(c^2 - x^2)^{\frac{3}{2}}}{c^3} \frac{d\zeta}{dx} \right\} = \frac{hx + xy}{a^2}$$

$$= \frac{h + b - \sqrt{c^2 - x^2}}{a^2} \cdot x, \quad (29)$$

by neglecting ζ in the value of y. Integrating between $x = 0$ and $x = x$,

$$\frac{x^2}{c} + \frac{x(c^2 - x^2)^{\frac{3}{2}}}{c^3} \frac{d\zeta}{dx} = \frac{(h + b) x^2}{2 a^2} + \frac{(c^2 - x^2)^{\frac{3}{2}}}{3 a^2} - \frac{c^3}{3 a^2}, \quad (30)$$

$$\therefore \frac{d\zeta}{dx} = \left(\frac{h + b}{2 a^2} - \frac{1}{c} \right) \frac{c^3 x}{(c^2 - x^2)^{\frac{3}{2}}}$$

$$+ \frac{c^3}{3 a^2 x} - \frac{c^6}{3 a^2} \cdot \frac{1}{x(c^2 - x^2)^{\frac{3}{2}}}, \quad (31)$$

$$\therefore \zeta = \left(\frac{h+b}{2a^2} - \frac{1}{c} - \frac{c}{3a^2}\right) \frac{c}{\sqrt{c^2-x^2}}$$

$$+ \frac{c^3}{3a^2} \log_e (c + \sqrt{c^2-x^2}) + C, \quad (32)$$

where C is a constant to be determined.

Now this equation would make $\zeta = \infty$ when $x = c$, which would, of course, be absurd. Hence we must have

$$\frac{h+b}{2a^2} - \frac{1}{c} - \frac{c}{3a^2} = 0. \quad \ldots \quad (33)$$

Again, $y = 0$ when $x = 0$, $\therefore b - c + \zeta = 0$ at V, and (32) becomes

$$\zeta = c - b + \frac{c^3}{3a^2} \log_e \frac{c + \sqrt{c^2-x^2}}{2c}, \quad \ldots \quad (34)$$

so that from (26)

$$y = c - \sqrt{c^2-x^2} + \frac{c^3}{3a^2} \log_e \frac{c + \sqrt{c^2-x^2}}{2c}, \quad \ldots \quad (35)$$

which is the approximate equation of the curve when c is known. Now we know that at the points B, B', of contact with the tube $\dfrac{dy}{dx} = \cot i$, and therefore if r is the radius of the tube,

$$\cot i = \frac{r}{\sqrt{c^2-r^2}} \left\{ 1 - \frac{c^3(c - \sqrt{c^2-r^2})}{3a^2 r^2} \right\} \quad \ldots \quad (36)$$

which determines c; and b is then known from (33).

As a first approximation, (36) gives $c = r \sec i$ from which, more accurately,

$$c = r \sec i \left\{ 1 - \frac{r^2}{3a^2} \frac{\sin^2 i (1 - \sin i)}{\cos^4 i} \right\}. \quad \ldots \quad (37)$$

Molecular Forces and Capillarity. 363

Finally, take the case in which liquid is contained between two vertical plates which make with each other a very small angle. Let the plates be $Ay\,Ox$, $A'y\,Ox'$ (Fig. 100), intersecting in the vertical line Oy, and making with each other the very small angle ϵ, or xOx'. Let the curves in which the liquid surface intersects the plates be $yPQx$, $yP'Q'x'$. It is required to find the nature of these curves.

Take any two indefinitely near points, P, Q, on one curve; let the corresponding points on the other be P', Q', the lines PP', QQ' being normal to the plates and in the liquid surface; draw the ordinates Pm, Qn, &c., and consider the separate equilibrium of the small prism $PQ'm$.

If $Pm = y$, $Om = x$, the weight of this prism is $\epsilon wxy\,dx$, where $mn = dx$, and it is balanced by surface tension round the contour $PQQ'P'$. Let T be the surface tension, and θ the inclination of the tangent to the curve PQ at P to Ox.

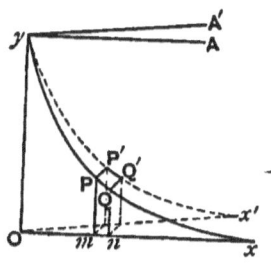

Fig. 100.

Then the amount of tension on PP' is $T.\epsilon x$, and its vertical component is $\epsilon Tx\sin\theta$; therefore the vertical component of the tension on QQ' is

$$\epsilon Tx\sin\theta + \epsilon T\frac{d(x\sin\theta)}{dx}dx.$$

Also the tension on PQ acts in the tangent plane to the liquid surface and at right angles to the line PQ. Now if i is the angle of contact (i.e., the angle between this tangent plane and the plate yOx), we easily find that the direction-cosines of the tangent plane are proportional to $\sin\theta$, $\cos\theta$, $\cot i$, while those of the line PQ are $\cos\theta$,

$-\sin\theta$, 0. Hence the direction-cosines of the line of action of the tension on PQ are $\sin\theta\cos i$, $\cos\theta\cos i$, $\sin i$; and if $PQ = ds$, the vertical component of the tension is $Tds \cdot \cos\theta\cos i$, or $T\cos i \cdot dx$; while the tension on $P'Q'$ gives the same amount. Hence for the equilibrium of the prism we have

$$2T\cos i = \epsilon w xy + \epsilon T \frac{d(x\sin\theta)}{dx}, \quad \ldots \quad (38)$$

which shows that the second term on the right-hand side is of the order ϵ^2.

Neglecting it in comparison with the first, we have for the approximate equation of the curve

$$xy = \frac{2T\cos i}{\epsilon w} = \frac{2a^2\cos i}{\epsilon}, \quad \ldots \quad (39)$$

where, as previously, $a^2 = \dfrac{T}{w}$.

The curve is, then, approximately a rectangular hyperbola —a result which is commonly assumed in virtue of the fact that the elevation of a liquid between two parallel close plates varies inversely as the distance between them.

81. Liquid Films. The forms which can be assumed by the surface of a liquid which is under the influence of none but molecular forces can be produced by means of thin films of liquid, such as soap-bubbles. Imagine a thin film of liquid in contact with air at both sides of its surface, the intensity of pressure of the air being, in general, different on these sides.

Let $ABCD$, Fig. 101, be a portion of such a film; let P be any point on its surface; let PQ, PS be elements of the arcs of the two principal sections of the surface at P; at Q and S draw the two principal sections QR and SR. Thus we determine a small area $PQRS$ on the surface.

Molecular Forces and Capillarity. 365

Let the normals to the surface at P and S intersect in C_2, and those at P and Q in C_1. Then $PC_1 = R_1$, $PC_2 = R_2$, where R_1, R_2 are the principal radii of curvature of the surface at P.

Let $PQ = ds_1$, $PS = ds_2$, p = intensity of air pressure on the lower or concave side of the surface at P, and p_0 = intensity of air pressure on the convex side. Then $(p-p_0)ds_1 ds_2$ is the resultant air pressure on the area $PQRS$ in the sense $C_1 P$; and for the equilibrium of the element this must be equal to the component of force in the sense PC, given by the surface tension exerted on the contour of the element, assuming that the film is so thin that the action of gravity is negligible. Now if T is the tension per unit length along PS, the whole tension on PS (which acts at its middle point perpendicularly to PS) is $T \cdot ds_2$, and the component of this along the normal to the surface is $Tds_2 \cdot \sin\dfrac{PC_1 Q}{2}$, or $Tds_2 \cdot \dfrac{ds_1}{2R_1}$. The tension on QR gives a component of the same magnitude; hence the sum of these is $\dfrac{T}{R_1} ds_1 ds_2$; similarly the sum of the normal components of the tensions acting on the sides PQ and SR is $\dfrac{T}{R_2} ds_1 ds_2$; so that the normal component of the tension acting on the whole contour $PQRS$ is

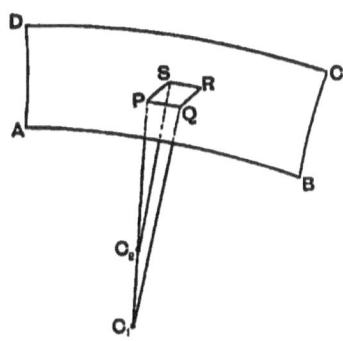

Fig. 101.

$$T\left(\frac{1}{R_1} + \frac{1}{R_2}\right) ds_1 ds_2. \quad \ldots \quad (1)$$

If the thickness of the film exceeds 2ϵ, where ϵ is the radius of molecular activity, there will be surface tension exerted both at the upper and at the under side of the surface, this action being confined (as explained in Art. 76) to a layer of thickness ϵ at each of these sides; so that we must replace T in (1) by $2T$, and the equation of equilibrium is

$$2T\left(\frac{1}{R_1} + \frac{1}{R_2}\right) = p - p_0. \quad \ldots \quad (2)$$

Hence, since p and p_0 are the same at all points of the film, the equation of its surface is

$$\frac{1}{R_1} + \frac{1}{R_2} = \text{const.}, \quad \ldots \quad (3)$$

and the forms of these films are the same as those of the surface of a liquid which is not acted upon by any external force, i.e., the shapes of thin films are the same as those of drops of oil in the water-alcohol mixture of Plateau (see p. 352).

The equation (2) can be otherwise deduced without considering the separate equilibrium of an element of the film. For, the intensity of pressure at any point inside the film (beyond the depth ϵ) due to the convex side is $p_0 + K + T\left(\frac{1}{R_1} + \frac{1}{R_2}\right)$; and the intensity of pressure at the same point due to the inner, or concave, side is

$$p + K - T\left(\frac{1}{R_1} + \frac{1}{R_2}\right).$$

Equating these, we have (2).

For a spherical bubble $R_1 = R_2 = r$,

$$\therefore 4T = (p - p_0)r, \quad \ldots \quad (4)$$

which shows that for all sizes of the bubbles the product $(p-p_0)r$ remains constant.

One possible shape of the bubble is that of a cylinder closed by two spherical ends. If r is the radius of the cylinder, r' that of each end, we have

$$\frac{2T}{r} = p-p_0,$$

$$\frac{4T}{r'} = p-p_0,$$

$$\therefore r' = 2r.$$

CHAPTER IX.

STEADY MOTION UNDER THE ACTION OF GRAVITY.

82. Steady Motion. When a fluid is in motion and we confine our attention to any point, P, in the space through which the fluid moves, it will be readily understood that the magnitude and direction of the velocity of the molecule which is passing through P at any instant may not be the same as the magnitude and direction of the velocity of the molecule which is passing through P at any other instant. If these should be the same at all instants, and if a like state of affairs prevails at all other points, the motion is said to be *steady*.

It is obvious, for example, that if a vessel is filled from a large reservoir of water, so that it is kept constantly full, while the liquid is allowed to flow out from an aperture made anywhere in the vessel, the motion at any fixed point in the vessel will be the same at all times.

83. Methods of Euler and Lagrange. It is at once obvious that the problem of the motion of a fluid acted upon by given forces may be attacked by two different methods. For, firstly, we may make it our aim to discover the condition of things—i. e., the magnitude and direction of the resultant velocity, and the pressure intensity—at each point, P, in space at any instant of time and at all times, and to do the same for all other points in the

Steady Motion under the Action of Gravity. 369

space through which the fluid moves, and thus, as it were, to obtain a map of the whole region—or a series of maps, if the motion is not steady—exhibiting the circumstances at each point as regards velocity and pressure.

Or, secondly, we may make it our aim to trace the path, and other circumstances, of each *individual molecule* throughout its whole motion.

The second object is much more difficult of attainment than the first, and, moreover, is not generally so desirable.

The first method is sometimes called the *statistical*, or the method of Euler; the second the *historical*, or the method of Lagrange.

84. Flow through a tube ; work of gravity. Suppose a column of water to occupy at any instant a length AB of a straight vertical tube of uniform cross-section, and let the end B of the tube be open.

If in a small element, Δt, of time a mass, Δm, of water flows out, what is the work done by gravity on the water during this interval?

Divide the tube by a series of very close horizontal planes, P, Q, R,... into sections such that the mass included between each adjacent pair is Δm. If Δs is the distance between the middle points of successive layers, PQ, QR, &c., while Δm flows out the middle point of each layer will fall through the height Δs, and the work done by the weight of this layer will be

$$\Delta m \cdot \Delta s \text{ or } g \Delta m \cdot \Delta s,$$

according as we use gravitation or absolute measure of

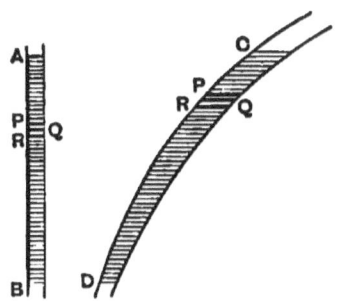

Fig. 102.

370 *Hydrostatics and Elementary Hydrokinetics.*

force. (If mass is measured in pounds and length in feet, the first expression gives the work in foot-pounds' weight, the second in foot-poundals.) If we use the first, the work is $\Sigma \Delta m \cdot \Delta s$, which can be written in either of the forms

$$\Delta m (\Delta s + \Delta s' + \Delta s'' + \ldots),$$

$$(\Delta m + \Delta m' + \Delta m'' + \ldots) \Delta s,$$

in which, of course, the successive distances $\Delta s, \Delta s', \Delta s'', \ldots$ are all equal, and the successive weights $\Delta m, \Delta m', \Delta m'', \ldots$ are also all equal.

Hence the work is

$$\Delta m \times AB,$$

or $M \times \Delta s,$

where M is the weight of the whole column.

The first expression shows that *the work done is the same as if the mass Δm which flows out at B fell through the height AB of the column.*

Precisely the same result holds if the shape of the tube is that represented in the right-hand figure. Let it be divided by close horizontal planes in the same manner as before. If now Δz is the *vertical* distance between the middle point of the layer, PQ, and the middle point of the next layer, QR, the work done in the descent of the first layer into the position of the second is $\Delta m \cdot \Delta z$, so that the work done by gravity on the whole tube of liquid while the quantity Δm flows out at D is

$$\Delta m \times z$$

in gravitation units, where z is the difference of level of the ends C and D.

This is again the same as the work of carrying Δm from C to D.

Steady Motion under the Action of Gravity. 371

85. Stream Lines. The actual path of a particle of a moving fluid is called a *stream line*. If at any point, A, Fig. 103, we describe a very small closed curve and at each point on the contour of this curve we draw the stream line, such as AP, and produce it indefinitely, we obtain a stream tube. When the fluid is a liquid, the mass contained between the normal sections of a tube at any two points, A, P, must always be the same; and therefore the same mass of fluid crosses every normal section of the tube per unit of time. Hence if v is the resultant velocity of the liquid at P and σ the area of the cross-section of the tube, the product

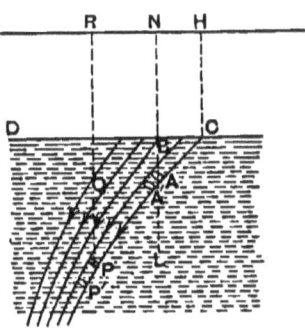

Fig. 103.

$$v\sigma$$

is constant all along the tube.

86. Theorem of Daniel Bernoulli. Consider at any instant the liquid contained in the stream tube between the normal sections at A and P, and suppose this liquid to occupy the volume $A'P'$ at the end of an infinitesimal element of time; let v_0, p_0, σ_0 be the velocity, pressure intensity, and cross-section of the tube at A; let v, p, σ be the same things at P; let z_0 and z be the depths of A and P below any fixed horizontal plane; let Δs_0 be the distance between the cross-sections at A and A', Δs being that between those at P and P'; and let $w =$ weight per unit volume of the liquid.

Now apply the equation of work and kinetic energy to the mass of liquid between A and P in the tube. The gain of kinetic energy in the small motion considered is

kinetic energy of $A'P'$ − kinetic energy of AP,

B b 2

in which, as the motion is steady, the kinetic energy of the portion $A'P$ is common to the two terms, and therefore disappears. Hence the gain of kinetic energy is

that of PP' — that of AA',

$$\text{or}\quad \Delta m \frac{v^2 - v_0^2}{2g}, \quad \ldots \ldots \quad (1)$$

where Δm = weight of PP' = weight of AA'.

The external forces doing work on the column of liquid considered are

gravity, the pressure at A, the pressure at P.

The work of gravity is

$$\Delta m \cdot (z - z_0), \quad \ldots \ldots \quad (2)$$

by Art. 84.

The pressure at A is $p_0 \sigma_0$, and its work $= p_0 \sigma_0 \cdot \Delta s_0$; the pressure at P is $p\sigma$, and its work $= -p\sigma \cdot \Delta s$. Hence the work of the pressure is

$$-\frac{\Delta m}{w} \cdot (p - p_0), \quad \ldots \ldots \quad (3)$$

since $\sigma \cdot \Delta s = \sigma_0 \cdot \Delta s_0$, i.e., the volume PP' = the volume $AA' = \dfrac{\Delta m}{w}$.

Equating (1) to the sum of (2) and (3), we have

$$\frac{v^2}{2g} + \frac{p}{w} - z = \frac{v_0^2}{2g} + \frac{p_0}{w} - z_0, \quad \ldots \quad (4)$$

in other words, since A and P are any two points along the stream line,

$$\frac{v^2}{2g} + \frac{p}{w} - z = C \quad \ldots \ldots \quad (5)$$

Steady Motion under the Action of Gravity. 373

at every point of the stream line, C being a constant for the stream line chosen; but this constant may have different values as we pass from one stream line to another.

This result is the theorem of D. Bernoulli.

If at P we draw a vertical line, PQ, of such length that

$$p = w \cdot PQ,$$

the height PQ is called the *pressure head* at P. If also QR is drawn vertically of such length that

$$v^2 = 2g \cdot QR,$$

QR is called the *velocity head* at P. Let AB be the pressure head and BN the velocity head at A. Then (4) gives
$$PR = AN + z - z_0$$
$$= LN,$$

where $AL = z - z_0$ and is the perpendicular from A on the horizontal plane through P.

Since PL is horizontal, it follows that RN is horizontal. Hence the theorem of Bernoulli may be expressed in these words: *if at each point along a stream line there be drawn a vertical line whose length = the pressure head + the velocity head at the point, the extremities of all these vertical lines lie in the same horizontal plane.*

If the liquid has a horizontal surface, CD, at rest at all points of which the intensity of pressure is constant (e. g., that of the atmosphere), the extremities of these lines drawn at *all* points of the liquid, and not merely along the same stream line, will all lie in the same horizontal plane. If CH is the pressure head on the surface CD (about 34 feet if CD supports the atmospheric pressure), the extremities $R,...$ of the vertical lines drawn at all points $P,...$ lie in the horizontal plane through H.

For a liquid in equilibrium, Q coincides with R, since $QR = 0$, and it has already been shown that the extremities of all vertical lines representing pressure heads lie in the same horizontal plane. The theorem of Bernoulli is the generalization of this result for a liquid in steady motion.

An approximate method of indicating the value of p, the pressure intensity at any point P in a moving liquid consists in inserting a vertical glass tube, open at both ends, into the liquid, one extremity of the tube being placed at P. The liquid will rise to a certain height in this tube and remain at rest. Thus, if the tube is so long that the upper end is above the free surface CD, the liquid would rise in it to the height PQ, the remainder of the tube being occupied by air. Such a tube is called a *pressure gauge*; but it is evident that it does not strictly measure the pressure, since the glass must, to some extent, alter the motion of the liquid.

87. Flow through a small orifice. Let Fig. 104 represent a vessel containing a liquid whose level is LM which flows out through a small aperture made anywhere in the side of the vessel, and let the thickness of the side be so small that the liquid touches the inner edge, AB, of the orifice and thence passes out without touching the outer edge or any intervening part of the aperture.

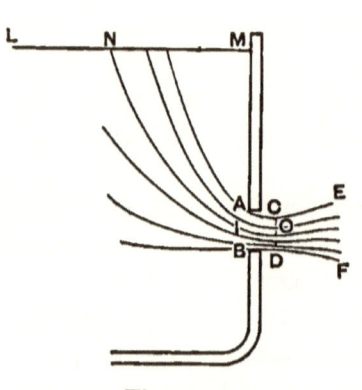

Fig. 104.

The curved lines in the figure represent the stream lines, or paths of particles, the forms and positions of which cannot, however, be determined mathematically.

We are obliged to have recourse to experiment for certain facts concerning the issuing jet. Firstly, it is found that after leaving the orifice AB, this jet contracts to a minimum cross-section, CD, beyond which, of course, the jet widens out again. This minimum cross-section is called the *vena contracta*.

The ratio of the area of the vena contracta to that of the orifice AB is called the *coefficient of contraction*.

For a circular orifice whose diameter is AB, if CD is the diameter of the vena contracta, it has been found experimentally that

$$\frac{CD}{AB} = \cdot 79,$$

so that if S is the area of the orifice and σ that of the vena contracta,

$$\frac{\sigma}{S} = \cdot 624.$$

It has also been found that the distance, IO, between the orifice and the vein is somewhere between $\cdot 39 \times AB$ and $\cdot 5 \times AB$, where, as before, AB is the diameter of the orifice; the uncertainty arising from the fact that in the neighbourhood of the minimum section the diameter of the jet varies very little. All the streams which pass through the vena contracta cut its plane perpendicularly. By consideration of the general equations of motion, it will follow from this fact that the intensity of pressure is the same at all points in the vena contracta. At all points on the outer surface, ACE, BDF, of the jet the pressure intensity is, of course, the same as that of the atmosphere, if the jet flows into the atmosphere; also the velocities at all points of the vein are equal; but in the interior of the jet this pressure intensity does not exist, except at points in the plane of the vena contracta.

376 *Hydrostatics and Elementary Hydrokinetics.*

Of course at the orifice AB the directions of motion are not all perpendicular to the cross-section of the jet, neither are the velocities all the same at points in this section.

88. Theorem of Torricelli. In the case of a jet escaping into the air, the velocities of particles in the vena contracta are expressed by a very simple formula.

In (4) of Art. 86, let p and z refer to a point, O, in the vena contracta while p_0, z_0 refer to the point, N, of the stream line through O which is on the free surface of the liquid. Then $p = p_0$, as we have said above, and as the velocities at the surface LNM are all very small, we may consider $v_0 = 0$.

Hence
$$v^2 = 2g(z - z_0)$$
$$= 2gh, \quad \ldots \ldots \ldots (a)$$

where h, or $z - z_0$, is the vertical depth of the vena contracta below the free surface LNM.

Hence when the particles reach the vena contracta, they have the same velocity as if they fell directly from the free surface. This is known as Torricelli's Theorem. Obviously it holds with considerable exactness in the case of a *small* orifice only.

EXAMPLES.

1. *The Syphon.* We now take a few common practical illustrations of the application of equation (4), Art. 86, which applies to the motion of a liquid acted upon by gravity. The first of the simple examples is furnished by the common syphon which is employed for the purpose of raising a liquid out of a vessel and lowering it into another vessel. The operation might, of course, in many cases be directly performed by taking the first vessel in the hand and pouring out the liquid bodily into the second; but if, as often happens, the liquid is Sulphuric or Nitric Acid, which it would be most undesirable

Steady Motion under the Action of Gravity. 377

to pour out with splashing, this method would not answer, and a syphon is used. The syphon is a bent tube (usually of glass) open at both ends, and with unequal branches.

Suppose M (Fig. 105) to be the vessel which it is desired to empty into another (not represented in the figure), and suppose the liquid to be water.

A bent tube, $DABC$, (the syphon) whose branch BC is longer than the branch BD is first filled with water, and the apertures at D and C held closely by the fingers. The end D is then inserted into the liquid in the vessel M, the fingers removed from D and C, and the tube held in the hand. The result will be a flow of the liquid through C until, if D is kept close to the bottom of the vessel, nearly all the liquid is removed.

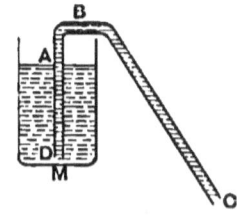

Fig. 105.

Let p_0 be the atmospheric intensity of pressure, which exists on the surface of the liquid at A and also at C; v_0, the velocity of liquid on the surface A, is nearly zero; if $v =$ velocity of efflux at C, $z =$ depth of C below A, equation (4) of Art. 86 gives, since ABC may be taken as a simple stream line,

$$\frac{v^2}{2g} + \frac{p_0}{w} - z = \frac{p_0}{w},$$

$$\therefore v = \sqrt{2gz},$$

so that the flow will continue all through if C is at a lower level than D. Of course there will be a small residue of liquid in M, because when nearly all has flowed out, air will enter the syphon at D.

If the liquid to be removed is an acid, as sulphuric or nitric, the syphon must be filled with it at the beginning by first inserting the end D into the vessel and then sucking the air out through C until the liquid rises in the syphon and falls in the leg BC to a lower level than A; and this suction may be effected by joining another tube to the end C by means of a short piece of indiarubber tubing which can be subsequently removed.

2. *Hero's Fountain.* Hero of Alexandria (120, B.C.) constructed a fountain, which is represented in Fig. 106. It consists of two

378 *Hydrostatics and Elementary Hydrokinetics.*

glass globes, M and N, and a dish, DD. Each globe is partly filled with water and partly with air at the atmospheric pressure. The globes are fitted with necks and are held together by two glass tubes, A, B, each open at both ends, which pass through necks fitted to the globes. The extremities of A are in the air in the globes; the lower extremity of B dips nearly to the bottom of the liquid in M, while its upper end barely projects into the dish DD. A third tube, C, open at both ends, passes through the neck of N, its lower end dipping nearly to the bottom of the liquid in N, while its upper end projects beyond the upper surface of the dish.

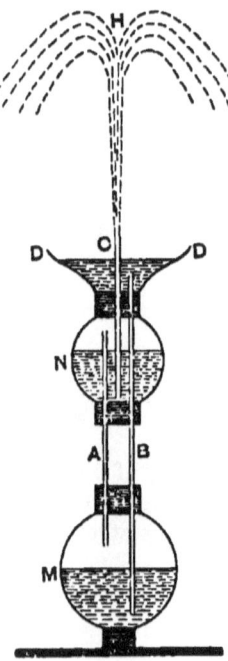

Fig. 106.

In this state of affairs the water is at rest in both vessels, the intensity of pressure on both water surfaces being p_0, that of the atmosphere.

If now water is poured into the dish, it will fall through B into M and drive some of the air into N where the surface pressure on the water becomes greater than p_0, and as a result the water from N is forced up through the tube C into the air.

To calculate the height to which it rises, let z be the difference of level between the water in M and that in N; let $c =$ difference of level between that in D and that in N; and let $h =$ the height of the top of the jet above the water in D.

Then, since the pressure intensity of the air in $N =$ that in $M = w(z+c)+p_0$; since the velocity of the water at the surface in N is nearly zero, and is also zero at H, where the pressure intensity is p_0, we have from (4) of Art. 86

$$\frac{p_0}{w} + z + c - (c+h) = \frac{p_0}{w},$$

$$\therefore h = z,$$

i. e., the height of the jet above the water in D is equal to the difference of level in M and N.

3. *Mariotte's Bottle.* It is sometimes desired to produce a narrow jet of water flowing for a considerable time with constant velocity. Of course a very large reservoir with a very small aperture made in the side would produce the result; but such a reservoir is not always at hand. The result can also be produced by means of a broad flask fitted with a stop-cock near the bottom. Fig. 107 represents the flask. The stop-cock (not represented) is fitted at C, and the aperture is supposed to be very small compared with the cross-section of the flask. The flask is first quite filled with water, the stop-cock being closed. In the top of the flask there is a neck fitted with a cork, and into this is inserted a tube, HD, open at both ends, the tube also being quite filled with water.

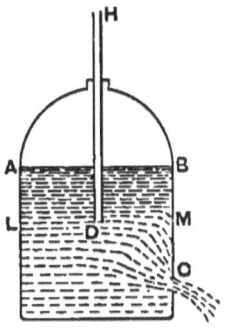

Fig. 107.

Now let the stop-cock be opened, and water will flow out, because the atmosphere presses at H and at the outside of C, and between C and H there is a column of water. The water that first flows out comes from the tube HD alone, the flask remaining filled to its upper surface; and, moreover, the velocity of efflux will be variable as the level sinks in HD. But when the tube is emptied of water, some air will be forced through D by superior atmospheric pressure, and it will rise to the upper part of the flask, and will begin to force down the water of the flask.

This being the case, the intensity of pressure at D in the water is p_0, the atmospheric intensity, and we may assume that p_0 is also the intensity all over the horizontal plane, LM, through D, because the motion of particles in this plane is very slow. The air at the top aided by the water above LM will keep the pressure intensity approximately equal to p_0 at points in LM other than D.

Now let $z = CM =$ vertical distance of orifice below D, the lower extremity of the tube; let v_0, p_0 in (4) of Art. 86 refer to D, while v is the velocity at C. Then, since p_0 is also the pressure intensity at C,

$$\frac{v^2}{2g} + \frac{p_0}{w} - z = \frac{p_0}{w},$$

$$\therefore v = \sqrt{2gz},$$

which shows that the velocity is constant whatever the position of the upper surface, AB, of the water in the flask.

The tube must, of course, have such a position that D is above the aperture. If the water, instead of escaping into the atmosphere, escapes into a medium (gaseous or liquid) in which the pressure intensity at C is p, we shall have

$$v^2 = 2g\left(z + \frac{p_0 - p}{w}\right).$$

This vessel is known as Mariotte's Bottle.

89. Discharge from a small orifice. If a liquid devoid of friction escapes from a small orifice in a vessel in which the free surface is maintained at a constant level, the velocity in the vena contracta is given by the equation (Art. 88)

$$v = \sqrt{2gh}. \quad \ldots \ldots \quad (1)$$

In the case of water, however, it is found that the velocity is not quite equal to this amount, but is very nearly a constant fraction, μ, of the value given by (1). The fraction μ is nearly equal to unity (about ·97). We may therefore put

$$v = \mu\sqrt{2gh}. \quad \ldots \ldots \quad (2)$$

Again, if S is the area of the aperture, and c denotes the coefficient of contraction (Art. 87), the area of the cross-section of the vena contracta is cS; so that the volume of water issuing from the vessel per unit of time is

$$c\mu S\sqrt{2gh}. \quad \ldots \ldots \quad (3)$$

If the unit of length is a foot and the unit of time

Steady Motion under the Action of Gravity.

a second, this will be the discharge in cubic feet per second, and multiplying it by w, the mass of the liquid per unit volume (in this case $62\frac{1}{2}$ pounds), we obtain the mass discharged per unit of time.

Practically the product $c\mu$ may be taken as ·62.

90. Flow through a large orifice. The determination of the discharge through a large orifice cannot be satisfactorily accomplished by theory.

Suppose, for example, that the orifice is a rectangle, $ABCD$, with vertical and horizontal sides, and that LM (Fig. 108) represents the level of the free surface in the vessel, the flow being supposed to take place through the orifice towards us as we look at the figure.

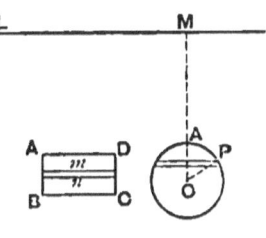

Fig. 108.

Divide the area of the aperture into an indefinitely great number of narrow horizontal strips, of which that between the horizontal lines m and n is the type.

Let the depths below LM of the lines AD and BC be h_1 and h_2, respectively, those of the lines m and n being z and $z+dz$. Let $AD = b$; then, supposing that the aperture between m and n alone existed, the volume of the discharge would be given by (3) of last Article, in which $S = bdz$. Denoting the product $c\mu$ by k, and by dQ the mass discharged per unit time through the strip, we have

$$dQ = kbw\sqrt{2gz}\,.\,dz. \quad . \quad . \quad . \quad . \quad (1)$$

Now the assumption that k is constant for all the strips enables us to find Q, the total discharge; but clearly this assumption cannot be strictly correct, for each strip does not discharge as if it alone existed as an aperture.

382 Hydrostatics and Elementary Hydrokinetics.

Assuming k to be constant, we integrate (1) from $z = h_1$ to $z = h_2$, and obtain

$$Q = \tfrac{2}{3} kbw \sqrt{2g}\,(h_2^{\frac{3}{2}} - h_1^{\frac{3}{2}}). \quad \ldots \quad (2)$$

To calculate the energy per second which flows through the orifice, if v is the velocity of the portion dQ, its kinetic energy is

$$\frac{v^2}{2g} \cdot dQ, \text{ i.e. } zdQ.$$

Energy per unit time is called *Power*. Hence if dP is the power of this flow,

$$dP = kbw\sqrt{2g} \cdot z^{\frac{3}{2}} dz,$$

$$\therefore P = \tfrac{2}{5} kbw \sqrt{2g}\,(h_2^{\frac{5}{2}} - h_1^{\frac{5}{2}}). \quad \ldots \quad (3)$$

If in this expression length is measured in feet, time in seconds, and w in pounds, since the unit of power called a Horse-Power is 550 foot-pounds' weight per second, we get the Horse-Power of the discharge equal to the right-hand side of (3) divided by 550.

If ABC is small compared with the depth h_1, and if h is the depth of the centre of area of the orifice, we can easily find from (2) that

$$Q = kwS\sqrt{2gh}, \quad \ldots \quad (a)$$

where $S = $ the area of the orifice. For if $AB = 2a$, we have $h_2 = h\left(1 + \dfrac{a}{h}\right)$, $h_1 = h\left(1 - \dfrac{a}{h}\right)$, and expanding in powers of $\dfrac{a}{h}$, we see that the term $\dfrac{a^2}{h^2}$ disappears, and (a) is true if we neglect the small fraction $\dfrac{a^3}{h^3}$.

As another example of the same kind, suppose the orifice to be circular (Fig. 108).

Steady Motion under the Action of Gravity.

Let h be the depth of O, the centre of the orifice, below the level, LM, of the water in the vessel or reservoir, let r be the radius of the circle, and break up the area by a series of indefinitely close horizontal lines. If P is any point on the circumference of the circle, OA the vertical diameter, and $\angle POA = \theta$, the area of the strip at P is $2r^2 \sin^2\theta \, d\theta$; therefore the discharge, dQ, through this strip is given by the equation

$$dQ = 2kr^2 w \sqrt{2g(h - r\cos\theta)} \cdot \sin^2\theta \, d\theta. \quad . \quad . \quad (4)$$

Supposing r to be small compared with h, it will be sufficient to expand the radical in powers of $\dfrac{r}{h}$ as far as the second. Then

$$dQ = 2kr^2 w \sqrt{2gh} \left(1 - \frac{r}{2h}\cos\theta - \frac{r^2}{8h^2}\cos^2\theta\right) \sin^2\theta \, d\theta. \quad (5)$$

Integrating from $\theta = 0$ to $\theta = \pi$, we have

$$Q = \pi k r^2 w \sqrt{2gh}\left(1 - \frac{r^2}{32h^2}\right). \quad . \quad . \quad . \quad (6)$$

91. Fluid revolving about vertical axis. If a vessel, represented in a vertical section by ACB, Fig. 109, and containing a fluid, is set rotating round a vertical axis, Cz, after a short time the fluid, owing to friction between its particles and against the surface of the vessel, will rotate like a rigid body with the angular velocity ω; each particle, P, will describe a horizontal circle with this angular velocity, so that if PN is the perpendicular from P on the axis of rotation, the resultant acceleration of the particle is directed along PN from P towards N, and is $\omega^2 \cdot NP$ in magnitude. Denote NP by r, and consider an indefinitely small particle of mass dm at P; then the resultant mass-acceleration of this particle is

$$\omega^2 r \, . \, dm, \quad \ldots \ldots \quad (1)$$

and this vector is directed from P towards N. [The *reversed* mass-acceleration, $-\omega^2 r\,dm$, is called the *force of inertia* of the particle, or its *resistance to acceleration*. It is most important to understand that this force of inertia is not a force acting *on* the particle, but one exerted *by* it on the surrounding medium, or, generally, on the agent or agents accelerating its motion. Thus, then, if a is the vector representing the resultant acceleration of a particle, dm, a force completely represented by

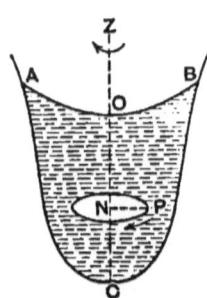

Fig. 109.

$-a \, . \, dm$ in absolute units, or

$-\dfrac{a}{g} \, . \, dm$ in gravitation units,

is the resultant force exerted *by* the particle *on* the agents acting upon it.

Now the fundamental principle of all Dynamics is this: for *each* particle of any material system (whether rigid body, natural solid, liquid, or gas) the resultant mass-acceleration is in magnitude and direction the exact resultant of *all* the forces acting upon the particle. These forces will, in general, consist partly of pressures from the surrounding particles, partly of attractions from these particles, and partly of attractions from bodies outside the system. If we consider the equivalence of the resultant mass-acceleration, $a\,dm$, to these forces under three separate heads, we deduce

Steady Motion under the Action of Gravity.

three great principles of Dynamics. Thus, if we consider that $a\,dm$ has—

1. The same virtual work for any imagined displacement of the particle,

2. The same moment about any axis,

3. The same component along any line,

as the whole system of forces acting on the particle, and that this is true for *every* particle of the system, we have at once the principles of—

1. Kinetic Energy and Work,

2. Time-rate of change of Moment of Momentum,

3. Motion of Centre of Mass,

for every material system. If the forces acting are measured in gravitation units, their complete equivalent is $\dfrac{a}{g}dm.$]

Suppose that at any point, P, Fig. 110, we take as the element dm a very short and thin cylinder, $abcd$, of the fluid having its axis along the tangent at P to any curve AB. Let the length $bc = ds$; let σ be the area of the cross-section, ab, of the cylinder; let F be the external force per unit mass exerted on the fluid at P, and therefore Fdm the force on the cylinder, not including the pressure exerted on its surface by the surrounding fluid; let a be the resultant acceleration of the particle; and let p be the intensity of pressure on the face ab, and therefore $p + \dfrac{dp}{ds}ds$ that on cd.

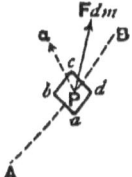

Fig. 110.

Then $dm = w\sigma ds$, if w is the mass of the fluid per unit volume at P; and if we resolve forces in the direction of the tangent at P to the curve, we see that $w\sigma ds \cdot \dfrac{a}{g}$ has the same component along this tangent, in the sense PB, as $F \cdot w\sigma ds$ and $-\dfrac{dp}{ds} ds \cdot \sigma$, the length of the arc s being measured from A. Hence if a_s is the component of a along the tangent at P and S is the component of F, we have

$$\frac{a_s}{g} = S - \frac{1}{w}\frac{dp}{ds}. \quad \ldots \ldots (2)$$

This equation connects the acceleration in any direction with the force intensity and the rate of change of pressure intensity in that direction.

Now suppose the external force to be gravity. Taking the right line NP (Fig. 109) as the direction of s, (2) becomes

$$w \cdot \frac{\omega^2 r}{g} = \frac{dp}{dr}; \quad \ldots \ldots (3)$$

and again, taking the vertical downward direction at P as that of s, (2) becomes

$$0 = w - \frac{dp}{dz}, \quad \ldots \ldots (4)$$

where z is the depth of P below any fixed horizontal plane.

Now p is a function of r and z only, so that

$$dp = \frac{dp}{dr} dr + \frac{dp}{dz} dz$$

$$= w \left(\frac{\omega^2 r}{g} dr + dz \right). \quad \ldots \ldots (5)$$

$$\therefore p = w \left(\frac{\omega^2 r^2}{2g} + z \right) + C,$$

where C is a constant, which may be determined from a knowledge of p at some one point. If p_0 is the value of p at O, the point in which the free surface cuts the axis of rotation, and if O is taken as origin, since $r = z = 0$ at O, we have $C = p_0$; hence

$$p = p_0 + w\left(\frac{\omega^2 r^2}{2g} + z\right). \quad \ldots \quad (6)$$

At all points on the free surface $p = p_0$, therefore the equation of this surface is

$$\frac{\omega^2 r^2}{2g} + z = 0, \quad \ldots \quad (7)$$

showing that the z of every point on it is negative, i.e., all these points are higher than O. This equation denotes a parabola whose latus rectum is $\dfrac{2g}{\omega^2}$, and the free surface is therefore a paraboloid generated by the revolution of this surface round Oz.

If the vertical upward line Oz is taken as axis of x, and a tangent at O as axis of y, the equation of the parabola is, in its usual form,

$$y^2 = \frac{2g}{\omega^2} \cdot x. \quad \ldots \quad (8)$$

If the fluid contained in the vessel is a gas, equations (2), (3), (4) still hold, and, in addition, $p = kw$ (Art. 48); hence (5) becomes

$$\frac{dp}{p} = \frac{1}{k}\left(\frac{\omega^2}{g} r\,dr + dz\right), \quad \ldots \quad (9)$$

$$\therefore\ p = A e^{\frac{1}{k}\left(\frac{\omega^2 r^2}{2g} + z\right)}, \quad \ldots \quad (10)$$

where A is a constant to be determined either from a

knowledge of p at some point or from the given mass of the fluid. Equation (10) shows that for a gas the free surface and the surfaces of constant pressure intensity are still paraboloids.

In the same way, if the vessel contains two fluids that do not mix, their surface of separation is a paraboloid of revolution. For if w, w' are their specific weights, we have (if they are liquids)

$$p = w\left(\frac{\omega^2 r^2}{2g} + z\right) + C \text{ for one,}$$

$$p' = w'\left(\frac{\omega^2 r^2}{2g} + z\right) + C' \text{ for the other,}$$

and since at all points on the surface of separation $p = p'$, we have the equation of a paraboloid of revolution, as before.

The equation of the free surface can also be deduced from the principle of Virtual Work thus. When the position of relative equilibrium has been assumed, imagine the bounding surface AOB to be very slightly displaced and to become $A'O'B'$.

Then the variation contemplated in the equation of Virtual Work may be confined to the surface particles occupying the volume between the surfaces AOB and $A'O'B'$. As in Fig. 86, p. 316, a part of this volume will be positive and a part negative. The principle of Virtual Work applied to any system of particles in motion is simply that the sum of the virtual works of the mass-accelerations of the particles is equal to the sum of the virtual works of the forces (in absolute measure) external and internal acting on the system. Now the resultant acceleration of a particle dm at a distance r from the axis of revolution is $\omega^2 r$, towards the axis, and the resultant mass-acceleration of this particle is $\omega^2 r\,dm$, whose virtual work is $-\omega^2 r\,dm\,\delta r$. The external force acting on the particle is its weight, $g\,dm$, whose virtual work is $g\,\delta z\,.\,dm$, if z is measured downwards from a fixed plane.

Hence
$$-\omega^2 \int r\,\delta r\,.\,dm = g\int \delta z\,.\,dm,$$

$$\int \left(\frac{\omega^2 r\,\delta r}{g} + \delta z\right)\delta n\,dS = 0,$$

where, as in Art. 72, δn is the arbitrary normal displacement of the element dS of the surface AOB. Hence as δn is arbitrary at all points, we have

$$\frac{\omega^2 r\,\delta r}{g} + \delta z = 0,$$

the integral of which gives the equation (6) of the free surface before obtained.

EXAMPLES.

1. A cylinder contains a given quantity of water. If it is rotated round its axis (held vertical), find the angular velocity at which the water begins to overflow.

Let AOB represent the surface of the rotating liquid, the points A and B being at the top of the cylinder; let r and h be the radius and height of the cylinder, and c the height to which the cylinder, when at rest, is filled.

Then since B is on the parabola, if ξ is the depth of O below AB,
$$r^2 = \frac{2g}{\omega^2}\,.\,\xi. \qquad \ldots \ldots \ldots (1)$$

But the volume of the water remains unchanged, and it is the volume of the cylinder minus the volume of the paraboloid AOB. This latter is $\dfrac{\pi g}{\omega^2}\,.\,\xi^2$. Hence

$$r^2 h - \frac{g}{\omega^2}\xi^2 = r^2 c, \qquad \ldots \ldots \ldots (2)$$

$$\therefore\ \omega = \frac{2\sqrt{g(h-c)}}{r}. \qquad \ldots \ldots (3)$$

The above is on the supposition that the water begins to overflow before the vertex, O, of the parabola reaches the base, C, of the cylinder. In this case, with any angular velocity, ω, if PQ is the level to which the water rises, and LM is the level at which the water stands when at rest, it is easily proved that

$$\text{depth of } O \text{ below } LM = \text{height of } PQ \text{ above } LM. \quad (4)$$

Take now the case in which c is so small in comparison with h that O reaches C (or the base begins to get dry) before the water begins to overflow. The angular velocity at which O reaches C is $\dfrac{2\sqrt{gc}}{r}$. Let $\omega = (1+n)\dfrac{2\sqrt{gc}}{r}$; then if O' (below C) is the vertex of the parabola, we have

$$CO' = 2n(1+n)c, \quad \ldots \ldots \quad (5)$$

the height of PQ (the water level) above the base is $2c(1+n)$; and if the free surface cuts the base in R, we have

$$CR = r\sqrt{\frac{n}{1+n}}, \quad \ldots \ldots \quad (6)$$

which is the radius of the dry circle on the base. The water will begin to overflow when the height of PQ above the base is h; i.e.,

$$\omega = \frac{h}{r}\sqrt{\frac{g}{c}}, \quad \ldots \ldots \quad (7)$$

which is quite different in form from (3); so that if c is infinitely small, i.e., if there is only an infinitely thin layer of water put originally into the cylinder, it will not begin to overflow until ω is infinitely great; and in this case $CR = r$, as it should be.

2. A heavy cylinder floats with its axis vertical in a liquid contained in a vessel which rotates uniformly round a vertical axis; find the length of the portion of the cylinder immersed.

Let PQ, Fig. 111, be the level of the liquid round the cylinder, and $PEDQ$ the immersed portion, the free surface being $APOQB$, and O the vertex of the parabola.

Now, by the same reasoning as that in Art. 22, it is obvious that the resultant action of the liquid on the cylinder is the same as that which the liquid would exert on the liquid which would fill the volume $POQDEP$; hence the weight, W, of the cylinder must be equal to the weight of this volume of the liquid.

Let r be the radius of the cylinder and Om the perpendicular from O on PQ; then $Om = \dfrac{\omega^2 r^2}{2g}$, and the volume of the displaced liquid

$$= \pi r^2 . PE - \text{vol.} POQ$$
$$= \pi r^2 \left(PE - \dfrac{\omega^2 r^2}{4g}\right);$$

hence if $w =$ specific weight of liquid,

$$PE = \dfrac{W}{\pi r^2 w} + \dfrac{w^2 r^2}{4g}.$$

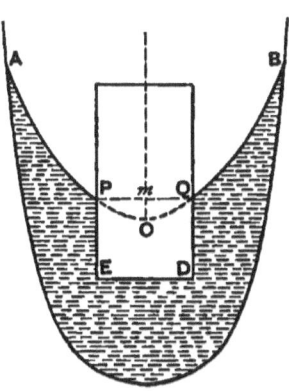

Fig. 111.

3. A vessel of given form containing water is set rotating round a vertical axis, the vessel and the liquid being in relative equilibrium; find the greatest angular velocity of the vessel which will allow all the water to escape through a small orifice at the lowest point of the vessel.

Let the vessel be ACB, Fig. 109, C being its lowest point; assume the free surface to pass through C, the latus rectum of the parabola being $\dfrac{2g}{\omega^2}$; then, taking C as origin, the tangent at C as axis of y and the vertical upward line as axis of x, express the condition that the parabola $y^2 = \dfrac{2g}{\omega^2} x$ intersects the curve ACB in no other point than C.

Thus, if the vessel is a sphere (with another small hole at the top) of radius a, the parabola will be completely outside the sphere if $\omega^2 = \dfrac{g}{a}$.

4. A cylinder whose axis is vertical is filled with a given mass of gas and set rotating round its axis; if the gas is assumed to move in relative equilibrium with the cylinder, find the intensity of pressure at any point.

We have, measuring z from the top of the cylinder

$$dp = wd\left(\frac{\omega^2 r^2}{2g} + z\right),$$

$$\therefore \quad p = A e^{\frac{1}{k}\left(\frac{\omega^2 r^2}{2g} + z\right)},$$

$$\therefore \quad w = \frac{A}{k} e^{\frac{1}{k}\left(\frac{\omega^2 r^2}{2g} + z\right)}.$$

Also if W is the weight of the gas put into the cylinder, we have
$$W = \int w d\Omega,$$
where $d\Omega$ = element of volume at any point, P (Fig. 109). Now if θ is the angle which the plane of P and the axis, Oz, of rotation makes with any fixed vertical plane,

$$d\Omega = r d\theta \, dr \, d\theta \, dz;$$

$$\therefore \quad W = \frac{2\pi A}{k} \iint e^{\frac{1}{k}\left(\frac{\omega^2 r^2}{2g} + z\right)} r \, dr \, dz.$$

Integrating with respect to r, the limits of r are 0 and a, where a is the radius of the cylinder, so that the integrations in r and z may be performed independently, the limits of z being 0 and h. We easily find

$$W = \frac{2\pi g k A}{\omega^2} \left(e^{\frac{\omega^2 a^2}{2gk}} - 1\right)\left(e^{\frac{h}{k}} - 1\right),$$

which determines A.

5. If the cylinder is replaced by a spherical shell rotating about a vertical diameter, solve the previous problem.

6. A hemispherical bowl containing a given quantity of water is set rotating about a vertical diameter, find the angular velocity at which the water begins to overflow.

Ans. If V is the volume of the water, a the radius of the bowl,
$$\omega^2 = \frac{4g}{\pi a^4}\left(\tfrac{2}{3}\pi a^3 - V\right).$$

7. If in the last case the angular velocity is increased beyond the value $\left(\frac{2g}{a}\right)^{\frac{1}{2}}$, find how much of the bowl is dry.

Ans. It is dry to a vertical height $a - \dfrac{2g}{\omega^2}$ above the lowest point.

8. If a hollow open cone with its axis vertical and vertex downwards containing a given quantity of water is made to revolve round a vertical axis, discuss the question as to the possibility of emptying the cone by increasing the angular velocity.

9. A narrow horizontal tube, BC, has two open vertical branches BA and CD, water being poured into the continuous tube, thus formed, to a given height. If this tube is set rotating round a vertical axis through a point O in BC, find the position of the liquid in its state of relative equilibrium.

Ans. If $BO = m$, $OC = n$, the difference of level in the two vertical branches is $\dfrac{\omega^2}{2g}(m^2 - n^2)$.

CHAPTER X.

WAVE MOTION UNDER GRAVITY (SIMPLE CASES).

92. General Equation of Motion. For a particle of any shape which forms part of any moving material system, we have the fundamental law of motion that the mass-acceleration of the particle is at each instant the exact resultant of all the forces acting on this particle; and from this it follows that, when the particle belongs to a perfect fluid,

$$\frac{a_s}{g} = S - \frac{1}{w}\frac{dp}{ds}, \quad \ldots \ldots \quad (1)$$

(p. 386) which may be considered as the equation of motion of the particle in any direction (tangent to the arbitrary curve AB at P, Fig. 110).

Now if the direction of the curve AB at P is such that the intensity of pressure, p, does not vary from point to point along it, $\dfrac{dp}{ds}$ will be zero, and we have

$$\frac{a_s}{g} = S, \quad \ldots \ldots \ldots \quad (2)$$

which gives a rule for finding the direction of the surface of constant pressure intensity at P, viz.,—*draw a vector PA, Fig. 112, representing the resultant acceleration, a, of the particle at P, and also a vector PF representing the force, $g \cdot F$, in absolute units, exerted on the fluid per unit mass at P by all causes other than the pressure of surrounding particles;*

Wave Motion under Gravity (Simple Cases). 395

then the plane through P perpendicular to the right line AF is the tangent plane at P to the surface of constant pressure intensity. For, if AF meets the perpendicular PT in r, the vector Pr is at once the component of a along PT and the component of $g.F$; hence $\dfrac{dp}{ds} = 0$ along PT at P.

If the external force is simply gravity, we take PA to represent a and PF to represent g.

The student may verify this construction in the case of a fluid revolving about a vertical axis (Art. 91).

Instead of using the acceleration, PA, of the particle we may use the reversed acceleration, PA', or force of inertia per unit mass at P (Art. 91); and then the result is that *the direction of the surface of constant pressure at P is at right angles to the resultant, Φ, of the external force and the force of inertia.* (By the *external force* at P is meant, as before explained, the resultant of all forces, excluding pressures, which act on the particle.)

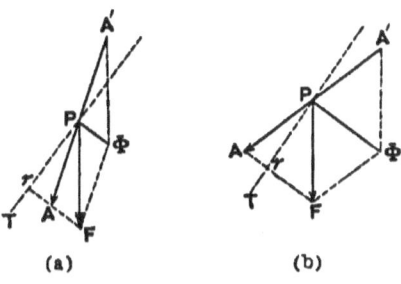

Fig. 112.

The following is also an important result: if in the fluid we describe a surface of constant pressure intensity, p_1, and also a very close surface of constant pressure intensity, p_2 (p_1 and p_2 differing by an infinitesimal amount), the normal distance between these close surfaces at any point P is inversely proportional to the magnitude of the force Φ at P.

For in Fig. 112, (a) or (b), take on the line PΦ a point Q indefinitely close to P, and let the surface of constant pressure through Q be described. Then taking the equation (1) with reference to the direction PQ, denoting PQ by Δn, we have

$$\frac{1}{w}\frac{p_2-p_1}{\Delta n}=gS-a_s, \quad \ldots \ldots \quad (3)$$

if p_2 and p_1 are measured in absolute units of force.

Now a_s means the component of PA along PQ in the sense $\overline{PQ}$, and hence in Fig. (a) the right-hand side of (3) is the projection of PF along $P\Phi$ minus the projection of PA, which is obviously $P\Phi$, which we have denoted simply by Φ. In Fig. (b) the projection of PA along $P\Phi$ in the sense $\overline{PQ}$ is negative, so that the right-hand side of (3) is the arithmetic sum of the projections of PA and PF, which is, again, Φ. Hence

$$\Delta n = \frac{1}{w}\cdot\frac{p_2-p_1}{\Phi}, \quad \ldots \ldots \quad (4)$$

and since at all points on the surface of constant pressure through P we have $p=p_1$, and at all points on that through Q, $p=p_2$, we see that Δn, or PQ, the normal distance between the two surfaces at any point, varies inversely as Φ.

The case of a fluid at rest is, of course, a particular instance. In this case Φ is simply the resultant external force per unit mass, and the normal distance between any two close surfaces of constant pressure varies inversely as this force.

93. Definition of a Wave. Any disturbance which is communicated from point to point of a body whereby each particle is displaced from its position of rest and the relative distances and directions of the particles are altered is called a *wave*. As a rule, in such disturbances the displacement of each particle from its position of rest in the body is small—or, at any rate, small compared with

Wave Motion under Gravity (Simple Cases). 397

other magnitudes which are involved in the particular case concerned. Moreover, the motions of the individual particles may be very complicated, or may be simple oscillatory motions in small circles or other closed curves. Thus, when in a long tube filled with air the air at one end is disturbed by a sudden impulse along the tube, the whole air column will be set in motion of a to-and-fro kind, and the disturbance will reach the far end of the tube, while no particle ever departs far from its position of rest.

So likewise in the case of an iron bar which is struck at one end; and so, again, in the well-known case of a long stretched string one end of which is fixed while the other end is agitated by the hand; or both ends may be fixed while the string is rubbed by a bow at any intermediate point. In all these cases the disturbance which travels by communication from particle to particle throughout the medium is called a wave.

94. Trochoids. If a circle rolls without sliding along a right line, any point 'carried' by the circle traces out a curve called a *trochoid*. If the carried point is one on the circumference of the rolling circle, the trochoid becomes the common cycloid; if the carried point lies outside the circumference, the trochoid is a looped curve; and if it lies inside the circumference, the locus is devoid of loops.

Thus, in Fig. 113, let the fixed line on which the circle rolls be LM; let E be the rolling circle having its centre at O and touching the line at B; let A be the carried point at a distance OA, or r, from O, while $OB = R$. Then as the circle rolls along LM, its centre describes the line Ox, parallel to LM. Let C be the position of O at any instant; then if θ is the angle rolled through, we have $OC = BI = R\theta$, where I is the point of contact of the rolling circle with LM. Now to find the position, P, occupied by the point A, we may measure off the arc Bi equal to BI; then BOi is the angle, θ, through which the circle—and every right line carried by the circle—has revolved, the sense of the rotation being denoted by the

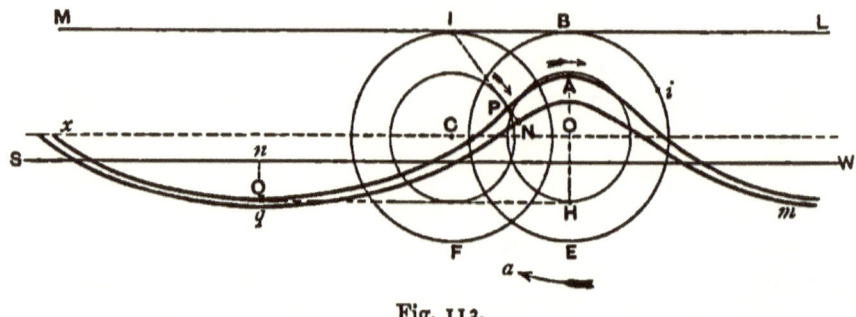

Fig. 113.

arrow a. Hence the line OA has revolved through this angle, and therefore if we draw CP parallel to Oi and equal to r, we obtain the position P to which A has come.

The trochoid traced out by A is the wavy curve APQ, symmetrical at both sides of the line OA; and obviously the trochoids described by all other points on the circle of radius OA will be merely the same curve in different positions.

This curve will have a series of crests, such as that at A, and a series of hollows or troughs, such as that at Q—at which point the moving point A reaches a maximum distance, $R+r$, from LM.

We may put the case in another way. Instead of imagining a single circle, E, to roll into successive positions and to carry a point A with it, let us imagine at one and the same instant a series of circles, E, F, ... each of radius R touching LM, and a series of points A, P, ... taken on circles of radius r with centres O, C, ... so that the angular deflection of the lines OA, CP, ... from the vertical are pro-

Wave Motion under Gravity (Simple Cases). 399

portional to the distances of the centres from O; then the points $A, P, \ldots$ all lie on a trochoid. If now these points $A, P, \ldots$ are all revolving with the same angular velocity, ω, they will all at each instant lie on a trochoid, which will be merely the trochoid APQ in the figure displaced in a direction parallel to LM. The trochoid will appear to travel towards the right or towards the left of the figure, so that there will no longer be a crest above O, until all the moving points $A, P, \ldots$ have completed revolutions in their circles, and then a complete wave length (distance between two consecutive crests or two consecutive troughs) of the curve will have travelled past O, and past every other fixed point.

The radius r is called the *tracing arm* of the trochoid.

Theorem I. If the radius of the rolling circle is taken equal to $\dfrac{g}{\omega^2}$, where ω is the angular velocity of the points $A, P, \ldots$ in their circles, the trochoid is a curve of constant intensity of pressure of a liquid whose surface particles when at rest under gravity are $A, P, \ldots$, and when set in motion revolve in the smaller circles with constant angular velocity ω, all these circles lying in the same vertical plane.

(We assume for the present that such a motion of the liquid particles is a possible one.)

For, assuming that the typical particle P moves with constant velocity in the circle round C, its resultant acceleration is $\omega^2 r$ and is directed in PC; also the direction of g is IC, therefore the direction of Φ is IP, because the reversed acceleration is $\omega^2 . CP$, and gravity is $\omega^2 . IC$, and the direction of their resultant is IP, which is the normal to the surface of constant pressure. But since I is the instantaneous centre of the rolling circle, IP is the normal to the trochoid; therefore, &c.

The trochoid APQ being that occupied at any instant by

the surface particles of the liquid, the constant intensity of pressure along it is p_0, the atmospheric intensity.

If BM and BO are taken as axes of x and y, respectively, the co-ordinates of P are

$$x = R\theta - r\sin\theta,$$
$$y = R - r\cos\theta.$$

Theorem II. As we descend vertically into the fluid, the curves of constant pressure are also trochoids.

To get the indefinitely near curve, qNm, of constant pressure p, we produce the normal IP through P and on it take a length given by (4) of Art. 92.

Let, then, $PN = \dfrac{p - p_0}{w\Phi}$, Fig. 114. Now $\Phi = \omega^2 . IP$, therefore

$$PN = \frac{p - p_0}{w\omega^2 IP}, \quad \dots \dots \quad (1)$$

We shall now show that the locus of N is a trochoid.

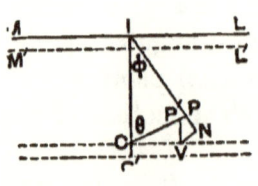

Fig. 114.

Let P' be a point on the radius CP, let $PP' = -dr$, and consider the trochoid which is traced out by P' as the circle of radius CI, or R, rolls along LM. Let this trochoid receive a motion of translation $P'V$ vertically downwards. It will then, in this new position, be the trochoid generated by the rolling of a circle of radius R along a horizontal line at a depth equal to $P'V$ below LM, and if $CC' = P'V$, the centre of the circle which generates this trochoid is C'.

We shall now show that the lengths PP' and $P'V$ can be determined so that the trochoid last mentioned shall pass through all the points N.

Let $\angle CIP = \phi$, and $\angle ICP = \theta$, as in Fig. 114. Then if PN = sum of projections of PP' and $P'V$ along IN, the trochoid will pass through N. Assume, then,

$$PN = -dr\cos(\theta+\phi) + P'V\cos\phi.$$

Putting $P'V = d\eta$, and observing that $\cos\phi = \dfrac{R - r\cos\theta}{IP}$ and $\sin\phi = \dfrac{r\sin\theta}{IP}$, we have

$$PN.IP = r\,dr + R\,d\eta - (R\,dr + r\,d\eta)\cos\theta, \quad \ldots \quad (2)$$

so that $PN.IP$ will be constant, as (1) requires, at all points P of the surface trochoid, provided that

$$R\,dr + r\,d\eta = 0, \quad \ldots \ldots \quad (3)$$

and we can, of course, choose dr and $d\eta$ so as to satisfy this condition.

As we descend vertically in the fluid the relation (3) is that which holds between each trochoid and the next, so that if we integrate (3), we obtain the relation between the tracing arm and the depth, η, below C, (or below the line of centres of surface particles,) of the centre of the corresponding circle. Denoting by r_0 the tracing arm of the surface trochoid, we have, therefore,

$$r = r_0\, e^{-\frac{\eta}{R}}. \quad \ldots \ldots \quad (4)$$

Equation (2) then gives

$$PN.IP = \frac{R^2 - r^2}{R}.d\eta, \quad \ldots \ldots \quad (5)$$

and (1) shows that, if $p - p_0$ is denoted by dp,

$$dp = w\omega^2\frac{R^2 - r^2}{R}.d\eta, \quad \ldots \ldots \quad (6)$$

from which and (4) the value of p is easily found in terms of η.

To show now that the assumed motion of the liquid is a

possible one, we shall adopt the method of Rankine (see his *Miscellaneous Scientific Papers*, p. 483).

If the motion of the particles in vertical circles, as described, is possible, it will be equally possible supposing the whole liquid to receive a horizontal motion of translation equal to $R\omega$, i.e., the velocity of the centre C of the rolling circle, in the sense LM; and conversely. Now this motion combined with the velocity $r\omega$ of P at right angles to CP will cause P to have a resultant velocity along the tangent to the trochoid at P; for, a velocity $\omega \cdot CP$ at right angles to CP combined with a velocity $\omega \cdot IC$ at right angles to IC will give a velocity $\omega \cdot IP$ at right angles to IP, by the proposition of the triangle of velocities.

Now supposing the liquid at P to be moving along the trochoid APQ and also the liquid at N to be moving along the consecutive trochoid, mNq, so that the space between the two trochoids is a channel of flow, the same quantity of liquid must flow across each normal section PN of this channel in the same time; and the fulfilment of this condition is not only necessary but sufficient for the possibility of the motion.

But the condition is obviously fulfilled; for, the velocity, v, of P along the trochoid is $\omega \cdot IP$, and $IP \cdot PN$ has been proved constant, so that we have

$$v \propto \frac{1}{PN},$$

and $v \cdot PN$ is proportional to the quantity which flows across PN per unit time; this is, therefore, the same at all points of the channel; and hence the motion is possible.

Observe that the motion here shown to be possible does not require the displacements of the particles from their original positions to be *small*: the circles, each of radius r, described by them may be of any magnitude, with the sole condition that the depth of the liquid must be very great, because at the bottom of the liquid mass the liquid must either be at rest or move along the bed; and since all through the liquid the particles describe circles, these circles must become infinitesimally small at the bottom; but since their radii are given by equation (4), if r becomes infinitesimal, η must be very great.

This wave is known as *Gerstner's Trochoidal Wave*, having been first discovered by Gerstner; it was afterwards discovered independently by Rankine.

The line, OCx, on which lie the centres of the circles described by the particles of liquid forming the surface, does not coincide with the surface of the liquid when it is at rest, but is vertically above this surface. Let SW be the surface when the liquid is at rest; let the tangent to the trochoid at the lowest point, Q, meet the vertical through A in H; let Ox be at a distance z above SW, and let the perpendicular from Q on SW be Qn. Then the volume $APQHA$ of disturbed liquid occupies the rectangular volume Hn when at rest. But the perpendicular from P on QH is $r(1+\cos\theta)$, while the distance of P from AH is $R\theta - r\sin\theta$; hence the area $APQHA$ is

$$r\int_0^\pi (1+\cos\theta)(R-r\cos\theta)\,d\theta, \text{ i.e., } \pi r\left(R - \frac{r}{2}\right),$$

while the area of the rectangle Hn is $\pi R(r-z)$; hence

$$z = \frac{r^2}{2R},$$

which is the height of the line of orbit centres above the surface of the still liquid.

It is easy to show that at any instant after the regular wave motion has been established all those particles which, when the liquid was at rest, lay on a vertical line will lie on a curve having this line for an asymptote. For, if we draw the vertical line IC and produce it downwards into the liquid, the position of the particle P is obtained by drawing CP equal to r_0 and making with IC the angle θ, or $\dfrac{OC}{R}$; and if we take any other point, C', on IC and draw $C'P'$ parallel to CP and equal to $r_0 e^{-\frac{\eta}{R}}$, where $\eta = CC'$, we obtain the position of a particle P' which originally lay on the line CI. As the point C' travels down IC, the radius $C'P'$ diminishes, according to the above law, and hence the locus of P' approaches IC asymptotically.

This curve possesses the simple property that if the tangent to it at any point, P', meets the vertical line IC in T, the length $C'T$ is constant and $= R$. From this the slope of the columns which were originally vertical is easily determined.

The inclination of the surface of the wave APQ to the horizon, being zero at A and at Q, is a maximum at some intermediate point. In Fig. 114 it is obvious that ϕ is the inclination of the surface at P, and that since $IC = R$, and $CP = r$, ϕ is a maximum when CP is perpendicular to PI. In general

$$\cot \phi = \frac{R - r \cos \theta}{r \sin \theta}.$$

It is obvious that when any particle, P, completes a revolution in its circle, the wave has travelled horizontally over one wave-length, so that if T is the time of travelling over a wave-length, $\omega = \dfrac{2\pi}{T}$; and if t is the time since the crest, A, of the wave has passed over the point C, we have

$$\theta = 2\pi \frac{t}{T}.$$

Wave Motion under Gravity (Simple Cases). 405

The wave-length, λ, is $2qH$, or $2\pi R$, and the velocity, v, of propagation of the wave is $R\omega$, or since $R = \dfrac{g}{\omega^2}$,

$$v = \sqrt{gR}, \text{ and } \lambda = 2\pi R.$$

If T is the time taken by the wave to travel a wave-length, $\lambda = vT$; therefore

$$T = 2\pi \sqrt{\dfrac{R}{g}}.$$

Momentum of the Wave. Consider the thin layer of liquid between two consecutive trochoidal surfaces, APQ, mNq, and take an element of this at any point P, the element being contained between two consecutive values of PN, the normal distance between the layers. If ds is the element of length of the trochoid at P, the volume of the element of liquid is $PN.ds$ multiplied by its length perpendicular to the plane of the figure. This last dimension we shall assume to be unity. Now $ds = IP.d\theta$, therefore the weight of the liquid element is

$$w.IP.PN.d\theta,$$

which at all points is simply proportional to $d\theta$, since $IP.PN$ is the same at all points on the surface of the wave.

Now the velocity of P is ωr, whose horizontal component is $\omega r \cos\theta$; hence the horizontal momentum of the element is proportional to $\cos\theta\, d\theta$, and the total horizontal component for the half wave-length qH, or for a whole wave-length, is zero.

This results from the fact that the particles near the crest, A, are moving backwards while those near the trough, Q, are moving in the reverse direction, as seen in the figure.

406 *Hydrostatics and Elementary Hydrokinetics.*

The horizontal momentum for a quarter wave is

$$r\omega.w.IP.PN\int_0^{\frac{\pi}{2}}\cos\theta\,d\theta,\text{ i. e.,}$$

$$r\omega.w.IP.PN.$$

Here substitute for $IP.PN$ from (5), and we have

$$w.\omega\frac{r}{R}(R^2-r^2)\,d\eta,\text{ or }-w.\omega(R^2-r^2)\,dr\,;$$

and if we integrate this from $r = r_0$ to $r = 0$, we obtain for the whole mass of liquid

$$w.\omega(R^2-\tfrac{1}{3}r_0^2)\,r_0.$$

Energy of the Wave. Let us calculate the amount of kinetic energy contained between the portion APQ of the trochoid and the corresponding portion of the next consecutive trochoid.

The velocity of the particle at P being ωr, the element of kinetic energy is

$$w.IP.PN.d\theta.\frac{\omega^2 r^2}{2g},\text{ or }-w\frac{\omega^2}{2g}d\theta.(R^2-r^2)\,r\,dr,$$

whose integral for the half wave APQ is

$$-w\pi\frac{\omega^2}{2g}(R^2-r^2)\,r\,dr\,;$$

and the integral of this from $r = r_0$ to $r = 0$ is

$$\frac{\pi\omega}{4R}(R^2-\tfrac{1}{2}r_0^2)\,r_0^2,$$

since $\frac{g}{\omega^2} = R$.

Calculate now the static energy of the layer APQ. The height of the element at P above the level, SW, of still liquid is $r\cos\theta + \frac{r^2}{2R}$, so that the element of static energy is $\omega \cdot IP \cdot PN\, d\theta \left(r\cos\theta + \frac{r^2}{2R}\right)$. Now the integral of the term in $\cos\theta$ for the half wave length vanishes, and since $\frac{r^2}{2R}$ is the same as $\frac{\omega^2 r^2}{2g}$, the static energy of the layer is equal to the kinetic energy above found. The energy of the wave, then, is half kinetic and half static.

95. Small Displacements. Assuming that the particle of liquid which, when at rest, occupies the point (x, y, z), or P, never moves very far from this position while the liquid is in motion, the co-ordinates, x', y', z', of its position P' at any instant may be denoted by

$$x + \xi, \quad y + \eta, \quad z + \zeta,$$

where ξ, η, ζ are small quantities each of which depends on the time, t, as well as on the values of x, y, z.

Confining our attention to the motion of a liquid under gravity, we shall assume the displacements of all particles confined to the vertical plane x, y—i.e., we consider the motion to take place in two dimensions only—the motion being the same in all planes parallel to the plane x, y. We shall take the origin of co-ordinates at a point on the bed of the liquid, the axis of x horizontal, and that of y vertically upwards. The displacement ζ is zero; and the components of acceleration of any particle are $\frac{d^2\xi}{dt^2}$ and $\frac{d^2\eta}{dt^2}$.
Hence for the motion of the particle the equations are, by (1) of Art. 92,

$$\frac{d^2\xi}{dt^2} = -\frac{g}{w}\frac{dp}{dx'}. \qquad (1)$$

$$\frac{d^2\eta}{dt^2} = -g - \frac{g}{w}\frac{dp}{dy'}. \qquad (2)$$

408 *Hydrostatics and Elementary Hydrokinetics.*

Again, $\quad \dfrac{d}{dx'} = \dfrac{dx}{dx'} \cdot \dfrac{d}{dx} = \left(1 + \dfrac{d\xi}{dx}\right)^{-1} \dfrac{d}{dx},$

and $\quad \dfrac{d}{dy'} = \left(1 + \dfrac{d\eta}{dy}\right)^{-1} \dfrac{d}{dy}.$

Also since the liquid is incompressible, ξ and η must satisfy the condition

$$\dfrac{d\xi}{dx} + \dfrac{d\eta}{dy} = 0. \quad \ldots \ldots \quad (3)$$

These equations contain the theory of small displacements of a liquid acted upon gravity only, irrespective of the particular kind of motion which may exist in any case.

We shall now proceed to consider some particular kinds of motion in which the displacements of all particles are small; and we assume, as above implied, that the axis of x is in the direction in which the disturbance travels along the surface of the liquid.

96. Oscillatory Waves. Assuming the displacements of the particles to be small, and also periodic, we shall have

$$\dfrac{d^2\xi}{dt^2} = -n^2 \xi, \quad \dfrac{d^2\eta}{dt^2} = -n^2 \eta, \quad \ldots \quad (1)$$

where n is a constant; for, it is well known that these are the equations of oscillatory motion, whose integrals give $\xi = A \cos(nt - a)$, where A and a are arbitrary constants, with a similar value of η; so that the values of ξ and η repeat themselves whenever t increases by $\dfrac{2\pi}{n}$.

For such motion, then, the equations are

$$n^2 \xi = \dfrac{g}{w} \dfrac{dp}{dx'}. \quad \ldots \ldots \quad (2)$$

Wave Motion under Gravity (Simple Cases). 409

$$n^2\eta = g + \frac{g}{w}\frac{dp}{dy'} \quad . \quad . \quad . \quad . \quad . \quad . \quad (3)$$

$$\frac{d\xi}{dx} + \frac{d\eta}{dy} = 0. \quad . \quad . \quad . \quad . \quad . \quad . \quad (4)$$

Hence $\dfrac{d\xi}{dy'} - \dfrac{d\eta}{dx'} = 0$; and since ξ and η are small, we can take $\dfrac{d}{dx'} = \dfrac{d}{dx}$, and $\dfrac{d}{dy'} = \dfrac{d}{dy}$. Therefore also

$$\frac{d\xi}{dy} - \frac{d\eta}{dy} = 0. \quad . \quad . \quad . \quad . \quad . \quad . \quad (5)$$

From (4) and (5) we have

$$\frac{d^2\xi}{dx^2} + \frac{d^2\xi}{dy^2} = 0, \text{ or } \nabla^2\xi = 0, \quad . \quad . \quad . \quad (6)$$

and $\quad \dfrac{d^2\eta}{dx^2} + \dfrac{d^2\eta}{dy^2} = 0, \text{ or } \nabla^2\eta = 0. \quad . \quad . \quad . \quad (7)$

Now from (1) the value of ξ is $A\cos(nt-a)$, where A and a are independent of t but may involve x and y. To satisfy (6) we shall assume this form for ξ, with the further assumption that the *amplitude* A is a function of y alone, i.e., it depends on the depth of the disturbed particle only, while the *phase*, a, is the same for all disturbed particles which originally lay on the same vertical line. Hence the general value of ξ will be given by

$$\xi = A\cos(nt - mx) + B\sin(nt - mx), \quad . \quad . \quad (8)$$

where A and B are functions of y only, and n and m are independent of x, y, and t.

A similar expression gives the value of η; but the latter is, of course, deducible from ξ by means of (4) and (5).

The periodic nature of a wave disturbance travelling

410 *Hydrostatics and Elementary Hydrokinetics.*

through a liquid, or through any disturbed medium, will be readily understood from Fig. 115, in which AB represents the surface of the liquid when at rest, and O is the origin.

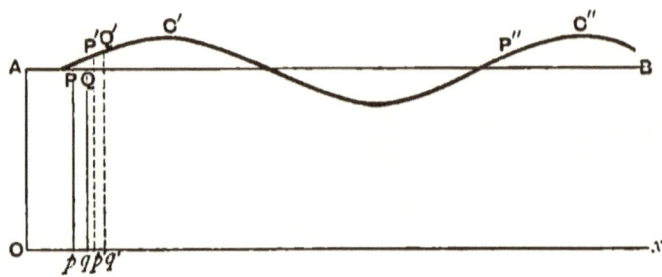

Fig. 115.

Let P be the position of a particle when the surface is at rest, AP being x, and P' be its position at any time, t. Then, considering the value of ξ only (since η runs through all its values in the same time as ξ), we see that when t is increased by the time $\dfrac{2\pi}{n}$, the position of P' will be exactly the same as at the time t. Denote this interval by T, so that
$$T = \frac{2\pi}{n}. \quad \ldots \ldots \ldots \quad (9)$$

Then *T is the time in which every particle completes its oscillation.*

Again, considering the displaced positions of all particles at the same time t, let P'' be the position occupied at this time by another particle, whose abscissa, along AB, was originally x'.

Then, if $nt - mx' = nt - mx - 2\pi$, the displacements ($\xi, \eta$) of Q will be exactly the same as those of P. This gives $x' - x = \dfrac{2\pi}{m}$, and this is therefore the distance, at each instant, between all particles which are in the same state, or *phase*, of disturbance. This distance is the wave-length of the disturbance. Denoting it by λ, we have

$$\lambda = \frac{2\pi}{m}. \quad \ldots \quad \ldots \quad (10)$$

Of course the wave-length is also the distance between any two successive crests, C', C''.

It is, then, obvious that *the time, T, taken by any particle, P', to complete its oscillation is also the time taken by the disturbance to travel over a wave-length.*

We now proceed to determine the values of A and B. Substituting the value (8) of ξ in (6) and expressing the fact that (6) is satisfied for all values of t, we have

$$\frac{d^2 A}{dy^2} - m^2 A = 0 \text{ and } \frac{d^2 B}{dy^2} - m^2 B = 0, \ldots \quad (11)$$

which give
$$A = Le^{my} + Me^{-my},$$
$$B = L'e^{my} + M'e^{-my},$$

where L, M, ... are constants, whose values must be determined from considerations not yet mentioned.

These values of A and B give

$$\xi = (Le^{my} + Me^{-my})\cos(nt - mx)$$
$$+ (L'e^{my} + M'e^{-my})\sin(nt - mx). \quad (12)$$

To determine η, equate $\dfrac{d\eta}{dx}$ to $\dfrac{d\xi}{dy}$, and then integrate; thus

$$\eta = -(Le^{my} - Me^{-my})\sin(nt - mx)$$
$$+ (L'e^{my} - M'e^{-my})\cos(nt - mx) + f(y), \quad (13)$$

where $f(y)$ is an unknown function of y, to determine which equate $\frac{d\eta}{dy}$ to $-\frac{d\xi}{dx}$, and we find that $f'(y) = 0$, ∴ $f(y)$ is a constant, C.

We shall now assume the bed of the liquid to be a horizontal plane, so that the undisturbed liquid is of uniform depth, h. Now the displacement of every particle at the bottom takes place along the bed of the channel; hence $\eta = 0$ when $y = 0$, whatever t may be.

This gives a zero value to the above constant C, and also makes
$$M = L, \quad M' = L',$$
so that we have

$$\xi = (e^{my} + e^{-my})\{L\cos(nt-mx) + L'\sin(nt-mx)\}, \quad (14)$$

$$\eta = (e^{my} - e^{-my})\{-L\sin(nt-mx) + L'\cos(nt-mx)\} \quad (15)$$

There is this further consideration, that on the free surface of the liquid p is at all times equal to p_0, the atmospheric intensity of pressure; so that

$$\frac{dp}{dt} \equiv 0 \text{ at the free surface.} \quad \ldots \quad (16)$$

Now since
$$dp = \frac{dp}{dx'} \cdot dx' + \frac{dp}{dy'} \cdot dy',$$

we have from (2) and (3)

$$\frac{g}{w}dp = n^2(\xi dx + \eta dy) - g\,dy', \quad \ldots \quad (17)$$

in which, of course, the term $\xi dx + \eta dy$ is a perfect differential; and (17) gives

$$\frac{g}{w}p = \frac{n^2}{m}(e^{my} + e^{-my})(-L\sin\phi + L'\cos\phi) - gy' + C, \quad (18)$$

where ϕ stands for $nt - mx$, and C is a constant.

Wave Motion under Gravity (Simple Cases).

Since $y' = y + \eta$, if we use the hyperbolic notation

$$e^{my} + e^{-my} = 2 \cosh my\,; \quad e^{my} - e^{-my} = 2 \sinh my, \quad (19)$$

we have

$$\frac{g}{w} p = 2\,(L \sin \phi - L' \cos \phi)\left(-\frac{n^2}{m} \cosh my + g \sinh my\right)$$
$$- gy + C. \quad (20)$$

Hence the value of p at the surface will be independent of t if

$$\frac{n^2}{m} \cosh mh - g \sinh mh = 0, \quad \ldots \quad (21)$$

since $y = h$ for surface points.

If v is the velocity of propagation of the wave

$$\lambda = vT, \quad \ldots \ldots \quad (22)$$

so that (21) becomes

$$v^2 = \frac{g\lambda}{2\pi} \tanh\left(\frac{2\pi h}{\lambda}\right). \quad \ldots \quad (23)$$

The velocity with which the wave is propagated will therefore depend on the depth of the liquid and on the length of the wave.

Long waves in shallow water. If the depth of the liquid is small compared with the length of the wave,

$$e^{\frac{2\pi h}{\lambda}} + e^{-\frac{2\pi h}{\lambda}} = 2, \text{ neglecting } \frac{h^2}{\lambda^2}, \text{ and } e^{\frac{2\pi h}{\lambda}} - e^{-\frac{2\pi h}{\lambda}} = 4\pi \frac{h}{\lambda};$$

and in this case

$$v = \sqrt{gh}, \quad \ldots \ldots \quad (24)$$

which gives the well-known result that in shallow water, all waves (provided their lengths are very much greater than the depth of the water) are propagated with the same velocity, which is that acquired by a body falling freely through a height equal to half the depth of the water.

In this case it also appears from (14) and (15) that the maximum horizontal displacements of all particles are much greater than the maximum vertical displacements. In general, these equations show that the orbit described by a disturbed particle about its position of rest is an ellipse

$$\frac{\xi^2}{\cosh^2\left(\frac{2\pi y}{\lambda}\right)} + \frac{\eta^2}{\sinh^2\left(\frac{2\pi y}{\lambda}\right)} = 4(L^2+L'^2), \quad (25)$$

whose horizontal axis in the case of long waves in shallow water is much greater than the vertical axis.

The value of ξ in this case (i.e., when $\frac{h}{\lambda}$ is small) shows also that, to the second order of small quantities, it is true that all particles which were originally in the same vertical plane perpendicular to the direction of propagation of the wave, will at all times be in a common vertical plane parallel to the original one: the vertical motions of these points are, of course, different; so that we can suppose the motion to be produced by parallel vertical planes of particles oscillating backwards and forwards horizontally, while the particles in these planes have small up and down motions in the planes. The problem of such waves is, indeed, often solved by starting with this assumption, and the equation (4) is then used in an *integrated form* as follows: let the vertical section Pp of the liquid at rest have abscissa x; and assume the particles in this section to occupy the plane $P'p'$ at any time when the liquid is in motion, the distance between the planes Pp and $P'p'$ being ξ.

Let Qq be a vertical section parallel to Pp, at distance Δx from Pp, and let the particles in this plane occupy the plane $Q'q'$ when the liquid is in motion; then, since $\xi = f(x)$, the displacement of $Q'q'$ from its original position is $f(x+\Delta x)$, i.e., $\xi + \frac{d\xi}{dx}\Delta x$; and the abscissae of $P'p'$ and $Q'q'$ being $x+\xi$ and $x+\Delta x+\xi+\frac{d\xi}{dx}\Delta x$, the distance be-

Wave Motion under Gravity (Simple Cases). 415

tween these planes is $\left(1 + \dfrac{d\xi}{dx}\right)\Delta x$. We have now to express the fact that the volume of liquid contained between the planes Pp and Qq is equal to that contained between $P'p'$ and $Q'q'$.

Let the elevation of P' above AB be ϵ; then $P'p' = h + \epsilon$, and if the sides of the channel containing the liquid are vertical and parallel, i.e., if the channel is a rectangular canal, its cross-section perpendicular to AB is of constant area, so that we have simply

$$h \Delta x = (h+\epsilon)\left(1 + \frac{d\xi}{dx}\right)\Delta x, \quad \ldots \quad (26)$$

$$\therefore \quad h\frac{d\xi}{dx} + \epsilon = 0. \quad \ldots \ldots \ldots (27)$$

Now this equation is the equivalent of (4), because if we integrate (4) with respect to y from the bottom to the top, i.e., from p to P, we have (since ξ does not sensibly involve y)

$$\frac{d\xi}{dx}\int_0^h dy + \int_0^\epsilon \frac{d\eta}{dy}dy = 0,$$

or $\quad h\dfrac{d\xi}{dx} + \epsilon = 0.$

Observe that this equation holds in *all* cases in which the vertical motions are small compared with the horizontal, and not merely in the case in which the motion is oscillatory, so that we are now dealing with the general equations (1), (2) of Art. 95 instead of the special equations (2), (3) of Art. 96, which are limited to oscillatory motion.

Again, (2) of Art. 95 shows that if $\dfrac{d^2\eta}{dt^2}$ is very small, the value of p at any point in the disturbed liquid is $p_0 + w(h + \epsilon - y')$ where p_0 is the value of p on the free surface; hence from (1) of Art. 95,

$$\frac{d^2\xi}{dt^2} = -g\frac{d\epsilon}{dx}$$

$$= gh\frac{d^2\xi}{dx^2}. \quad \ldots \quad \ldots \quad (28)$$

Now if $v = \sqrt{gh}$, the integral of this equation is well known to be

$$\xi = \phi(x-vt) + \psi(x+vt), \quad \ldots \quad (29)$$

where ϕ and ψ are any symbols of functionality whatever.

If ϕ and ψ are circular functions (sines or cosines) the disturbance is periodic, or oscillatory, because the values of ξ are reproduced after a constant interval, and then the case becomes that which we have just discussed—viz., *oscillatory* motion in which the vertical displacement is very small with respect to the horizontal.

In the general case now supposed—viz., motion in which the vertical displacement is very small with respect to the horizontal, while the motion is not necessarily of the oscillatory or periodic kind, the function $\phi(x-vt)$ denotes a disturbance travelling in the positive sense of x, while $\psi(x+vt)$ denotes one travelling in the opposite sense; and the velocity of each is v.

Waves in very deep water. Applying now the results obtained for periodic or oscillatory waves to the case in which the depth, h, of the liquid is vastly greater than λ, the wave-length—as in the case of waves in the ocean—we have $e^{-\frac{2\pi h}{\lambda}}$ negligible, so that $\tanh\frac{2\pi h}{\lambda} = 1$, and (23) gives

$$v = \sqrt{\frac{g\lambda}{2\pi}}, \quad \ldots \quad \ldots \quad (30)$$

so that the velocity depends on the wave-length.

Also the ellipses described by the particles become circles.

Wave Motion under Gravity (Simple Cases). 417

Channel of any uniform cross-section. In obtaining equation (26) we have assumed the cross-section of the channel to be rectangular. But if the cross-section has any form, the corresponding equation is easily obtained. Let A be the area of the cross-section; then, still expressing the fact that the volume between the planes $P'p'$, $Q'q'$ is equal to that between Pp, Qq, if ϵ is the elevation of P' above the surface AB and b is the breadth of the channel at the level AB, the area of the cross-section $P'p'$ is $A + b\epsilon$, and we have

$$A \Delta x = (A + b\epsilon)\left(1 + \frac{d\xi}{dx}\right) \Delta x, \quad \ldots \quad (31)$$

$$\therefore A \frac{d\xi}{dx} + b\epsilon = 0, \quad \ldots \ldots \quad (32)$$

which shows that in the previous results for a rectangular channel we have merely to replace h by $\frac{A}{b}$, since the dynamical equations, (1), (2) of Art. 95, still hold. Hence, in particular, the velocity of the disturbance is $\left(\frac{gA}{b}\right)^{\frac{1}{2}}$ for *long* waves.

Channel with sloping sides. The determination of the displacements of water contained in a channel of any cross-section—even when this cross-section is the same at all points along the channel—has not yet been effected.

Moreover, the problem has not been solved even in the simple case in which the cross-section is a triangle whose two sides are equally inclined to the horizon, except when the inclination is 45° or 30°. The solution for the former case was obtained by Kelland on the assumption that the motion of the water is *irrotational*, i. e., that the components of displacement are the differential coefficients of a function of x, y, z. Thus, let the cross-section be an isosceles triangle whose inclined sides are AB and AC, its base BC being horizontal; let the horizontal line through A be taken as axis of x, the vertical upward line through A as axis of z, and the axis of y parallel to the breadth of the channel. Let the small components of displacement of the particle which at rest occupied the point (x, y, z) be ξ, η, ζ. Then the equations of motion are

$$\left.\begin{aligned}\frac{d^2\xi}{dt^2} &= -\frac{g}{w}\frac{dp}{dx}, \\ \frac{d^2\eta}{dt^2} &= -\frac{g}{w}\frac{dp}{dy}, \\ \frac{d^2\zeta}{dt^2} &= -g - \frac{g}{w}\frac{dp}{dz}.\end{aligned}\right\} \quad \ldots \ldots (33)$$

With these must be combined the equation

$$\frac{d\xi}{dx} + \frac{d\eta}{dy} + \frac{d\zeta}{dz} = 0, \ldots \ldots (34)$$

which implies the incompressibility of the liquid.

Now whatever values we determine for ξ, η, ζ must satisfy the boundary condition, viz., that the displacement of every particle which, when at rest, lay on the side AB must be along AB, and the displacement of every particle which lay on AC must be along AC; i.e., the values of ξ, η, ζ must be such that

$$\left.\begin{aligned}\text{when}\quad y - z &= 0 \text{ we have } \eta - \zeta = 0, \\ \text{and when}\quad y + z &= 0 \text{ we have } \eta + \zeta = 0.\end{aligned}\right\} \quad \ldots (a)$$

Suppose ξ, η, ζ to be the differential coefficients with respect to x, y, z, respectively, of some function ϕ, and assume

$$\phi \equiv f(y, z) \cdot \cos(nt - mx). \ldots \ldots (35)$$

Then (34) gives

$$\frac{d^2 f}{dy^2} + \frac{d^2 f}{dz^2} - m^2 f = 0. \ldots \ldots (36)$$

Wave Motion under Gravity (Simple Cases). 419

To satisfy this assume
$$f(y, z) \equiv A(e^{ky}+e^{-ky})(e^{lz}+e^{-lz}), \quad \ldots \quad (37)$$
where A is a constant.

Then (36) gives
$$k^2 + l^2 = m^2, \quad \ldots \quad (38)$$
which will be satisfied by assuming
$$k = m \cos a, \; l = m \sin a.$$

Calculating η and ζ from (37) with these values of k and l, we find that the conditions (a) will be satisfied if $a = \dfrac{\pi}{4}$.

Hence the required function ϕ whose differential coefficients give the displacements is given by
$$\phi = A(e^{\frac{my}{\sqrt{2}}}+e^{-\frac{my}{\sqrt{2}}})(e^{\frac{mz}{\sqrt{2}}}+e^{-\frac{mz}{\sqrt{2}}})\cos(nt-mx). \quad (39)$$

The subject of Wave Motion will be found treated at length in Basset's *Hydrodynamics*, vol. ii., Greenhill's *Wave Motion in Hydrodynamics* (reprinted from the American Journal of Mathematics), Airy's Article on *Tides and Waves* in the Encyclopædia Metropolitana, and the Mathematical Papers of Stokes and of Green.

INDEX

ABSOLUTE temperature, 191, 257.
Adiabatic transformation, 246.
  work done in, 249.
Air thermometer, 197.
Air, moist, weight of, 234.
Andrews's experiments on liquefaction, 269.
Angle of contact of liquid and solid, 324.
Aqueous vapour, pressure of at boiling point, 236.
  present in air, 267.
Archimedes, principle of, 115.
  screw of, 278.
Area, plane, pressure on, 45, 102.
  not influenced by molecular forces, 311.
Atmosphere, any layer of liquid may be treated as, 42.
  pressure at any height in, 80.
Avogadro, law of, 196.

Balloon, pressure not uniform on, 117.
Barometric formula, 209, 238.
Bellows, hydrostatic, 20.
  pump, 275.
Bernoulli, theorem of, 371.
Boiling point of water at various pressures, 237.
Boyle and Mariotte, law of, 21, 186, 187, 189.

Bramah's press, 19.
Buoyancy, principle of, 111, 113, 122, 179.
  centre of, 113.

Capillary surface in front of vertical plane, 353.
  between two parallel planes, 357.
  in cylindrical tube, 360.
  between inclined planes, 363.
  tubes, rise or fall of liquids in, 333.
Carnot's cycle, 246.
  view of heat, 253.
Catenoid, 350.
Centre of pressure, 52.
  position of on plane area, 53.
  „ „ on parallelogram, 54.
  „ „ on triangle, 55, 56, 57.
  „ „ in general, 100.
  „ „ referred to principal axes, 102.
  change of by rotation of area, 104.
Compressibility, cubical, 20.
Compression of water in ocean, 107.
Condensing air pump, 286.
Co-ordinates, polar and cylindrical, 93.
Critical temperature of gas, 269.
Curl of force and vector, 84, 87, 181.

Dalton and Gay-Lussac, law of, 190.
related to Boyle's, 193.
Density, 24.
laws of, which allow equilibrium, 88, 89, &c.
of water at any depth in ocean, 107.
Displacement, constant, theorem on, 153.
work done in, 175.
Diving bell, principle of, 201.

Efficiency of reversible engine, 254.
Energy of a gas, function of temperature alone, 241.
Equilibrium, general equations of, 82.
necessary condition of, 84.
Equipotential surfaces, 85.
Expansion, coefficient of for gas, 190.

Films, liquid, 364.
Fire engine, 276.
Floating body, positions of, 145.
Fluid, perfect, definition of, 3.
Forces, bodily and surface, 12.
fictitious, introduction of, 117.
gravitation and absolute measure of, 78.
Free surface of a liquid, 39.
everywhere at same level, 39.
French Commissioners' formula, 236.

Gas, perfect, definition of, 21, 185.
pressure at top of house apparently greater than at bottom, 80.
general equation of transformation of, 195, 207.
weight of, 205.
pressure of, on kinetic theory, 217.
Gases, mixture of, 226.
mixture with vapours, 233.

Gases, liquefaction of, 268.
Gay-Lussac's law, 190.
Geissler air pump, 287.
Gerstner's trochoidal wave, 403.
Green's equation, 177.

Hero's fountain, 377.
Heterogeneous fluid, equilibrium of body in, 130, 132.
buoyancy in, 179.
Hydraulic Press, 17.
Screw, 277.
Screw, differential, 281.
Ram, 282.
Hydrometers, 294.
Hygrometer, 265.

Irrotational motion, definition of, 417.
Isothermal transformation, 186.
work done in, 200.

Joule's equivalent, 239, 240.
experiment on energy of gas, 241.

Kinetic theory of gases, 216.

Laplace's formula for molecular pressure, 309.
Latent heat of steam, 261.
Level surfaces, 85.
Level of free surface of water, 39.
Line and surface integrals, 181.
Lines of resistance in masonry dams, 63.
Liquefaction of gases, 268, 272.

Manometers, 292.
Mariotte's bottle, 379.
law, 21, 186, 187, 189.
Mass moments, theorem of, 30.
McCay's proof of theorem on centre of pressure, 60.

Index. 423

etacentre, 154.
experimental determination of, 159.
etacentric evolute, 165.
oist air, weight of, 234.
olecular forces, 298.
oment of stability, 156.
otion of fluid, equation of, 386, 394.
ountain, height of, by barometer, 209.
height of, by boiling point, 238.

odoid, 351.

rifices, flow through, 380.
scillatory waves, 408.

arallel forces, centre of, 28.
ascal's principle, 13, 42.
ressure, intensity of, constancy round a point, 6, 11.
intensity of, due to gravity, 38, 77.
uniform on closed surface has no resultant, 113.
molecular, 305.
intrinsic, of liquid, 311.
gauge in a stream, 374.
of water vapour at boiling points, 237.
rism, floating, positions of, 149.
umps, water, 273.
air, 284.

uincke, on range of molecular forces, 302.

egnault's formula for total heat of steam, 261.
elative volume of steam, 215.
evolving fluid, 383.

aturated vapour, 231.

Screw of Archimedes, 278.
Separate equilibrium, principle of, 5.
'Skin' of a liquid, 339.
Specific weight and gravity, 22.
heat of a gas, 244.
Sprengel pump, 291.
Stability of floating body, 139, 158.
in two fluids, 166.
of floating vessel containing liquid, 168.
in heterogeneous fluid, 170.
Steam, total heat of, 260.
Strain and stress, intensity of, 1, 4.
Stream lines, 371.
Superposed liquids, 40.
Surface, closed, subject to uniform pressure, 113.
unclosed, resultant pressure on, 118.
of floatation, of buoyancy, 146.
tension of a liquid, 336, 342.
of constant pressure in moving fluid, 395.
Syphon, 376.

Temperature, critical, of a gas, 269.
Tension in a liquid surface, 336, 342.
Thermal unit, 239.
Torricelli, theorem of, 376.
Transformation, reversible, 254.
Triangle of surface-tensions, 348.
Trochoids, 397.

Unduloid, 351.

Vapours, 228.
saturated, 231.
Vector, curl of, 181.
Velocity of mean square for a gas, 225.
Vena contracta, 375.
Virtual Work, 136.

Waterfall, pressure due to, 218.
Waves, velocity of, in shallow water, 413.
  deep water, 416.
Wet and Dry Bulb Thermometer, 265.

Work done in lateral displacement of floating body, 175.
  expansion of gas, 199, 249.
  external and internal in evaporation of water, 263.

THE END.

Clarendon Press Series.

ENGLISH LANGUAGE AND LITERATURE . . pp. 1-6
HISTORY AND GEOGRAPHY . . . . . . p. 6 .
MATHEMATICS AND PHYSICAL SCIENCE . . p. 7
MISCELLANEOUS . . . . p. 8

The English Language and Literature.

HELPS TO THE STUDY OF THE LANGUAGE.

1. DICTIONARIES.

A NEW ENGLISH DICTIONARY ON HISTORICAL PRIN-CIPLES, founded mainly on the materials collected by the Philological Society Imperial 4to. Parts I-IV, price 12s. 6d. each.
    Vol. I (A and B), half-morocco, 2l. 12s. 6d.
    Vol. II (C and D). *In the Press.*
        Part IV, Section 2, **C—CASS**, beginning Vol. II, price 5s.
        Part V, **CAST—CLIVY**, price 12s. 6d.
        Part VI, **CLO—CONSIGNER**, price 12s. 6d.
    Edited by JAMES A. H. MURRAY, LL.D., sometime President of the Philological Society; with the assistance of many Scholars and Men of Science.
    Vol. III (**E, F, G,**) Part I, **E—EVERY**, Edited by HENRY BRADLEY, M.A., price 12s. 6d.

Bosworth and **Toller**. *An Anglo-Saxon Dictionary*, based on the MS. Collections of the late JOSEPH BOSWORTH, D.D. Edited and enlarged by Prof. T. N. TOLLER, M.A. Parts I-III, A-SAR. . . . [4to, 15s. each.
    Part IV, Section I, SÁR—SWÍÐRIAN. . . . [4to, 8s. 6d.

Mayhew and **Skeat**. *A Concise Dictionary of Middle English*, from A. D. 1150 to 1580. By A. L. MAYHEW, M.A., and W. W. SKEAT, Litt.D.
                                            [Crown 8vo, half-roan, 7s. 6d.

Skeat. *A Concise Etymological Dictionary of the English Language.* By W. W. SKEAT, Litt.D. *Fourth Edition* . . . [Crown 8vo, 5s. 6d.

2. GRAMMARS, READING BOOKS, &c.

Earle. *The Philology of the English Tongue.* By J. EARLE, M.A., Professor of Anglo-Saxon. *Fifth Edition.* . . [Extra fcap. 8vo, 8s. 6d.

—— *A Book for the Beginner in Anglo-Saxon.* By J. EARLE, M.A., Professor of Anglo-Saxon. *Third Edition.* . . [Extra fcap. 8vo, 2s. 6d.

Mayhew. *Synopsis of Old-English Phonology.* By A. L. MAYHEW, M.A. . . . . . [Extra fcap. 8vo, bevelled boards, 8s. 6d.

Morris and **Skeat.** *Specimens of Early English.* A New and Revised Edition. With Introduction, Notes, and Glossarial Index:—

    Part I. From Old English Homilies to King Horn (A.D. 1150 to A.D. 1300). By R. MORRIS, LL.D. *Second Edition.* . . [Extra fcap. 8vo, 9s.

    Part II. From Robert of Gloucester to Gower (A.D. 1298 to A.D. 1393). By R. MORRIS, LL.D., and W. W. SKEAT, Litt.D. *Third Edition.* 7s. 6d.

Skeat. *Specimens of English Literature*, from the 'Ploughmans Crede' to the 'Shepheardes Calender' (A.D. 1394 to A.D. 1579). By W. W. SKEAT, Litt.D. *Fifth Edition.* . . . [Extra fcap. 8vo, 7s. 6d.

—— *The Principles of English Etymology:*
    First Series. The Native Element. . . . . [Crown 8vo, 9s.
    Second Series. The Foreign Element. . . [Crown 8vo, 10s. 6d.

Sweet. *An Anglo-Saxon Primer, with Grammar, Notes, and Glossary.* By HENRY SWEET, M.A. *Third Edition.* . . [Extra fcap. 8vo, 2s. 6d.

—— *An Anglo-Saxon Reader.* In Prose and Verse. With Grammatical Introduction, Notes, and Glossary. By the same Author. *Sixth Edition, Revised and Enlarged.* . . . [Extra fcap. 8vo, 8s. 6d.

—— *A Second Anglo-Saxon Reader.* By the same Author. [4s. 6d.

—— *Old English Reading Primers.* By the same Author:—
    I. Selected Homilies of Ælfric. [Extra fcap. 8vo, *stiff covers*, 1s. 6d.
    II. Extracts from Alfred's Orosius. [Extra fcap. 8vo, *stiff covers*, 1s. 6d.

—— *First Middle English Primer,* with Grammar and Glossary. By the same Author. *Second Edition.* . . . . [Extra fcap. 8vo, 2s.

—— *Second Middle English Primer.* Extracts from Chaucer, with Grammar and Glossary. By the same Author. . . [Extra fcap. 8vo, 2s.

—— *A Primer of Spoken English.* . . [Extra fcap. 8vo, 3s. 6d.

—— *A Primer of Phonetics.* . . . [Extra fcap. 8vo, 3s. 6d.

Tancock. *An Elementary English Grammar and Exercise Book.* By O. W. TANCOCK, M.A. *Second Edition.* . [Extra fcap. 8vo, 1s. 6d.

—— *An English Grammar and Reading Book,* for Lower Forms in Classical Schools. By O. W. TANCOCK, M.A. *Fourth Edition.* 3s. 6d.

Wright. *A Primer of the Gothic Language.* With Grammar, Notes, and Glossary. By JOSEPH WRIGHT, Ph. D. . . [Extra fcap. 8vo, 4s. 6d.

A SERIES OF ENGLISH CLASSICS.

(CHRONOLOGICALLY ARRANGED.)

Chaucer. I. *The Prologue to the Canterbury Tales.* (*School Edition.*) Edited by W. W. SKEAT, Litt.D. . . [Extra fcap. 8vo, *stiff covers*, 1s.

—— II. *The Prologue; The Knightes Tale; The Nonne Prestes Tale.* Edited by R. MORRIS, LL.D. *A New Edition, with Collations and Additional Notes,* by W. W. SKEAT, Litt.D. . . [Extra fcap. 8vo, 2s. 6d.

—— III. *The Prioresses Tale; Sir Thopas; The Monkes Tale; The Clerkes Tale; The Squieres Tale, &c.* Edited by W. W. SKEAT, Litt.D. *Fourth Edition.* . . . . [Extra fcap. 8vo, 4s. 6d.

—— IV. *The Tale of the Man of Lawe; The Pardoneres Tale; The Second Nonnes Tale; The Chanouns Yemannes Tale.* By the same Editor. *New Edition, Revised.* . . . . [Extra fcap. 8vo, 4s. 6d.

—— V. *Minor Poems.* By the same Editor. [Crown 8vo, 10s. 6d.

—— VI. *The Legend of Good Women.* By the same Editor.
[Crown 8vo, 6s.

Langland. *The Vision of William concerning Piers the Plowman,* by WILLIAM LANGLAND. Edited by W. W. SKEAT, Litt.D. *Fourth Edition.*
[Extra fcap. 8vo, 4s. 6d.

Gamelyn, The Tale of. Edited by W. W. SKEAT, Litt.D.
[Extra fcap. 8vo, *stiff covers*, 1s. 6d.

Wycliffe. *The New Testament in English,* according to the Version by JOHN WYCLIFFE, about A.D. 1380, and Revised by JOHN PURVEY, about A.D. 1388. With Introduction and Glossary by W. W. SKEAT, Litt.D.
[Extra fcap. 8vo, 6s.

—— *The Books of Job, Psalms, Proverbs, Ecclesiastes, and the Song of Solomon*: according to the Wycliffite Version made by NICHOLAS DE HEREFORD, about A.D. 1381, and Revised by JOHN PURVEY, about A.D. 1388. With Introduction and Glossary by W.W. SKEAT, Litt.D. [Extra fcap. 8vo, 3s. 6d.

Minot. *The Poems of Laurence Minot.* Edited, with Introduction and Notes, by JOSEPH HALL, M.A. . . . . [Extra fcap. 8vo, 4s. 6d.

Spenser. *The Faery Queene.* Books I and II. Edited by G. W. KITCHIN, D.D., with Glossary by A. L. MAYHEW, M.A.
Book I. *Tenth Edition.* . . . . . [Extra fcap. 8vo, 2s. 6d.
Book II. *Sixth Edition.* . . . . [Extra fcap. 8vo, 2s. 6d.

Hooker. *Ecclesiastical Polity,* Book I. Edited by R. W. CHURCH, M.A., Dean of St. Paul's. *Second Edition.* . . . [Extra fcap. 8vo, 2s.

Marlowe and Greene. MARLOWE'S *Tragical History of Dr. Faustus,* and GREENE'S *Honourable History of Friar Bacon and Friar Bungay.* Edited by A. W. WARD, Litt.D. *New Edition.* . [Extra fcap. 8vo, 6s. 6d.

Marlowe. *Edward II.* Edited by O. W. TANCOCK, M.A. *Second Edition.* . . . . . [Extra fcap. 8vo. *Paper covers*, 2s.; *cloth*, 3s.

Shakespeare. Select Plays. Edited by W. G. CLARK, M.A., and W. ALDIS WRIGHT, D.C.L. . . . . . [Extra fcap. 8vo, *stiff covers*.
   The Merchant of Venice. 1s.   *Macbeth.* 1s. 6d.
   Richard the Second. 1s. 6d.   *Hamlet.* 2s.
            Edited by W. ALDIS WRIGHT, D.C.L.
   The Tempest. 1s. 6d.   *Coriolanus.* 2s. 6d.
   As You Like It. 1s. 6d.   *Richard the Third.* 2s. 6d.
   A Midsummer Night's Dream. 1s. 6d.   *Henry the Fifth.* 2s.
   Twelfth Night. 1s. 6d.   *King John.* 1s. 6d.
   Julius Cæsar. 2s.   *King Lear.* 1s. 6d.
            Henry the Eighth. 2s.

Shakespeare as a Dramatic Artist; *a popular Illustration of the Principles of Scientific Criticism.* By R. G. MOULTON, M.A. Second Edition, Enlarged. . . . . [Crown 8vo, 6s.

Bacon. *Advancement of Learning.* Edited by W. ALDIS WRIGHT, D.C.L. *Third Edition.* . . . . . . [Extra fcap. 8vo, 4s. 6d.

Bacon. *The Essays.* Edited, with Introduction and Illustrative Notes, by S. H. REYNOLDS, M.A. . . . . [Demy 8vo, *half-bound*, 12s. 6d.

Milton. I. *Areopagitica.* With Introduction and Notes. By JOHN W. HALES, M.A. *Third Edition.* . . . . [Extra fcap. 8vo, 3s.

——— II. *Poems.* Edited by R. C. BROWNE, M.A. In two Volumes. *Fifth Edition.*
        [Extra fcap. 8vo, 6s. 6d. Sold separately, Vol. I. 4s., Vol. II. 3s.
        In paper covers :—
   Lycidas, 3d.   *L'Allegro*, 3d.   *Il Penseroso*, 4d.   *Comus*, 6d.

——— III. *Paradise Lost.* Book I. Edited with Notes, by H. C. BEECHING, M.A. . . [Extra fcap. 8vo, 1s. 6d. *In Parchment*, 3s. 6d.

——— IV. *Samson Agonistes.* Edited, with Introduction and Notes, by JOHN CHURTON COLLINS, M.A. . . [Extra fcap. 8vo, *stiff covers*, 1s.

Bunyan. *The Pilgrim's Progress, Grace Abounding, Relation of the Imprisonment of Mr. John Bunyan.* Edited by E. VENABLES, M.A.
               [Extra fcap. 8vo, 5s. *In Parchment*, 6s.

Clarendon. I. *History of the Rebellion.* Book VI. Edited with Introduction and Notes by T. ARNOLD, M.A. . . . [Extra fcap. 8vo, 4s. 6d.

——— II. *Selections.* Edited by G. BOYLE, M.A., Dean of Salisbury.
               [Crown 8vo, 7s. 6d.

Dryden. *Select Poems.* (*Stanzas on the Death of Oliver Cromwell; Astræa Redux; Annus Mirabilis; Absalom and Achitophel; Religio Laici; The Hind and the Panther.*) Edited by W. D. CHRISTIE, M.A.
               [Extra fcap. 8vo, 3s. 6d.

——— *Essay of Dramatic Poesy.* Edited, with Notes, by T. ARNOLD, M.A. . . . . [Extra fcap. 8vo, 3s. 6d.

Locke. *Conduct of the Understanding.* Edited, with Introduction, Notes, &c., by T. FOWLER, D.D. *Third Edition.* . [Extra fcap. 8vo, 2s. 6d.

Addison. *Selections from Papers in the 'Spectator.'* By T. ARNOLD, M.A. *Sixteenth Thousand.* . [Extra fcap. 8vo, 4s. 6d. *In Parchment*, 6s.

ENGLISH LITERATURE. 5

Steele. *Selected Essays from the Tatler, Spectator, and Guardian.* By AUSTIN DOBSON. . . . [Extra fcap. 8vo, 5s. *In Parchment,* 7s. 6d.
Pope. I. *Essay on Man.* Edited by MARK PATTISON, B.D. *Sixth Edition.* . . . . [Extra fcap. 8vo, 1s. 6d.
—— II. *Satires and Epistles.* By the same Editor. *Second Edition.*
[Extra fcap. 8vo, 2s.
Parnell. *The Hermit.* . . . . . [*Paper covers,* 2d.
Thomson. *The Seasons, and the Castle of Indolence.* Edited by J. LOGIE ROBERTSON, M.A. . . . . . [Extra fcap. 8vo, 4s. 6d.
Berkeley. *Selections.* With Introduction and Notes. By A. C. FRASER, LL.D. *Fourth Edition.* . . . . . . [Crown 8vo, 8s. 6d.
Johnson. I. *Rasselas.* Edited, with Introduction and Notes, by G. BIRKBECK HILL, D.C.L.
[Extra fcap. 8vo, *limp,* 2s.; *Bevelled boards,* 3s. 6d.; *in Parchment,* 4s. 6d.
—— II. *Rasselas; Lives of Dryden and Pope.* Edited by ALFRED MILNES, M.A. . . . . . . [Extra fcap. 8vo, 4s. 6d.
—— *Lives of Dryden and Pope.* By the same Editor.
[*Stiff covers,* 2s. 6d.
—— III. *Life of Milton.* Edited, with Notes, &c., by C. H. FIRTH, M.A. . . . [Extra fcap. 8vo, *stiff covers,* 1s. 6d.; *cloth,* 2s. 6d.
—— IV. *Vanity of Human Wishes.* With Notes, by E. J. PAYNE, M.A. . . . . [*Paper covers,* 4d.
Gray. *Selected Poems.* Edited by EDMUND GOSSE, M.A.
[*In Parchment,* 3s.
—— *The same,* together with Supplementary Notes for Schools. By FOSTER WATSON, M.A. . . . . [Extra fcap. 8vo, *stiff covers,* 1s. 6d.
—— *Elegy, and Ode on Eton College.* . . . [*Paper covers,* 2d.
Goldsmith. *Selected Poems.* Edited, with Introduction and Notes, by AUSTIN DOBSON . . . [Extra fcap. 8vo, 3s. 6d. *In Parchment,* 4s. 6d.
—— *The Traveller.* Edited by G. BIRKBECK HILL, D.C.L.
[Extra fcap. 8vo, *stiff covers,* 1s.
—— *The Deserted Village.* . . . . . [*Paper covers,* 2d.
Cowper. I. *The Didactic Poems of* 1782, with Selections from the Minor Pieces, A.D. 1779-1783. Edited by H. T. GRIFFITH, B.A.
[Extra fcap. 8vo, 3s
—— II. *The Task, with Tirocinium,* and Selections from the Minor Poems, A.D. 1784-1799. By the same Editor. *Second Edition.*
[Extra fcap. 8vo, 3s.
Burke. I. *Thoughts on the Present Discontents; the two Speeches on America.* Edited by E. J. PAYNE, M.A. *Second Edition.*
[Extra fcap. 8vo, 4s. 6d.
—— II. *Reflections on the French Revolution.* By the same Editor. *Second Edition.* . . . . . . [Extra fcap. 8vo, 5s.
—— III. *Four Letters on the Proposals for Peace with the Regicide Directory of France.* By the same Editor. *Second Edition.*
[Extra fcap. 8vo, 5s.

Burns. *Selected Poems.* Edited by J. LOGIE ROBERTSON, M.A.
[Crown 8vo, 6s.
Keats. *Hyperion,* Book I. With Notes, by W. T. ARNOLD, B.A. 4d.
Byron. *Childe Harold.* With Introduction and Notes, by H. F. TOZER, M.A. . . . . . [Extra fcap. 8vo, 3s. 6d. *In Parchment*, 5s.
Shelley. *Adonais.* With Introduction and Notes. By W. M. ROSSETTI. . . . . [Crown 8vo, 5s.
Scott. *Lady of the Lake.* Edited, with Preface and Notes, by W. MINTO, M.A. With Map. . . . [Extra fcap. 8vo, 3s. 6d.
—— *Lay of the Last Minstrel.* Edited by W. MINTO, M.A. With Map. . . . [Extra fcap. 8vo, *stiff covers*, 2s. *In Parchment*, 3s. 6d.
—— *Lay of the Last Minstrel.* Introduction and Canto I, with Preface and Notes, by W. MINTO, M.A. . . . . [*Paper covers*, 6d.
—— *Marmion.* Edited by T. BAYNE. . [Extra fcap. 8vo, 3s. 6d.
Campbell. *Gertrude of Wyoming.* Edited, with Introduction and Notes, by H. MACAULAY FITZGIBBON, M.A. *Second Edition* . [Extra fcap. 8vo, 1s.
Wordsworth. *The White Doe of Rylstone.* Edited by WILLIAM KNIGHT, LL.D., University of St. Andrews. . . [Extra fcap. 8vo, 2s. 6d.

Typical Selections *from the best English Writers.* Second Edition. In Two Volumes. . . . . . [Extra fcap. 8vo, 3s. 6d. each.

HISTORY AND GEOGRAPHY, &c.

Freeman. *A Short History of the Norman Conquest of England.* By E. A. FREEMAN, M.A. *Second Edition.* . . [Extra fcap. 8vo, 2s. 6d.
George. *Genealogical Tables illustrative of Modern History.* By H. B. GEORGE, M.A. *Third Edition, Revised and Enlarged.* [Small 4to, 12s.
Greswell. *History of the Dominion of Canada.* By W. PARR GRESWELL, M.A. . . . . [Crown 8vo, 7s. 6d.
—— *Geography of the Dominion of Canada and Newfoundland.* By the same Author. . . . . [Crown 8vo, 6s.
Hughes (Alfred). *Geography for Schools.* Part I, *Practical Geography.* With Diagrams. . . . . . . [Extra fcap. 8vo, 2s. 6d.
Kitchin. *A History of France.* With Numerous Maps, Plans, and Tables. By G. W. KITCHIN, D.D., Dean of Winchester. *Second Edition.*
Vol. I. To 1453. Vol. II. 1453-1624. Vol. III. 1624-1793. Each 10s. 6d.
Lucas. *Introduction to a Historical Geography of the British Colonies.* By C. P. LUCAS, B.A. . . . [Crown 8vo, with 8 maps, 4s. 6d.
—— *Historical Geography of the British Colonies:*—
 I. *The Mediterranean and Eastern Colonies* (exclusive of India).
[Crown 8vo, with 11 maps, 5s.
 II. *The West Indian Dependencies.* With Twelve Maps.
[Crown 8vo, 7s. 6d.

MATHEMATICS AND PHYSICAL SCIENCE.

Aldis. *A Text Book of Algebra (with Answers to the Examples).* By W. STEADMAN ALDIS, M.A. . . . . . . [Crown 8vo, 7s. 6d.

Combination Chemical Labels. In Two Parts, gummed ready for use. Part I, Basic Radicles and Names of Elements. Part II, Acid Radicles.
[Price 3s. 6d.

Hamilton and **Ball.** *Book-keeping.* By Sir R. G. C. HAMILTON, K.C.B., and JOHN BALL (of the firm of Quilter, Ball, & Co.). *New and Enlarged Edition.* . . . . [Extra fcap. 8vo, 2s.

*** *Ruled Exercise Books adapted to the above;* fcap. folio, 1s. 6d. *Ruled Book adapted to the Preliminary Course;* small 4to, 4d.

Hensley. *Figures made Easy: a first Arithmetic Book.* By LEWIS HENSLEY, M.A. . . . . [Crown 8vo, 6d.

—— *Answers to the Examples in Figures made Easy,* together with 2000 additional Examples formed from the Tables in the same, with Answers. By the same Author. . . . . . . [Crown 8vo, 1s.

—— *The Scholar's Arithmetic.* By the same Author.
[Crown 8vo, 2s. 6d.

—— *Answers to the Examples in the Scholar's Arithmetic.* By the same Author. . . . . [Crown 8vo, 1s. 6d.

—— *The Scholar's Algebra.* An Introductory work on Algebra. By the same Author. . . . . [Crown 8vo, 2s. 6d.

Nixon. *Euclid Revised.* Containing the essentials of the Elements of Plane Geometry as given by Euclid in his First Six Books. Edited by R. C. J. NIXON, M.A. *Second Edition.* . . . . . . [Crown 8vo, 6s.
May likewise be had in parts as follows :—
Book I, 1s. Books I, II, 1s. 6d. Books I-IV, 3s. Books V, VI, 3s.

—— *Supplement to Euclid Revised.* By the same Author.
[Stiff covers, 6d.

—— *Geometry in Space.* Containing parts of Euclid's Eleventh and Twelfth Books. By the same Editor. . . . [Crown 8vo, 3s. 6d.

Fisher. *Class-Book of Chemistry.* By W. W. FISHER, M.A., F.C.S. *Second Edition* . . . . [Crown 8vo, 4s. 6d.

Harcourt and **Madan.** *Exercises in Practical Chemistry.* Vol. I. *Elementary Exercises.* By A. G. VERNON HARCOURT, M.A., and H. G. MADAN, M.A. *Fourth Edition.* Revised by H. G. MADAN, M.A.
[Crown 8vo, 10s. 6d.

Williamson. *Chemistry for Students.* By A. W. WILLIAMSON, Phil. Doc., F.R.S. . . . . [Extra fcap. 8vo, 8s. 6d.

Hullah. *The Cultivation of the Speaking Voice.* By JOHN HULLAH.
[Extra fcap. 8vo, 2s. 6d.

Maclaren. *A System of Physical Education: Theoretical and Practical.* With 346 Illustrations drawn by A. MACDONALD, of the Oxford School of Art. By ARCHIBALD MACLAREN, the Gymnasium, Oxford. *Second Edition.*
[Extra fcap. 8vo, 7s. 6d.

Troutbeck and **Dale.** *A Music Primer for Schools.* By J. TROUTBECK, D.D., formerly Music Master in Westminster School, and R. F. DALE, M.A., B.Mus., late Assistant Master in Westminster School. [Crown 8vo, 1s. 6d.

Tyrwhitt. *A Handbook of Pictorial Art.* By R. St. J. TYRWHITT, M.A. With coloured Illustrations, Photographs, and a chapter on Perspective, by A. MACDONALD. *Second Edition.* . . . [8vo, *half-morocco*, 18s.

Upcott. *An Introduction to Greek Sculpture.* By L. E. UPCOTT, M.A. . . . . [Crown 8vo, 4s. 6d.

Student's Handbook to the University and Colleges of Oxford. *Eleventh Edition.* . . . . [Crown 8vo, 2s. 6d.

Helps to the Study of the Bible, taken from the *Oxford Bible for Teachers,* comprising Summaries of the several Books, with copious Explanatory Notes and Tables illustrative of Scripture History and the Characteristics of Bible Lands ; with a complete Index of Subjects, a Concordance, a Dictionary of Proper Names, and a series of Maps. . . . . [Crown 8vo, 3s. 6d.

⁎ A READING ROOM *has been opened at the* CLARENDON PRESS WAREHOUSE, AMEN CORNER, *where visitors will find every facility for examining old and new works issued from the Press, and for consulting all official publications.*

☞ *All communications on Literary Matters and suggestions of new Books or new Editions, should be addressed to*

THE SECRETARY TO THE DELEGATES,
CLARENDON PRESS,
OXFORD.

London: HENRY FROWDE,
OXFORD UNIVERSITY PRESS WAREHOUSE, AMEN CORNER.
Edinburgh: 12 FREDERICK STREET.
Oxford: CLARENDON PRESS DEPOSITORY,
116 HIGH STREET.

www.ingramcontent.com/pod-product-compliance
Lightning Source LLC
Chambersburg PA
CBHW032137010526
44111CB00035B/603